Next-Gen Manufacturing 2020

Topical Conference at the 2020 AIChE Annual Meeting

Online
16 – 20 November 2020

ISBN: 978-1-7138-2291-2

Printed from e-media with permission by:

Curran Associates, Inc.
57 Morehouse Lane
Red Hook, NY 12571

CURRAN ASSOCIATES INC.
proceedings.com

Some format issues inherent in the e-media version may also appear in this print version.

AIChE
120 Wall Street, FL 23
New York, NY 10005-4020

Phone: (800) 242-4363
Fax: (203) 775-5177

www.aiche.org

Additional copies of this publication are available from:

Curran Associates, Inc.
57 Morehouse Lane
Red Hook, NY 12571 USA
Phone: 845-758-0400
Fax: 845-758-2633
Email: curran@proceedings.com
Web: www.proceedings.com

TABLE OF CONTENTS

596258 Detection of Cyber-Attacks and Resilient Operation of Nonlinear Processes Under Economic Model Predictive Control

Abstract:

In recent years, the cyber-security of chemical plants has become increasingly important as more communication networks are replaced or complemented by wireless networks in addition to point-to-point communications. Stable and secure operation of chemical plants require accurate information and reliable communication technologies. In the last few years, increasing research efforts have been dedicated to developing system and control designs to address cyber-attacks using machine-learning tools for cyber-attack detection [1,2,3]. Despite this progress, the integration of artificial intelligence and advanced control algorithms to address the problem of cyber-security is an emerging field of research interest with many open issues.

This work presents a resilient operating scheme for nonlinear processes subject to standard types of cyber-attacks, and proposes detection and handling strategies of these cyber-attacks. In particular, considering a general class of nonlinear systems, a modified Lyapunov-based Economic Model Predictive Controller (LEMPC) using combined closed-loop and open-loop control action implementation schemes is proposed with the control objective of optimizing economic benefits while maintaining closed-loop process stability. Process stability and resiliency against particular types of targeted destabilizing attacks will be ensured under the proposed controller modification and operation strategy. Moreover, machine-learning methods are used to develop sensor-data-based cyber-attack detectors, which are periodically activated online during process operation. Simulation results based on a continuously stirred tank reactor example demonstrate the effectiveness of the resilient LEMPC control strategy in the presence of various cyber-attacks, as well as the capability of the detection algorithm in reporting and differentiating the different types of cyber-attacks.

[1] Cárdenas, A., Amin, S., Lin, Z., Huang, Y., Huang, C., Sastry, S., 2011. Attacks against process control systems: risk assessment, detection, and response, in: Proceedings of the 6th ACM Symposium on Information, Computer and Communications Security, pp. 355–366.

[2] Chen, S., Wu, Z., Christofides, P.D., 2020. A cyber-secure control-detector architecture for nonlinear processes. AIChE Journal 66, e16907.

[3] Durand, H., 2018. A nonlinear systems framework for cyberattack prevention for chemical process control systems. Mathematics 6, 169.

608617 A Framework for the Identification and Mitigation of Cyber-Attacks in Networked Process Control Systems

Abstract:

Process operations have become increasingly reliant on networked control system architectures in which dedicated sensor-controller and controller-actuator links are replaced by real-time shared communication networks (wired or wireless). The integration of a shared communication network in the control system can substantially improve the operational flexibility and fault-tolerance capabilities of an industrial control system, in addition to reducing the installation, reconfiguration and maintenance times and costs.

Despite the economic and operational benefits of networked control systems, the increased reliance on a shard communication network comes with a host of fundamental challenges. Some of these challenges are tied to the inherent limitations on the transmission and processing capabilities of the communication medium. For example, issues such as network resource constraints, data losses, communication delays and real-time scheduling constraints can cause operational instabilities or closed-loop performance deterioration if left unaddressed. These challenges have motivated a significant body of research work in this area (e.g., [1], [2], [3], [4]).

Another challenge that networked control systems face, especially in the context of wireless networks, is the issue of network cyber-security. Owing to the deployment of large-scale sensor and actuator networks and the open nature of wireless networks, networked control systems have become more vulnerable to cyber-attacks. Cyber-attacks generally aim to alter process inputs and cause them to deviate from their normal operating values. One form of such attacks is the falsification of sensor measurements sent to the feedback controller, resulting in incorrect signals to the control actuators, or manipulating stored process data [5]. If left unchecked, these cyber-security risks can potentially lead to injury, death or physical damage, and are therefore a critical problem to address.

An approach to mitigate the risk of sensor measurement falsification cyber-attacks in networked control systems has been proposed in an earlier study [6]. In that study, a model-based networked control structure that enforces closed-loop stability was developed wherein a set of predictive models were embedded within the local control systems. Utilizing local state measurements local control actions were generated when communication between the sensor and network was suspended, and the model states were eventually updated when communication was re-established, at discrete times. In doing this, a minimum communication rate that guarantees closed-loop stability was established. To mitigate the effect of a cyber-attack, an assessment of the robustness of the networked control system to the cyber-attack was performed by modeling the cyber-attack in the closed-loop system formulation thus making it possible to explicitly characterize the stability of the closed-loop system in terms of the measurement error resulting from the cyber-attack as well as the communication rate and the controller design parameters. Using this characterization, the range of feasible operating conditions within which robust stability is guaranteed under cyber-attacks can be obtained. This characterization also reveals the parameters that play a critical role in the mitigation of the cyber-attack.

While the results of this study are promising in that they confirm the feasibility of the proposed cyber-attack mitigation strategy, the study does not provide a mechanism by which these cyber-attacks can be detected. Furthermore, in order to characterize the stability region, knowledge of the magnitude of the sensor measurement error is required, thus the magnitude of the cyber-attack must be identified before the proposed mitigation strategy can be implemented.

Motivated by these considerations, the objective of this contribution is to integrate a data-based method for the detection of a cyber-attack and the identification of the size of the attack in order to determine the magnitude of the sensor measurement error. To this end we utilize machine learning methods (e.g., [7], [8]) to build a neural network (NN) based detection system. To achieve our objective of detecting both the occurrence and the magnitude of the cyber-attack, the NN training data set is constructed with data for the system under normal operation and the system under different cyber-attack magnitudes, all subject to process disturbances. While this NN is utilized as a classification tool, training the NN with different cyber-attack magnitudes allows for the classification of both the existence of a cyber-attack and the approximate magnitude of this attack. The choice of the range and resolution of the training data set under different cyber-attack magnitudes affects the accuracy of the classification of the size of the cyber-attack. The NN is trained off-line and then used on-line to identify the occurrence and magnitude of the cyber-attack. The implementation and efficacy of the integrated cyber-attack identification and mitigation strategies are demonstrated using a chemical process example.

References:

[1] Hespanha JP, Naghshtabrizi P, Xu Y. A survey of recent results in networked control systems. Proceedings of the IEEE. 2007; 95:138–162.

[2] Sun Y, El-Farra NH. Quasi-decentralized model-based networked control of process systems. Computers & Chemical Engineering. 2008; 32(9):2016–2029.

[3] Sun Y, El-Farra NH. Resource-aware quasi-decentralized control of networked process systems over wireless sensor networks. Chemical Engineering Science. 2012; 69:93-106.

[4] You KY, Xie LH. Survey of recent progress in networked control systems. Acta Automatica Sinica. 2013; 39:101–117.

[5] Khorrami F, Krishnamurthy P, Karri R. Cybersecurity for control systems: A process-aware perspective. IEEE Design and Test. 2016; 33:75–83.

[6] Zedan A, El-Farra NH. A model-based approach for the analysis and mitigation of cyber-attacks in networked process control systems. In: AIChE Annual Meeting.2019.

[7] Wu Z, Albalawi F, Zhang J, Zhang Z, Durand H, Christofides PD. Detecting and handling cyber-attacks in model predictive control of chemical processes. Mathematics. 2018; 6(10).

[8] Wu Z, Tran A, Rincon D, Christofides PD. Machine-learning-based predictive control of nonlinear processes. Part II: Computational implementation. AIChE Journal. 2019; 65 (11).

611782 Relating State Estimation, Data-Driven Modeling, and Computing to Control System Cybersecurity

Abstract:

Cybersecurity of process control systems is receiving increasing interest in the process systems engineering community due to the potential for cyberattacks to impact both safety and profits in the chemical and pharmaceutical industry. An important question is defining the conditions under which it can be guaranteed that even if a cyberattack is performed on an aspect of a control loop (e.g., by falsifying the sensor measurements provided to a controller, or by falsifying a signal received by an actuator), the process does not experience closed-loop stability issues. Directions which look like they may have promise for handling this issue include cyberattack detection methods using neural networks (e.g., [1]), or methods which attempt to reconstruct the state when only a limited number of sensor measurements are attacked in a linear system [2]. However, methods for guaranteeing that closed-loop stability is not lost under a cyberattack on a nonlinear system are still needed.

In this work, we explore several concepts for preventing cyberattack success for nonlinear control systems. First, we explore how state estimation for nonlinear systems can be utilized to develop mechanisms for preventing closed-loop stability issues when a limited number of sensors is attacked. In addition, we develop a methodology for probing for cyberattacks using Lyapunov-based economic model predictive control (LEMPC) [3] that randomly develops constraints based on state-space regions in which the manner in which a Lyapunov function changes over the next sampling period is known in the absence of an attack, so that if the expected trend is not observed over the next sampling period, a cyberattack may be occurring. We also explore how data-driven models used in optimization-based control designs could be impacted by falsified sensor measurements, and how such modeling errors could impact closed-loop stability. Finally, we explore how the computing and communication strategies with optimization-based control may impact cybersecurity considerations.

[1] Wu, Z., F. Albalawi, J. Zhang, Z. Zhang, H. Durand and P. D. Christofides, "Detecting and Handling Cyber-Attacks in Model Predictive Control of Chemical Processes," *Mathematics*, **6**, 173, 22 pages, 2018.

[2] Fawzi, H., P. Tabuada and S. Diggavi. “Secure estimation and control for cyber-physical systems under adversarial attacks.” *IEEE Transactions on Automatic Control*. **59**, 1454-1467, 2014.

[3] M. Heidarinejad, J. Liu and P. D. Christofides. Economic model predictive control of nonlinear process systems using Lyapunov techniques, *AIChE Journal*, 58:855-870, 2012.

596262 Post Cyber-Attack Secure State Reconstruction for Nonlinear Processes Using Machine Learning

Abstract:

In recent years, several works have been dedicated to investigating various aspects of securing complex chemical plants, e.g., by ensuring the networking security amongst cyber devices, the prompt detection of faults and attacks, as well as accurate estimation of states using corrupted state measurements [1,3]. As it is important to develop accurate detectors to promptly report the intrusion of a cyber-attack as well as building robust frameworks to mitigate the impact of cyber-attacks before the detector is activated, it is equally important to have recuperating measures in place to maintain controllability of the system in the absence of reliable sensors.

This work investigates state-reconstruction strategies to maintain the controllability of the system effectively following the detection of cyber-attacks on sensor measurements. Considering a general class of nonlinear systems of which the sensor measurements are vulnerable to malicious cyber-attacks, there have been previous works on robust control frameworks that effectively maintain the stability of the process in the presence of cyber-attacks, within which the detection and differentiation of various types of cyber-attacks have been successfully carried out by machine-learning-based detection algorithms [2,4]. In this work, we further explore recuperation measures as a response plan after the detection of cyber-attacks to minimize or eliminate its impact. A machine-learning-based state reconstruction approach is presented to provide estimated state measurements based on falsified state measurements, and it ensures stable operation of the process before reliable sensor measurements can be re-installed.

[1] Ao, W., Song, Y., Wen, C., 2016. Adaptive cyber-physical system attack detection and reconstruction with application to power systems. IET Control Theory & Applications 10, 1458–1468.

[2] Chen, S., Wu, Z., Christofides, P.D., 2020a. Cyber-attack detection and resilient operation of nonlinear processes under lyapunov-based economic model predictive control. Computers & ChemicalEngineering 136, 106806.

[3] Hu, Q., Fooladivanda, D., Chang, Y.H., Tomlin, C.J., 2017. Secure state estimation for nonlinearpower systems under cyber attacks, in: Proceedings of the American Control Conference, Seattle,Washington. pp. 2779–2784.

[4] Wu, Z., Albalawi, F., Zhang, J., Zhang, Z., Durand, H., Christofides, P.D., 2018. Detecting andhandling cyber-attacks in model predictive control of chemical processes. Mathematics 6, 173, 22pages.

610274 Dynamical-Systems-Guided Learning of PDEs from Data

Abstract:

Traditionally, a Partial Differential Equation (PDE) is derived by applying basic laws of physics (e.g. Newton's second law, conservation of energy) to one or several continuum variables. However, for many systems of current interest, like fluid systems sampled sparsely in space, we do not know physical laws that lead to analytical PDE forms. Yet, it is suspected that there may exist PDEs that could (approximately) predict the evolution of target variables in those systems. Recent data-driven methods have shown great promise for discovering such PDEs from experimental and computational data. These methods have potential drawbacks, such as dependence on training data, lack of guarantees for stability, and high computational cost compared to classical computational schemes. In this talk, we discuss some concepts from dynamical systems theory that underpin such data-driven approaches; through those concepts, we connect data-driven methods to a special class of physics-informed discretization schemes that can eliminate some of those drawbacks.

First, we investigate the problems where the PDE trajectories are attracted to a global low-dimensional manifold in the state space. For these problems, if the dimension of the underlying manifold is n then knowledge of the solution at $2n+1$ spatial locations is enough to predict the evolution everywhere. This observation — justified by the well-known Whitney embedding theorem — allows us to learn multiple forms of PDEs that explain the evolution of the system. We demonstrate the application of this idea by learning a few PDEs for a reaction-diffusion system from data.

It is known that PDEs with strongly nonlinear behavior (e.g. turbulent fluid flows) do not possess a global low-dimensional attracting manifold. However, recent works on data-driven learning of coarse-grained PDEs suggest that post-transient low-dimensional behavior may still prevail locally. In particular, the neural-net-based architecture by Bar-Sinai et al. [1] learns the evolution of the solution profile restricted to a small spatial subdomain by using a smaller number of local measurements compared to, say, classical finite volume schemes. We justify this observation by appealing to a dynamical-systems coarse-graining of nonlinear PDEs known as "holistic discretization" [2]. This methodology, based on center manifold theory, gives a systematic approximation of the manifold that attracts the local solution profiles and hence provides relatively sparse but accurate discretizations for time stepping of the solution. Previous work, e.g. [3], has shown the effectiveness of this approach in guiding equation-free modeling of multi-scale systems.

Through examples from a linear advection-diffusion PDE and the nonlinear Burgers equation, we show that indeed the low-dimensional manifold exploited in data-driven methods coincides with the manifold approximated by the holistic discretization. We compare the performance of the holistic method to recent neural-net architectures, and explore the behavior of both approaches for a wide range of system parameters (Peclet/Reynolds number) and coarse-graining factors. An important feature of the holistic approach is that its stability and accuracy does not depend on the choice of the training data.

[1] Y. Bar-Sinai, S. Hoyer, J. Hickey, and M. P. Brenner, "Learning data-driven discretizations for partial differential equations," Proceedings of the National Academy of Sciences 116, 15344–15349 (2019).

[2] A. Roberts, "Holistic discretization ensures fidelity to Burgers' equation," Applied numerical mathematics 37, 371–396 (2001).

[3] A. Roberts and I.G. Kevrekidis, ``General tooth boundary conditions for equation-free modeling", SIAM Journal of Scientific Computing 29.4, 1495-1510 (2007)

611103 A New Method for Building Surrogate Models of Stochastic Simulations

Abstract:

With the increase in computational power and our understanding of fundamental phenomena, we can now develop high-fidelity simulation models to study, design, and optimize engineering systems (Eaton et al., 2017; Ookawara et al., 2007; Szilágyi et al., 2018). Despite these advances, some of these simulation models are still computationally expensive to utilize in sensitivity analysis or optimization studies where many simulation runs may be necessary (Liu et al., 2016; Peherstorfer et al., 2017). Different approaches have been recommended in the literature for handling this computational expense. One commonly-used approach is building a surrogate/meta-model model of the simulation and employing the surrogate model in the analysis when an evaluation of the high-fidelity simulation is required (Han and Zhang, 2012). Surrogate models or meta-models, are representative of the original models, but cheaper to evaluate computationally. The surrogate models approximate the output of the high-fidelity simulation models based on the inputs (Quirante et al., 2015).

When high-fidelity simulation models have uncertain parameters and/or uncertain inputs, the traditional techniques for building surrogate models fail to accurately represent the original model outputs (Staum, 2009). Current approaches for building surrogate models of stochastic simulations, in general, fall into one of two groups. The first group considers a nominal value for the uncertain parameter(s) and builds surrogate models at these values of uncertain parameters (Hüllen et al., 2019) or defines an uncertainty set for the stochastic parameter, evaluates the high-fidelity simulation model using the values of the elements of these sets and carries out the surrogate modeling (Hüllen et al., 2019). The second group, which has been quite popular in recent years, applies stochastic kriging (Ankenman et al., 2008). By utilizing the methods under the first group, more uncertainty may be added to the surrogate model because part of the information is being ignored by fixing the value of uncertain parameters (Hüllen et al., 2019). Stochastic kriging has been shown to exhibit good performance in predicting the expected output value (Ankenman et al., 2008). However, by using this method, the surrogate model type is predetermined because it is built based on kriging meta-models, while other surrogate modeling techniques may be more appropriate for representing the data (Williams and Cremaschi, 2019). Developing approaches for building surrogate models to estimate the outputs of high-fidelity stochastic simulation models accurately can overcome these shortcomings.

In this study, we propose that accurate surrogate models can be built for estimating the outputs of high-fidelity stochastic simulations by considering each uncertain parameter of the stochastic simulation as an uncertain input to the simulation. Isolating the uncertain stochastic parameters of the high-fidelity simulation converts the simulation to a deterministic one, for which most of the popular surrogate modeling techniques can be used to train a meta-model. The uncertain parameter(s) are still part of the meta-model, which paves the way for easier analysis of the uncertainty. Assume we have a high-fidelity simulation model g(X;K), where X is the input vector with dimension d_1, where some of the inputs may be uncertain, and K is the vector of system uncertain parameters with d_2 components. The new simulation model is defined as g'(X*), where X* is a d dimensional vector of inputs with $d = d_1 + d_2$, which contains all inputs and the uncertain parameters of the stochastic simulation (g(X;K)). It is assumed that the distributions of the uncertain parameters (K) are known, and the parameters of the distributions are constant. With these definitions, the new simulation model (g'(X*)) becomes deterministic, and any of the surrogate modeling techniques can be utilized to train a model representing this simulation (Fig. 1).

With these definitions, the new simulation model (g'(X*)) becomes deterministic, and any of the surrogate modeling techniques can be utilized to train a model representing this simulation.

$g'(X^*) \approx F'(X^*)$, where $F'(X^*)$ is the trained meta-model representing the deterministic formulation. With this new method, the high fidelity stochastic simulation is approximated using the existing surrogate modeling techniques for deterministic models.

There are various machine learning techniques practiced for surrogate modeling. Seven different techniques are implemented in this study to train the meta-models of deterministic formulation, g'(X*). The performance of the new method is evaluated computationally using these seven surrogate modeling methods for test functions with uncertain parameters. The test functions are from the Virtual Library of Simulation Experiments (Surjanovic and Bingham, 2013) with a different number of inputs and uncertain parameters. The quality of the predictions is evaluated for test functions with various features. The number of inputs of the function (dimension of X), number of uncertain parameters (cardinality of K), and nonlinearity of the uncertain parameters are considered as features of the test functions. Root mean square error and the mean absolute error are the metrics of the quality of the performance. One thousand data points generated for each of the test functions were used as the training data set. The test set was 10^5 new data points for comparison of the model performances. The results suggest that the prediction accuracy depends on the specifications of the test functions. The number of uncertain parameters, as well as the number of input parameters have a significant effect on the quality of the predictions.

References

Ankenman, B., Nelson, B.L., Staum, J., 2008. Stochastic kriging for simulation metamodeling. Proc. - Winter Simul. Conf. 362–370. https://doi.org/10.1109/WSC.2008.4736089

Eaton, A.N., Beal, L.D.R., Thorpe, S.D., Hubbell, C.B., Hedengren, J.D., Nybø, R., Aghito, M., 2017. Real time model identification using multi-fidelity models in managed pressure drilling. Comput. Chem. Eng. 97, 76–84. https://doi.org/10.1016/j.compchemeng.2016.11.008

Han, Z.-H., Zhang, K.-S., 2012. Surrogate-based optimization. Real-world Appl. Genet. algorithms 343–362.

Hüllen, G., Zhai, J., Kim, S.H., Sinha, A., Realff, M.J., Boukouvala, F., 2019. Managing Uncertainty in Data-Driven Simulation-Based Optimization. Comput. Chem. Eng. 106519. https://doi.org/10.1016/j.compchemeng.2019.106519

Liu, B., Koziel, S., Zhang, Q., 2016. A multi-fidelity surrogate-model-assisted evolutionary algorithm for computationally expensive optimization problems. J. Comput. Sci. 12, 28–37. https://doi.org/10.1016/j.jocs.2015.11.004

Ookawara, S., Kuroki, M., Street, D., Ogawa, K., 2007. High-fidelity DEM-CFD modeling of packed bed reactors for process intensification. Chem. Eng. 16–20.

Peherstorfer, B., Kramer, B., Willcox, K., 2017. Combining multiple surrogate models to accelerate failure probability estimation with expensive high-fidelity models. J. Comput. Phys. 341, 61–75. https://doi.org/10.1016/j.jcp.2017.04.012

Quirante, N., Javaloyes, J., Ruiz-Femenia, R., Caballero, J.A., 2015. Optimization of chemical processes using surrogate models based on a kriging interpolation, in: Computer Aided Chemical Engineering. Elsevier, pp. 179–184.

Staum, J., 2009. Better Simulation Metamodeling: The why, what, and how of Stochastic Kriging 119–133.

Surjanovic, S., Bingham, D., 2013. Virtual Library of Simulation Experiments: Test Functions and Datasets.

Szilágyi, B., Agachi, P.Ş., Nagy, Z.K., 2018. Chord Length Distribution Based Modeling and Adaptive Model Predictive Control of Batch Crystallization Processes Using High Fidelity Full Population Balance Models. Ind. Eng. Chem. Res. 57, 3320–3332. https://doi.org/10.1021/acs.iecr.7b03964

Williams, B.A., Cremaschi, S., 2019. Surrogate Model Selection for Design Space Approximation And Surrogatebased Optimization, in: Computer Aided Chemical Engineering. Elsevier, pp. 353–358.

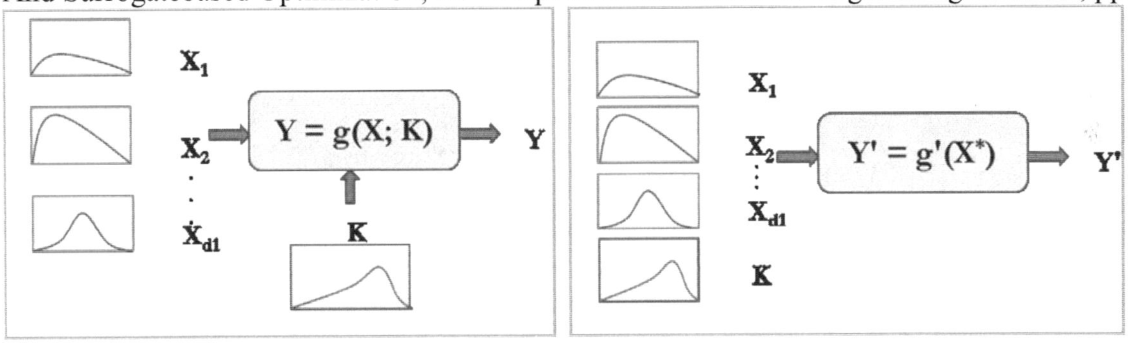

602617 Machine Learning-Based Prediction of Liquid Wettability of iCVD Polymers

Abstract:

Initiated chemical vapor deposition (iCVD) is a reactive process that creates polymeric materials by adsorbing vapor phase monomers, which then polymerize on a surface via free-radical additive propagation. Although contiguous polymer thin films are usually formed, the iCVD process can sometimes lead to discrete micro- and nano-structures. For example, our iCVD synthesis of poly(perfluorodecyl acrylate) (PPFDA) has resulted in the growth of micro- and nano-worms, which tend to grow normal to the surface [1]. Due to the low surface energy of the fluorinated polymer and the heterogeneous worm-like surface, liquid wetting can be substantially reduced by entering the Cassie-Baxter state. The micro- and nano-structures of the worm-like features directly depend on iCVD operating conditions. They in turn influence bulk properties such as repellency of oils, like heptane. With the contact angle being dependent on iCVD operating conditions, it would be advantageous to create a model that captures the dependence.

Using conventional reaction rate equations and parameters, the reaction kinetics can be modeled with enough accuracy [2]. However, the mechanisms that determine the growth of the worm-like structures are currently not well understood. Given this currently-inadequate chemical and physical understanding, an alternative to first-principles modeling to relate product (polymer) properties such as contact angle to iCVD processing conditions is to use an empirical modeling approach such as neural networks. A typical neural network model can contain a number of fitting parameters, many times greater than the number of observations available to fit the model with. This relatively large number of fitting parameters provides a lot of flexibility for shaping the response surface of the target (output) variable(s).

In this work, we model the contact angle of heptane on PPFDA as a function of iCVD operating conditions using a neural network architecture. We then use the developed model to determine the optimal iCVD operating conditions that maximizes the heptane contact angle on the worm-like surface. To verify model predictions, we independently perform iCVD laboratory experiments with the model-predicted optimal operating conditions, perform microscopy to elucidate the polymer structure, and measure the resulting heptane contact angles.

References

1. Lau, K.K.S.; Chen, Z.; Nguyen, T. Directed Vapor Deposition and Assembly of Polymer Micro- and Nanostructures. *S. Patent Application* 62941086; November 27, 2019.
2. Lau, K.K.S. Growth Mechanism, Kinetics, and Molecular Weight. In *CVD Polymers: Fabrication of Organic Surfaces and Devices* (ed. Gleason, K.K.); Wiley 2015.

611261 Approximating Process Safety Metrics Using Artificial Neural Networks

Abstract:

Safety of a chemical plant and its personnel is always a priority. During expansion projects and unit upgrades, often changes in one equipment may result in upset conditions for a downstream equipment. At the most benign level, this can cause the relief valves to open, losing valuable products to the flare thereby affecting the economics of that unit. A more severe case can damage equipment or result in loss of life. As a result, it is paramount that process safety and consideration of system-wide consequences are at the forefront of any decision making process. One challenge in accounting for process safety incidents is that accurate modeling is complex, time-intensive, and requires many inputs [1]. Heat exchangers are a great example of this complexity. Heat exchanger modeling using first principles can be complicated [2]. At the same time, setting up experiments is not always practical [3]. The advent of machine learning and Artificial Neural Network (ANN)-based models have shown significant promises in addressing complex problems in many areas in recent times, heat exchangers [3,4]. This work expands on these by proposing how ANN can be used to predict process safety metrics. Specifically, we apply ANN to predict heat exchanger safety ratings. This metric is used to measure the severity of a heat exchanger tube rupture, an event that is listed under API 521 as a credible overpressure scenario [5-7]. In the event of a tube rupture, the tube side can quickly overpressure the shell side. Upon the shell side pressure increasing beyond the hydrotest pressure, the shell material is prone to fail, potentially resulting in a catastrophic outcome [8]. By training ANN to a set of tube rupture simulation data, we are able to bypasses the need for tedious non-steady state equations. Lastly, we illustrate how these predictions can be used to develop tube-rupture resilient heat exchanger networks.

References:

[1] Crowl, D. A., & Louvar, J. F. (2001). *Chemical process safety: fundamentals with applications*. Pearson Education.

[2] Shah, R. K., & Sekulic, D. P. (2003). *Fundamentals of heat exchanger design*. John Wiley & Sons.

[3] Wang, Q., Xie, G., Zeng, M., & Luo, L. (2006). Prediction of heat transfer rates for shell-and-tube heat exchangers by artificial neural networks approach. *Journal of Thermal Science*, *15*(3), 257-262.

[4] Mohanraj, M., Jayaraj, S., & Muraleedharan, C. (2015). Applications of artificial neural networks for thermal analysis of heat exchangers–a review. *International Journal of Thermal Sciences*, *90*, 150-172.

[5] Harhara, A., & Hasan, M. F. (2020). Dynamic modeling of heat exchanger tube rupture. *BMC Chemical Engineering*, *2*(1), 1-20.

[6] Harhara, A., & Hasan, M. F. (2019). Incorporating Process Safety into Heat Exchanger Network Synthesis and Operation. In *Computer Aided Chemical Engineering* (Vol. 47, pp. 221-226). Elsevier.

[7] API Standard 521. (2014). Pressure-Relieving and Depressuring Systems.

[8] Hellemans, M. (2009). *The safety relief valve handbook: design and use of process safety valves to ASME and International codes and standards*. Elsevier.

603315 Long-Term Hybrid AI-Expert Combustion Optimization System for Coal-Fired Electricity Generation NOx Reduction

Abstract:

An ever-increasing concern for the environmental impact of fossil-fuel electricity generation has amplified efforts by academia and industry to identify methods to mitigate this issue [1]–[5]. An attractive option has been the development and installation of optimization systems which utilize machine learning models to represent the combustion process and improve system performance. Consisting of primarily software, these systems can potentially provide significant improvements with little capital expenditure [6]. Numerous potential models and systems have been discussed within the literature, but each of these represent entirely offline or only short duration online performance studies [7], [8]. This work aims to satisfy the need for the long-term evaluation and characterization of the performance and effects of a hybrid AI-expert combustion optimization system (COS) on NOx emission rates as well as the secondary effects on the remainder of the system. The COS uses artificial neural networks to model the combustion process, particle swarm optimization (PSO) to optimize the 80+ manipulated variables, and expert logic to address adverse conditions in real-time. Over a two-year operational period, the hybrid AI-expert system was able to aid in the reduction of the NOx emission rate by more than 22.5% compared to baseline, while improving unit temperature management and realizing high operator acceptance leading to a system service factor greater than 86%.

[1] G. H. Richards, C. Q. Maney, J. L. Marion, R. Lewis, and C. Smith, "Ultra-Low NOx integrated system for coal fired power plants," *Fuel Energy Abstr.*, vol. 43, no. 3, pp. 218–219, 2003.

[2] J. F. Tuttle and K. M. Powell, "Analysis of a thermal generator's participation in the Western Energy Imbalance Market and the resulting effects on overall performance and emissions," *Electr. J.*, vol. 32, no. 5, pp. 38–46, 2019.

[3] F. Wang, S. Ma, H. Wang, Y. Li, Z. Qin, and J. Zhang, "A hybrid model integrating improved flower pollination algorithm-based feature selection and improved random forest for NO X emission estimation of coal-fired power plants," *Meas. J. Int. Meas. Confed.*, vol. 125, no. 1, pp. 303–312, 2018.

[4] P. Tan *et al.*, "Dynamic modeling of NOX emission in a 660 MW coal-fired boiler with long short-term memory," *Energy*, vol. 176, no. X, pp. 429–436, 2019.

[5] J. F. Tuttle, R. Vesel, S. Alagarsamy, L. D. Blackburn, and K. Powell, "Sustainable NOx emission reduction at a coal-fired power station through the use of online neural network modeling and particle swarm optimization," *Control Eng. Pract.*, vol. 93, no. September, 2019.

[6] D. Hasler and W. Rosenquist, "Coal-Fired Power Plant Heat Rate Reductions," 2009.

[7] J. Song, C. E. Romero, Z. Yao, and B. He, "A globally enhanced general regression neural network for on-line multiple emissions prediction of utility boiler," *Knowledge-Based Syst.*, vol. 118, no. X, pp. 4–14, 2017.

[8] J. Smrekar, P. Potočnik, and A. Senegačnik, "Multi-step-ahead prediction of NOx emissions for a coal-based boiler," *Appl. Energy*, vol. 106, no. 2013, pp. 89–99, 2013.

602401 Reinforcement Learning Augmented Model Predictive Control with Fuzzy Subtractive Clustering and Fuzzy Cognitive Mapping

Abstract:

Reinforcement learning (RL) is a model-free machine learning framework in which an agent learns from a system by direct interaction, as opposed to learning from labeled sets of data that occurs with unsupervised learning. RL can be applied to process control [1], but there are considerable opportunities for its improvement as well as expansion of application areas [2]. RL has been shown to outperform traditional controllers by accounting for plant-model mismatch [3] especially for systems with complex dynamics [4]. The potential of RL to improve the approximation of uncertain system dynamics has also been illustrated, and was shown to converge to a solution with small optimality gap for the discrete linear case [5]. Applications for decision making and process optimization have also been shown [6], [7]. Recent work has shown the application of deep neural networks to represent the Hamilton-Jacobi-Bellman equation in a large state space for an optimal control problem [8]. As can be seen, the research on RL for practical application to process control applications is diverse and growing, but significant work remains in solving key issues and exploiting all of the features of RL.

In this work, a novel adaptive model predictive controller (MPC) augmented with RL is proposed, where a supervisory RL agent learns an optimal tuning policy from the performance of the MPC while improving its performance through exploration and exploitation. However, for a general class of nonlinear systems, the state space is infinite dimensional thus leading to computational intractability due to the aggregation of infinite states.

Here we propose to reduce the size of the state space via a clustering approach. First, an updated value function approximation is considered, using a Bayesian Smoothing Spline (BSS)-ANOVA framework where eigenfunctions of the Karhunen-Loéve (KL) expansion are used as basis functions. Fuzzy Subtractive Clustering is then used to partition the state space in a control-centric way [9]. Clustering is performed in a state-optimal action space where the optimality of the action is evaluated with respect to cluster-specific basis functions. The state of the system is determined using a Fuzzy Cognitive Map (FCM) based on the existing cluster centers. This fuzzy clustering can be considered as an adaptive mapping of the states and actions where cluster aggregation is minimized using an information theoretic criterion. The FCM exploits stored memory for computing the optimal move. In this algorithm, policy improvement is performed using a TD(λ) update [10].

For fast and efficient learning, a two-stage learning approach is proposed where exploration in the first stage, where the agent learns on a reduced model, is followed by exploitation of the control space. The approach enables use of the RL technique for large-scale systems nonlinear with acceptable optimality gap. Performance of the proposed controller is benchmarked against the static underlying MPC by applying it to several applications. Of particular importance is the application of the proposed approach to a time-varying time-delay system.

[1] J. Shin, T. A. Badgwell, K.-H. Liu, and J. H. Lee, "Reinforcement Learning – Overview of recent progress and implications for process control," *Comput. Chem. Eng.*, vol. 127, pp. 282–294, Aug. 2019, doi: 10.1016/j.compchemeng.2019.05.029.

[2] J. B. Rawlings and C. T. Maravelias, "Bringing new technologies and approaches to the operation

and control of chemical process systems," *AIChE J.*, vol. 65, no. 6, Jun. 2019, doi: 10.1002/aic.16615.

[3] L. A. Brujeni, J. M. Lee, and S. L. Shah, "Dynamic tuning of PI-controllers based on model-free Reinforcement Learning methods," in *ICCAS 2010*, Oct. 2010, pp. 453–458, doi: 10.1109/ICCAS.2010.5669655.

[4] I. Carlucho, M. De Paula, S. A. Villar, and G. G. Acosta, "Incremental Q-learning strategy for adaptive PID control of mobile robots," *Expert Syst. Appl.*, vol. 80, pp. 183–199, Sep. 2017, doi: 10.1016/j.eswa.2017.03.002.

[5] J. E. Morinelly and B. E. Ydstie, "Dual MPC with Reinforcement Learning," *11th IFAC Symp. Dyn. Control Process Syst. Biosyst. DYCOPS-CAB 2016*, vol. 49, no. 7, pp. 266–271, Jan. 2016, doi: 10.1016/j.ifacol.2016.07.276.

[6] J. Shin and J. H. Lee, "Multi-timescale, multi-period decision-making model development by combining reinforcement learning and mathematical programming," *Comput. Chem. Eng.*, vol. 121, pp. 556–573, Feb. 2019, doi: 10.1016/j.compchemeng.2018.11.020.

[7] P. Petsagkourakis, I. O. Sandoval, E. Bradford, D. Zhang, and E. A. del Rio-Chanona, "Reinforcement learning for batch bioprocess optimization," *Comput. Chem. Eng.*, vol. 133, p. 106649, Feb. 2020, doi: 10.1016/j.compchemeng.2019.106649.

[8] J. W. Kim, B. J. Park, H. Yoo, T. H. Oh, J. H. Lee, and J. M. Lee, "A model-based deep reinforcement learning method applied to finite-horizon optimal control of nonlinear control-affine system," *J. Process Control*, vol. 87, pp. 166–178, Mar. 2020, doi: 10.1016/j.jprocont.2020.02.003.

[9] Q. Gao, G. Feng, Z. Xi, Y. Wang, and J. Qiu, "A New Design of Robust ${\rm H}_{\infty}$ Sliding Mode Control for Uncertain Stochastic T-S Fuzzy Time-Delay Systems," *IEEE Trans. Cybern.*, vol. 44, no. 9, pp. 1556–1566, Sep. 2014, doi: 10.1109/TCYB.2013.2289923.

[10] R. S. Sutton and A. G. Barto, *Reinforcement Learning - An Introduction*, 2nd ed. United States of America: Westchester Publishing Services, 2018.

613569 *Batch-to-Continuous Transition in the Specialty Chemicals Industry: A Case Study for Intensification of Dispersants Production*

Abstract:

Production of specialty chemicals to-date is conducted almost exclusively in batch or semi-batch reactors. Yet, transition to continuous processing promises strong improvements over the century-old batch processing technology—demonstrated widely in the pharmaceutical industry but barely explored for the much larger scale production of specialty chemicals. Among the anticipated improvements are lower unit energy consumption, less waste generation, reduced exposure hazards, smaller process footprint, lower capital investment and more consistent product quality. Continuous processing thus offers the prospect of truly revolutionizing the manufacture of specialty chemicals via "process intensification" (PI).

However, several key factors hamper adoption of this technology. Among those are: (i) The deep historic roots of this industry in batch processing with decades of industrial experience make a transition to "unproven" process technology risky in an overall risk-averse industry. A first-of-its-kind demonstration of such a transition by an industry leader could have a transformative impact on this industry. (ii) While batch reactors are highly tolerant to incomplete knowledge (in particular regarding kinetics), the design and operation of continuous processes requires availability of robust kinetics. However, the detailed chemistry and kinetics underlying current processes are often poorly understood (if not completely unknown). Transition to continuous processing hence requires collection of robust and accurate kinetics as a key prerequisite, including impact of impurities and contaminants which play a significant role in many specialty chemicals processes.

In this presentation, we will present the application of process intensification via batch-to-continuous transition for dispersant production, a large class of specialty chemicals that constitute a ~$20B global market and are well-suited for demonstration of PI principles due to their relatively simple chemistry and broad range of production volumes. The (on-going) project is a close collaboration between the University of Pittsburgh and Lubrizol Corporation , a global leader in dispersant technology, and aims to develop a methodology for the batch-to-continuous transition for the specialty chemicals industry and demonstrate the drastic advances in efficiency that are attainable during that transition. This collaboration was able to move a first continuous process for dispersant production within ~2 years from initial laboratory studies to deployment of a first fully operational commercial skid. We will highlight new insights into the reaction mechanisms and kinetics for the underlying chemistry, efficiency advances enabled by this transition, as well a first insights gained from operating the commercial scale unit.

606035 Sodium Tungstate-Promoted Perovskite Oxides As Effective Redox Catalysts for Redox Oxidative Cracking of Naphtha

Abstract:

Light olefins such as ethylene and propylene are important chemical building blocks produced primarily via steam cracking of naphtha and light alkanes; however, commercial cracking processes suffer from high energy consumption and significant CO_2 intensity (up to 2 tons CO_2/ton C_2H_4). This is largely due to the endothermic cracking reactions, equilibrium-limited single-pass conversion, and complex product separations. While oxidative cracking of naphtha using co-fed O_2 represents a potentially advantageous approach, the need to co-feed gaseous oxygen introduces safety concerns and leads to significant CO_x formation.

We propose a redox oxidative cracking (ROC) approach as an alternative scheme to intensify the production of light olefins from naphtha. ROC avoids direct contact between gaseous oxygen and naphtha by cracking the vaporized naphtha feed, forming light olefins and hydrogen. Meanwhile, the hydrogen is selectively combusted by the lattice oxygen of a solid metal oxide redox catalyst to form water. The reduced redox catalyst is then re-oxidized with air in a separate regeneration step, completing the redox cycle. Here, we report Na_2WO_4-promoted perovskites, e.g. $CaMnO_3$, as effective redox catalysts for ROC of naphtha. Using *n*-hexane and cyclohexane model compounds, over three-fold increase in olefin yield was achieved in the presence of the redox catalyst compared to thermal cracking. The redox reaction pathway, phase transition behavior of the perovskite oxides, and the role of Na_2WO_4 promoter were characterized in detail, providing important mechanistic insights for further optimization.

$CaMnO_3$ was prepared from metal nitrates, citric acid, and ethylene glycol via a modified Pechini method. 20 wt.% Na_2WO_4 was added using wet impregnation. *N*-hexane and cyclohexane ROC tests used a fixed-bed quartz U-tube reactor and an automated gas delivery panel with downstream gas chromatography (GC) and mass spectrometry (MS). Samples were characterized via oxygen temperature-programmed desorption (O_2-TPD), X-ray diffraction (XRD), *in-situ* XRD, X-ray photoelectron spectroscopy (XPS), low-energy ion scattering (LEIS), transition electron microscopy (TEM) with EDS mapping, FTIR spectroscopy, differential scanning calorimetry (DSC), and electrochemical impedance spectroscopy (EIS).

A 20 wt.% $Na_2WO_4/CaMnO_3$ redox catalyst demonstrated 64% and 58% olefin yields from *n*-hexane and cyclohexane feeds, respectively, at T = 750°C and GHSV = 4500–5400 h^{-1}, representing absolute single-pass olefin yield increases of 12% and 40% over thermal cracking of each compound at identical conditions (**Figure 1**). CO_x yield as low as 1.7% was achieved along with >80% H_2 combustion over 25 redox cycles. Multiple characterizations including TEM-EDS and LEIS (**Figure 2**) show the existence of a Na_2WO_4 surface layer and a Na- and W-enriched shell around the $CaMnO_3$ core. Moreover, *in situ* XRD and TGA-DSC indicate that the Na_2WO_4 surface layer forms a molten shell covering the $CaMnO_3$ core under the working state of the redox catalyst. Na_2WO_4 played a key bifunctional role as (i) a catalytically active surface layer for hexane conversion and (ii)
2- -

an O /e conductor facilitating *in-situ* combustion of H_2 by $CaMnO_3$ lattice oxygen at the gas-molten shell interface.

Core-shell structured $Na_2WO_4/CaMnO_3$ was an effective redox catalyst for enhanced olefin production. Aspen Plus simulation of naphtha ROC demonstrated 52% energy savings and 50% reduction in CO_2 emissions on a per-ton ethylene basis compared to naphtha steam cracking.

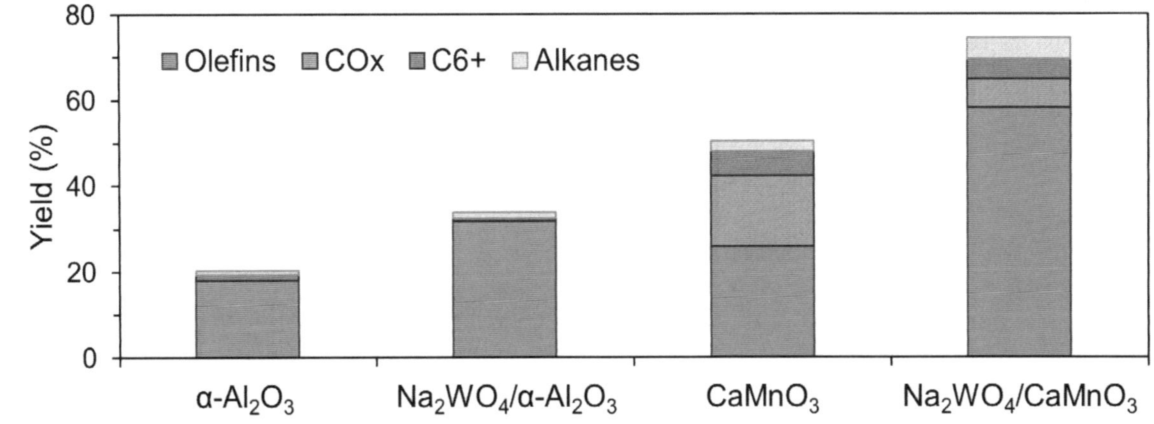

Figure 1. Reaction performance in cyclohexane ROC over four samples at 750°C.

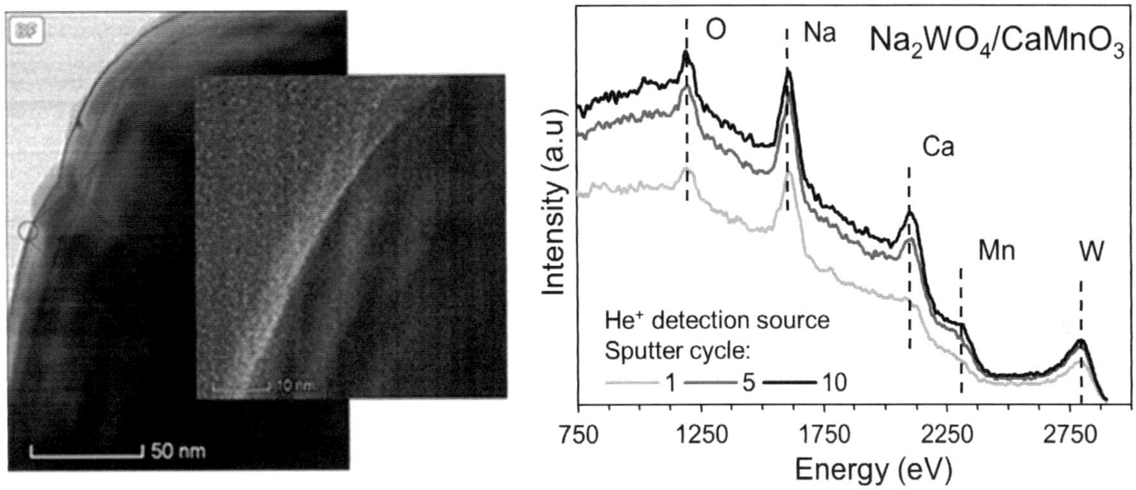

Figure 2. (a) BF-TEM and HR-TEM images of the $Na_2WO_4/CaMnO_3$ redox catalyst, showing a 4-5 nm surface layer, and; (b) LEIS spectra of same demonstrating Na and W surface enrichment.

606187 A Demonstration of Membranes for the Separation of Light Olefins and Paraffins

Abstract:

The separation of light olefins and paraffins is one of the most energy-intensive separations in petrochemical and oil and gas operations. The current state of the art technology is distillation, a decades-old capital and energy-intensive process. Compact Membrane Systems (CMS) has developed a membrane to preferentially permeate light olefins over light paraffins to allow for efficient and modular separation of these compounds. This membrane is a facilitated transport membrane where silver has been incorporated into a fluoropolymer combining the selective interaction of the silver with the olefin and the chemical resistance of the fluoropolymer for membrane longevity.

Previous testing in collaboration with Dow Chemical in their laboratory facility has demonstrated successful separation of light olefins and paraffins. This talk will discuss the latest updates on the field installation of this technology at the Freeport, Texas Dow Chemical facility. This installation will upgrade a feed stream of ~60% propylene to 90% or greater propylene. In addition to updates on the field installation, we will discuss the efforts to scale up membrane production to greater than 20ft2. These larger modules are being tested on a recycle rig at CMS and results from that testing will be shown. Thus far, 8 days of continuous run at operating conditions of ~160 psig and 45C have been demonstrated.

609129 Compact Delivery of UHP Hydrogen from Ammonia Decomposition Using Catalytic Membrane Reactors

Abstract:

The deployment of fuel cell electric vehicles is constrained by the paucity of hydrogen fueling stations and price, which is dominated by the costs of hydrogen storage and transportation. With more hydrogen per volume than liquid H_2 and an extensive distribution infrastructure in place, ammonia is a promising vector for efficient hydrogen distribution. In this talk, we describe the development of catalytic membrane reactor (CMR) technology for process-intensified delivery of high purity H_2 from ammonia cracking. The CMR integrates state-of-the-art catalysts with thin Pd –based metal membranes. We have developed a detailed axisymmetric model that captures the reactor performance with high fidelity, and this is used for the analysis and design of these systems. High volumetric productivity is obtained by combining highly active catalysts with an elevated transmembrane driving force to maximize conversion and hydrogen recovery, respectively. Ammonia decomposition rate is inhibited by both H_2 adsorption and reverse reaction, but the efficient removal of H_2 in a CMR enables temperature reductions by up to 200 °C and decreases catalyst loading requirements by a factor of 2. Commercial zeolite adsorbents are identified that reduce the ammonia in the hydrogen permeate stream from ~1000 ppm to below detectable limits (sub ppm) as required by commercial fuel cells.

611239 Ethylene Production Using Oxidative Dehydrogenation: Effects of Combined Membrane-Based & Alternative Distillation Separation Technologies on Process Safety & Economics

Abstract:

Previous work from this group utilized a HYSYS simulator to predict the performance of an ethylene oxidative dehydrogenation (ODH) production facility using alternative diluents and membrane separation technology not currently used in the standard industrial process [1-3]. An ODH process system has been previously modeled with ethane being fed along with air into a reactor with simultaneous mixing with pure methane to lower the product gas temperature. The methane feed was later separated from the other product gases and eventually fed into a recycle stream. This process was fairly expensive due to the use of a demethanizer within the process recycle stream. Safety and operational improvements for the ODH process was another aspect of this analysis that yielded interesting results. By the inclusion of storage vessels before and after the ODH reactor, the reactor unit can operate even in the event of an upstream or downstream failure, resulting in less downtime for the process.

The current work investigated the development of membrane-based separation in combination with alternative distillation technologies, such as bottom-flash, heat-pump assisted and low-temperature distillation, for the production of ethylene for the oxidative dehydrogenation (ODH) reaction of ethane using a HYSYS simulator with MEMCAL extension software in concert with economic and safety analyses using Dow Fire & Explosion Index, Mond Index and the Inherent Safety Index methods. The application of combinations of polymer-membrane separation with alternative distillation technologies were simulated and evaluated individually and in-tandem with membrane separation. The effects of alternative diluents, process flammability issues, and economic analysis results will also be discussed.

Literature Cited:

1.Maffia GJ, Gaffney AM, Mason OM, Techno-economic analysis of oxidative dehydrogenation options, Top. Catal. (2016), http://dx.doi.org/10.1007/ s11244-016-0677-9 (2016, August 16).

2.Baroi C, Gaffney AM, Fushimi R, Process economics and safety considerations for the oxidative dehydrogenation of ethane using M1 catalyst, Catal. Today (2017), http://dx.doi.org/10.1016/j.cattod.2017.05.041 R (2017, May 11).

3.Gao, Y., Neal, L., Ding, D., Wu, W., Baroi, C., Gaffney, A.M., Li, F., Recent Advances in Intensified Ethylene Production: A Review, ACS Catalysis (2019), http://10.1021/acscatal.9b02922 (2019, August 12).

604021 Microfluidics-Based Biocatalyst-Product Separation for Continuous Enzymatic Processes. a Case Study on β-Lactam Antibiotics Manufacturing

Abstract:

Immobilizing an enzyme on a support often is essential for downstream separation of the biocatalyst from a product and retaining the valuable enzyme in many enzymatic processes. This is particularly true in the case of continuous enzymatic manufacturing processes where the enzyme needs either to be retained in the reactor or separated from the product downstream and recycled back to the reaction vessel. This becomes challenging when the product is in solid form, e.g., crystals, necessitating the separation of two types of solid particles, biocatalyst and the crystalline product. This is the case for the continuous synthesis and crystallization of β-lactam antibiotics using the enzyme Penicillin G acylase (PGA) [1].

In this contribution, we propose a 3D-printed microfluidics-based device for downstream separation of the biocatalyst from the crystalline antibiotic product. Microfluidics is a technique characterized by the precise manipulation of fluid dynamics at the microscale and has shown a great promise for industrial processing applications. Among microfluidic devices, inertial microfluidics has experienced massive growth due to their high throughput, automation, operational simplicity, and low cost [2]. In such a device, particles with different sizes are separated by taking advantage of inherent inertial lift and drag forces. Inertial microfluidics can be employed by various channel shapes and structures; spiral channels are frequently used for the aim of high-throughput particle focusing by reduction of lateral migration positions and they can be an ideal candidate for this particular application. Here, we focus on the separation of high aspect ratio crystals from spherical biocatalyst particles. The efficiency of the separation for different product and catalyst sizes, and potentials/challenges for scaling up the proposed separation strategy will be discussed.

[1] McDonald, M. A., Bommarius, A. S., Grover, M. A., & Rousseau, R. W. (2019). Continuous reactive crystallization of β-lactam antibiotics catalyzed by penicillin G acylase. Part II: Case study on ampicillin and product purity. *Computers & Chemical Engineering*, *126*, 332-341.

[2] Moloudi, R., Oh, S., Yang, C., Teo, K. L., Lam, A. T. L., Warkiani, M. E., & Naing, M. W. (2018). Inertial-based filtration method for removal of microcarriers from mesenchymal stem cell suspensions. *Scientific reports*, *8*(1), 1-10

605697 Novel Integration of Continuous Crystallization Platforms for Control of Crystal Size Distribution and Polymorphic Form

Abstract:

Continuous crystallization has been in the focus of pharmaceutical industry as an appealing method to obtain drug substances of consistent critical quality attributes (CQAs), which meet strict regulatory guidelines. Traditionally, continuous crystallization is done in mixed suspension mixed product removal (MSMPR) crystallizers, but one main disadvantage of MSMPR crystallizers is the broad crystal size distribution (CSD) generated due to broad residence time distribution (RTD).[1] On the other hand, the new emerging continuous oscillatory baffled crystallizers (COBCs) offer plug-flow conditions with narrow RTD properties to generate product with narrow CSD.[2] Unlike traditional plug flow crystallizers that use high volumetric flow rate to achieve plug flow conditions, the COBCs use an oscillating piston to generate turbulent mixing within its concentric baffles without a high volumetric flow rate. However, a main disadvantage of COBCs is fouling and encrustation due to nucleation on the crystallizer walls which can lead to clogging and process failure.[3] To remedy fouling and prolong the operating time of the COBC, seeding is a common methodology to not only reduce the supersaturation in the system for nucleation but also provide polymorphic form control. Although seeding is very common method in the pharmaceutical industry, seeding for continuous crystallization has proven to be challenging, especially in the case of the COBC. Consistent manual preparation of seeds is labor intensive and economically infeasible, and, furthermore, any fluctuations of seed quality can impact startup dynamics.[4] In addition, seeds can be washed away after a few residence times, which no longer provide the benefits of seeding.

In this work, a novel integration of the MSMPR crystallizer with a COBC demonstrated in-situ seed generation capabilities while demonstrating CSD and polymorphic control in continuous combined cooling antisolvent crystallization (CCAC). Firstly, the 3D solubility surfaces of the metastable and stable form were generated to provide an operating regime for the continuous CCAC. It was discovered that the continuous generation of the metastable form required higher concentration of solvent than that of the stable form. Secondly, the MSMPR crystallizer was integrated into the COBC as a separate temperature zone to generate seeds via primary nucleation. The seeds generated were then fed into the COBC without pump transfer to allow for further growth. While compared to a single stage MSMPR crystallizer, the integrated system showed little signs of fouling and encrustation while generating product of uniform CSD. While both platforms are capable of polymorphic control using a temperature switch, the integrated system was able to maintain the narrower CSD while compared to the MSMPR, which makes this combined system a novel next generation continuous crystallization platform for systems with polymorphism and seeding requirements.

References:

(1) Ferguson, S.; Morris, G.; Hao, H.; Barrett, M.; Glennon, B. Characterization of the Anti-Solvent Batch, Plug Flow and MSMPR Crystallization of Benzoic Acid. *Chem. Eng. Sci.* **2013**, *104*, 44–54. https://doi.org/10.1016/j.ces.2013.09.006.

(2) Oliva, J. A.; Pal, K.; Barton, A.; Firth, P.; Nagy, Z. K. Experimental Investigation of the Effect of Scale-up on Mixing Efficiency in Oscillatory Flow Baffled Reactors (OFBR) Using Principal

Component Based Image Analysis as a Novel Noninvasive Residence Time Distribution Measurement Approach. *Chem. Eng. J.* **2018**, *351* (June), 498–505. https://doi.org/10.1016/j.cej.2018.06.029.

(3) Pena, R.; Oliva, J. A.; Burcham, C. L.; Jarmer, D. J.; Nagy, Z. K. Process Intensification through Continuous Spherical Crystallization Using an Oscillatory Flow Baffled Crystallizer. *Cryst. Growth Des.* **2017**, *17* (9), 4776–4784. https://doi.org/10.1021/acs.cgd.7b00731.

(4) Wood, B.; Girard, K. P.; Polster, C. S.; Croker, D. M. Progress to Date in the Design and Operation of Continuous Crystallization Processes for Pharmaceutical Applications. *Org. Process Res. Dev.* **2019**, *23*, 122–144. https://doi.org/10.1021/acs.oprd.8b00319.

606175 A Scalable Membrane Pervaporation Approach for Continuous Flow Ring Closing Metathesis

Abstract:

Pervaporation membranes can be used to avoid azeotropic distillation, enable flow chemistry technology, degas organic solvents, dehydrate solvents, drive reactions towards product formation, and more. Compact Membrane Systems (CMS) has demonstrated the effectiveness of their fluoropolymer pervaporation membranes to degas and dewater organic solvents for the pharmaceutical and specialty chemical industries in recent years. These membranes are applicable to a wide variety of solvents and operating conditions including difficult solvents such as NMP, DMSO or THF. In this talk, CMS will describe their newest pervaporation case study, performed in collaboration with Snapdragon Chemistry. A CMS membrane was used to selectively remove ethylene from a ring closing metathesis reaction in toluene to drive product formation and enable the use of flow chemistry.

Using traditional batch techniques, implementing olefin metathesis reactions on the commercial scale has proven challenging, if not impossible. Ring closing reaction mechanisms are commonly used during the preparation of small molecules for the pharmaceutical and fragrance industries. Membranes are a modular, scalable solution that can continuously remove the ethylene byproduct and enable the use of commercial scale continuous ring closing metathesis. This case study details the advantage of using a sheet-in-frame membrane reactor as opposed to a stainless steel tubular reactor. The fluoropolymer membrane has excellent chemical and thermal stability which allows for an extended range of thermal conditions without sacrificing selectivity or flux. Pure ethylene flux of the membrane was also investigated to ensure that the mass transport across the membrane would always be greater than the reaction's ethylene production rate. Scale up considerations including residence time, required ethylene removal rate and module design will be addressed. The feasibility of using a membrane reactor to remove byproducts shown in this case study is only one example of the reactions that could benefit from the implementation of membrane reactors.

604827 Development of Continuous Homogeneous Hydrogenation Process for Pharmaceuticals Using a Flow Reactor

Abstract:

In past decade there has been significant interest in the development of continuous processes in the pharmaceutical industry. Key drivers for continuous processing are product quality improvement, yield and process robustness, footprint reduction of processing area and increased process safety. Hydrogenations are commonly used in synthesis of pharmaceutical intermediates and active pharmaceutical ingredients. Types of hydrogenations include removal of protecting groups such as benzyl groups, reduction of nitro compounds and asymmetric hydrogenations. In this work we have demonstrated:

- Application of a laboratory scale tubular reactor for development and scale-up of a continuous asymmetric hydrogenations process.
- The effect of operating parameters such as pressure, gas to liquid ratio, temperature, catalyst loading and residence time of the substrate on enantioselectivity and product quality, are studied to develop a scalable and robust process.

A 'Fit for purpose' model was developed and found to be very effective in efficiently identifying the optimal reaction conditions, assessing risk to product CQAs, and increasing overall robustness.

Disclosures:

All authors are employees of AbbVie and may own AbbVie stock. AbbVie sponsored and funded the study; contributed to the design; participated in the collection, analysis, and interpretation of data, and in writing, reviewing, and approval of the final publication.

593483 Liquid Phase Peptide Synthesis Via Nanostar-Sieving

Abstract:

Peptides, an important class of polymers, mainly used for therapeutic purposes, are generally synthesised using solid phase methods. While the rapid synthesis characteristic of solid phase peptide synthesis (SPPS) benefits the drug discovery, this technology faces challenges for large scale synthesis, where its non-quantitative coupling leads to error sequences. In contrast liquid phase peptide synthesis (LPPS), offers high peptide purity and scalability but its development is hampered by inefficient intermediate separations.

Liquid phase peptide synthesis via nanostar-sieving (LPPS-NS) is a platform which synthesises peptides in solution with facile intermediate separations. Amino acids (AA) are coupled iteratively onto a 3-armed, star-shaped macromolecule, forming peptide-nanostar intermediates. After coupling, the unreacted AA is quenched and subsequently proceeded to N-terminal deprotection. The bulky intermediates are then 'sieved' out from the debris and quenched AA all together via organic solvent nanofiltration (OSN) thus omitting the post-coupling isolation step. This synthetic cycle is repeated until the desired peptide length is achieved, as shown in Figure 1. Standard Fmoc peptide chemistry is applied throughout the synthetic cycle. The use of nanostar greatly enhances the molecular sieving efficiency by the >3-fold mass difference between the nanostar and the unreacted building blocks. Most importantly, real-time reaction monitoring can be undertaken by LC-MS with high accuracy because of the monodispersity of the nanostar. OSN plays a pivotal role in synthesis efficiency. A solvent resistant membrane made of crosslinked polybenzimidazole (PBI) polymer was chosen for OSN. The polymer-based membrane was proven to be durable and to have a high separation factor which remained consistent throughout many synthesis cycles. We speculate that this new concept of LPPS is a truly one-pot synthesis where no transfer of liquids between cycles occurs and has a good potential for full-automation.

In this work we demonstrate the successful synthesis of Enkephalin-type model peptides (~5-10 mers) via nanostar-sieving technology. The products are of higher purity than, or at least comparable to, peptides produced by a reliable solid phase vendor, while using less equivalents of AA during coupling. To further validate this technology, high purity linear Octreotide (8AA) was synthesised. Our ambition is to develop this platform into a robust technology for large-scale, fully automated and high purity peptide synthesis.

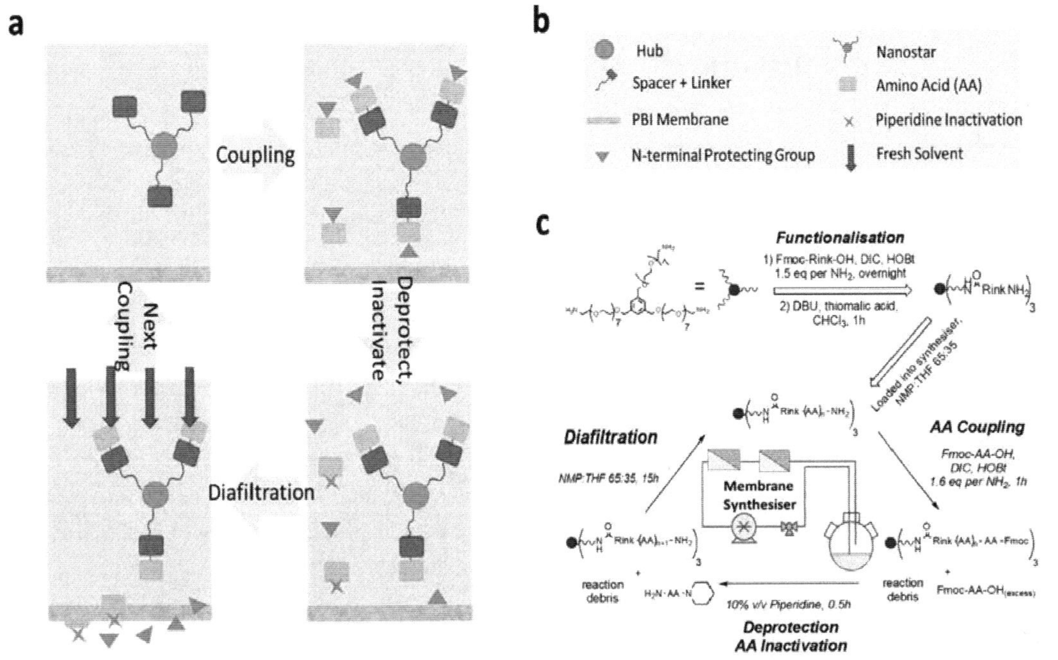

Figure 1: ***(a)*** *Flow diagram of Liquid Phase Peptide Synthesis via Nanostar-sieving (LPPS-NS) and* ***(b)*** *diagram legend.* ***(c)*** *The synthetic cycle of LPPS-NS. Prior to peptide synthesis, the star-shaped and 3-armed nanostar is functionalised with rink-amide linker. The functionalised nanostar then undergoes repeated cycle of coupling, deprotection / AA inactivation and diafiltration in the membrane synthesiser until target peptide length is reached. No liquid transfers between cycles ie. genuine one-pot synthesis.*

600743 Liquid-Liquid Extractions in API Manufacturing Processes: Is Continuous the Way to Go?

Abstract:

Extraction and separation operations are present in most API processes, in different stages, and with different purposes. Both unit operations are traditionally carried out under batch mode in multi-purpose reactors and consist in two/three sequential steps of mixing immiscible liquids with different densities and then separate them by gravity, which results in a time-consuming stage. Moreover, sometimes emulsions are obtained leading to difficulties in phase separation with impact in cycle time, yield and, potentially, product quality. Another bottleneck from these operations is related to the reactor occupancy at this stage, considering that usually one phase is carried forward and the other is discarded, which results in a limitation of batch size and production capacity.

To tackle all these operation deficiencies, continuous manufacturing is an attractive approach. The main drivers associated when going from batch to continuous highlight the opportunity to improve reproducibility, capacity, faster development, among others. For extraction and separation processes, several alternatives are available in the market with different characteristics and advantages.

In this work, we demonstrate our approach in selecting technologies for continuous process of extraction and separation, and case studies with different purposes such as pH adjustments, product/API transfer between phases and washings. The examples will show characteristics of the selected technologies (membrane based separator and centrifugal extractor) and types of liquid-liquid extraction and separation process evaluated along with the impacts resulting from going from batch to continuous manufacturing.

608996 Development and Process Intensification of a Three Step Continuous Flow Process for Telescoped High Energy Reactions

Abstract:

Continuous manufacturing (CM) offers significant advantages over traditional batch manufacturing for chemistries that generate or require high energy (usually in the form of heat). The comparatively large surface-area-to-volume ratio of flow reactors makes them particularly attractive for chemical reactions such as lithiation-carboxylation, nitro reduction, debenzylation, and thermal cyclization. Continuous processing also allows the reaction rates for these transformations to be orders of magnitude higher than would be possible with batch manufacturing. Additionally, performing these reactions in flow allows them to be telescoped to reduce operator handling and processing downtime.

This presentation will focus on development and process intensification of a three step continuous flow process toward synthesis of an active pharmaceutical ingredient (API). It will detail how construction of fit-for-purpose continuous reactors improves safety, cost, and process mass intensity while enabling use of preferred reagents and unlocking otherwise unattainable operating conditions.

601987 Producing Predictive Chemical Process Models from a State of Incomplete Mechanistic Knowledge and Small Data Sets

Abstract:

ABSTRACT: In this work, we explore the possibility of developing a quantitative and predictive model for chemical processes, without having to first gain a complete understanding of the underlying mechanisms. We find that this can be done efficiently combining: (1) expectations for the functional expression of the process—extracted from a general, *qualitative* understanding on how the chemical process ought to behave—and, (2) small data sets providing examples of the process outputs resulting from varying process inputs. The strategy discussed is labeled knowledge-constrained machine learning. The main example we take is on reaction kinetic models—and, here, the expectations are written in terms of constraints on the model predictions over time, and the data is a series of time-point measurements for the reaction outcome under different conditions. But, the intent of the presentation is to discuss a general methodology that may be used to develop predictive models across the range of unit operations found in the processes used to produce synthetic drug substance API's.

594384 Identifying the Processing Space of Continuous Granulation: A Case Study of Extended Release Tablets

Abstract:

While continuous manufacturing (CM) has been employed for directly compressible tablets, the continuous wet granulation has also been used as an alternative in continuous tableting lines. To date, no matrix-based extended release tablets manufactured by a CM technology has been approved. Compared to batch granulation, continuous twin screw granulation exerts higher shear to produce denser granules with broader size distribution within very short time (seconds). Changing the screw configuration, process, and formulation parameters can alter the physical properties of the granules, e.g., shape, flow, compressibility, density, size distribution, etc. In twin screw granulation, complex interactions among these parameters manage the granule properties. This study aimed at identifying the critical material attributes and process parameters and their interactions to be controlled for switching a batch process to a CM granulation process.

Using batch high shear granulation, we evaluated the formulation and process parameters that can produce metoprolol succinate extended release tablets. These formulation and process parameters were successfully transferred to continuous twin screw granulation to produce the same dissolution properties of the finished product. The effects of the screw configuration parameters on the characteristics of the granules were further investigated. The interaction between the processing, e.g., liquid to solid ratio, screw speed, barrel fill volume, etc., and the significant screw parameters was studies to unveil the processing space and to identify the failure modes.

Similar granule quality attributes were obtained for most of the formulations; although, the conditioned bulk density and compressibility of the granules determined the dissolution properties of the tablets. The number of kneading and sizing elements had the most significant impact on granule characteristics. The effects of the staggering angle and distance between the kneading zone of the screws were negligible.

This presentation will then highlight the processing space of screw granulation to produce extended release tablets. Some of the failure modes will be discussed along with the feasibility of implementing PAT tools to monitor the granulation process and granules properties.

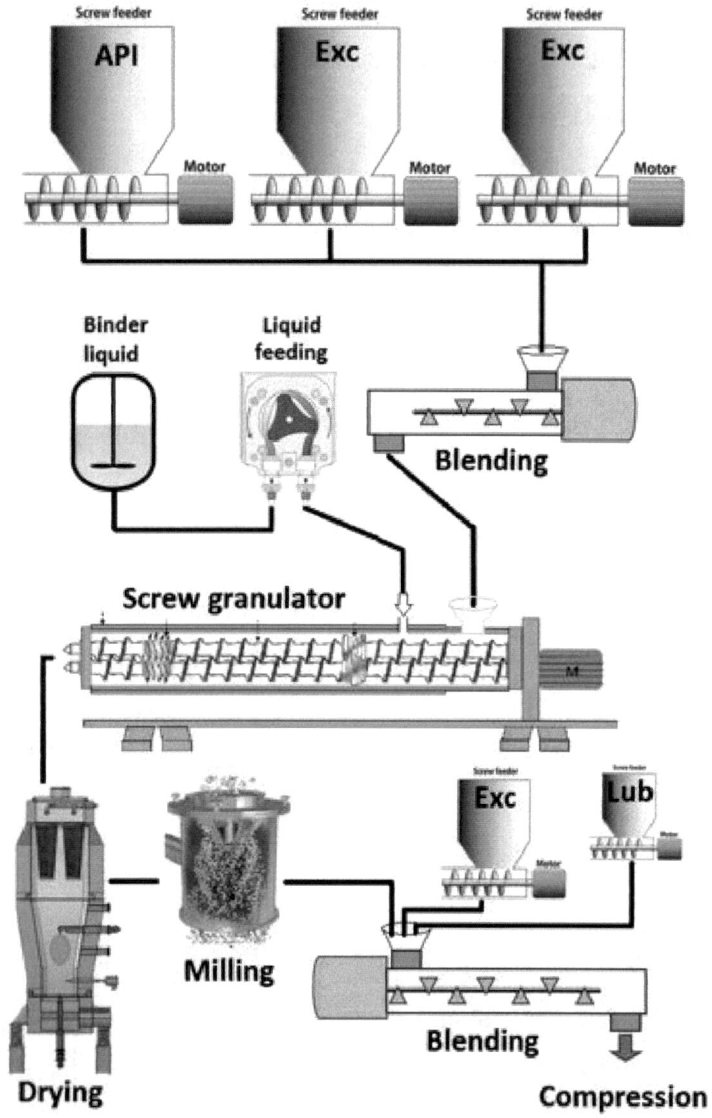
Screw feeder
API
Screw feeder
Exc
Screw feeder
Exc
Motor
Motor
Motor
Binder liquid
Liquid feeding
Blending
Screw granulator
M
Exc
Lub
Milling
Blending
Drying
Compression

608665 Dynamic Optimization Strategies Towards Addressing the Challenges in Protein Therapeutics

Abstract:

Protein therapeutics is an important field of the pharmaceutical industry. Current challenges in this field include meeting the increasing market demands, handling the entrance of biosimilar products and complying with quality by design (QbD) principles [1-3]. To tackle these challenges effectively, efficient use of existing tools, methods and solution approaches together with the development of novel ones is necessitated. This work provides an overview of our recent efforts to address some of the aforementioned challenges using model-based dynamic optimization strategies.

Model-based approaches are powerful tools towards facilitating process development and advancing process performance. They offer a promising alternative to costly and time-consuming experimentation [1], and can have a profound effect on process sustainability and profitability. Recent advances in cell culture modeling (with cell cultures representing a typical production platform for many of the products of this class) have enabled significant insight to the system's dynamics and the underlying production mechanisms, and have paved the way towards developing dynamic optimization strategies for process intensification (cf., [4-6]).

Existing numerical methods and optimization software for solving the resulting dynamic optimization problems rely almost exclusively on local techniques. Due to the noncovexity of the described complex biological systems, the vast majority of these problems exhibits the presence of multiple local minima. Nevertheless, deriving suboptimal local solutions can have direct quality and cost impacts. Despite the significant progress in deterministic global optimization [7], it is not yet tractable for the considered systems. Hence, alternative methods to overcome the hazards of suboptimal local optimality are required. Stochastic and heuristic approaches can be advantageous to this end. In our previous work [8], we have shown how incorporating biological process knowledge into the optimization problem formulation can facilitate the derivation of superior local solutions. In a similar direction, we illustrate how proper decomposition of the optimization problem can outweigh certain advantages [9].

Beyond increasing product yield, monitoring and controlling product quality is essential. Regulatory authorities including FDA are increasingly recommending consideration of QbD to develop processes that consistently meet quality specifications [3, 10]. This requires further attempts in modeling, as well as reconsiderations in the formulation of the dynamic optimization problem to effectively account for direct quality attributes. Yet, current methods in this direction are still in an early stage. The recent work in [11] demonstrates a successful example of model-based optimization to maximize galactosylation (a glycosylation related metric, with glycosylation presenting a critical quality attribute for many therapeutic proteins [12, 13]) content for an antibody case study. In our recent study [14], we have combined model simulations and design of experiments to investigate the effects of nutrients and nucleotide sugars feeding in protein glycosylation. Building on this basis, we have developed an optimization framework to maximize process performance while meeting quality specifications, and thus provided a successful example for utilizing dynamic optimization to apply the QbD paradigm in protein therapeutics [15].

Acknowledgements:
This work has received funding from the European Union's Horizon 2020 research and innovation programme under the Marie Skłodowska-Curie grant agreement no.675251.

References:

[1] Farzan, P., Mistry, B., & Ierapetritou, M. G. (2017). Review of the important challenges and opportunities related to modeling of mammalian cell bioreactors. AIChE Journal, 63(2), 398-408.
[2] Shukla, A. A., Wolfe, L. S., Mostafa, S. S., & Norman, C. (2017). Evolving trends in mAb production processes. Bioengineering & Translational Medicine, 2(1), 58-69.
[3] Collins, P. C. (2018). Chemical engineering and the culmination of quality by design in pharmaceuticals. AIChE Journal, 64(5), 1502-1510.
[4] Dhir, S., Morrow Jr, K. J., Rhinehart, R. R., & Wiesner, T. (2000). Dynamic optimization of hybridoma growth in a fed-batch bioreactor. Biotechnology and Bioengineering, 67(2), 197-205.
[5] Kiparissides, A., Pistikopoulos, E. N., & Mantalaris, A. (2015). On the model-based optimization of secreting mammalian cell (GS-NS0) cultures. Biotechnology and Bioengineering, 112(3), 536-548.
[6] Kappatou, C. D., Mhamdi, A., Campano, A. Q., Mantalaris, A., & Mitsos, A. (2018). Model-based dynamic optimization of monoclonal antibodies production in semibatch operation—Use of reformulation techniques. Industrial & Engineering Chemistry Research, 57(30), 9915-9924.
[7] Chachuat, B., Singer, A. B., & Barton, P. I. (2006). Global methods for dynamic optimization and mixed-integer dynamic optimization. Industrial & Engineering Chemistry Research, 45(25):8373–8392.
[8] Kappatou, C. D., Mhamdi, A., Campano, A. Q., Mantalaris, A., & Mitsos, A. (2017). Dynamic optimization of the production of monoclonal antibodies in semi-batch operation. In Computer Aided Chemical Engineering (Vol. 40, pp. 2161-2166). Elsevier.
[9] Kappatou, C. D., Altunok, O., Mhamdi, A., Mantalaris, A., & Mitsos, A. (2019). Sequential and Simultaneous Optimization Strategies for Increased Production of Monoclonal Antibodies. In Computer Aided Chemical Engineering (Vol. 46, pp. 1021-1026). Elsevier.
[10] Pais, D. A., Carrondo, M. J., Alves, P. M., & Teixeira, A. P. (2014). Towards real-time monitoring of therapeutic protein quality in mammalian cell processes. Current Opinion in Biotechnology, 30, 161-167.
[11] Kotidis, P., Jedrzejewski, P., Sou, S. N., Sellick, C., Polizzi, K., del Val, I. J., & Kontoravdi, C. (2019). Model-based optimization of antibody galactosylation in CHO cell culture. Biotechnology and Bioengineering, 116(7), 1612-1626.
[12] Hossler, P., Khattak, S. F., & Li, Z. J. (2009). Optimal and consistent protein glycosylation in mammalian cell culture. Glycobiology, 19(9), 936-949.
[13] Blondeel, E. J., & Aucoin, M. G. (2018). Supplementing glycosylation: A review of applying nucleotide-sugar precursors to growth medium to affect therapeutic recombinant protein glycoform distributions. Biotechnology Advances, 36(5), 1505-1523.
[14] Ehsani, A., Kappatou, C. D., Mhamdi, A., Mitsos, A., Schuppert, A., & Niedenfuehr, S. (2019). Towards model-based optimization for quality by design in biotherapeutics production. In Computer Aided Chemical Engineering (Vol. 46, pp. 25-30). Elsevier.
[15] Kappatou, C. D., Ehsani, A., Niedenführ, S., Mhamdi, A., Schuppert, A., & Mitsos, A. Quality-targeting dynamic optimization of monoclonal antibody production. (Submitted)

608965 Integrated Supply Chain Network Design and Inventory Management for Autologous Cell Therapy.

Abstract:

Personalized medicine is gaining a lot of attention in the pharmaceutical industry as it can treat diseases that earlier had no cure. In the past few years, we witnessed exponential growth in the development of cell and gene therapy products. Cancer Research Institute reported that 753 cancer cell therapies are currently under development all over the world, of which 375 are in clinical trials [1]. There has been an increase in the growth and demand for autologous therapy due to the recent approval from the FDA [2]. Therefore, autologous therapy is undergoing the transformation from laboratory to commercial scale.

The autologous therapy starts with a collection of cells from the patient, which are cryopreserved and sent to the manufacturing site. The cells are then formulated and enriched. The formulated cells are again cryopreserved and transferred back to the patient, where they are injected back. Unlike traditional pharmaceutical drugs that are produced in bulk, these drugs are patient-centric and have to be manufactured for each patient separately. In addition, the process also deals with living cells that go around different nodes in the supply chain. Therefore, the cells have to be cryopreserved in storage device and transported across the supply chain.

An integrated approach is considered for supply chain network design and inventory management of the raw materials. We consider a network with a set of hospitals where the patients arrive. The facilities to be set up are distribution centers and manufacturing centers. The manufacturing center receives the samples from the patient and engineers the cells to create the drug. The distribution centers have the inventory for the storage containers and are installed for the intermediate halt for the sample transportation between the hospital and manufacturing center. To account for the transit times for the material transfer between facilities, we define the parameter accounting for both transportation time and waiting time for different transportation modes as well as the processing times at each of the node. We also simultaneously optimize inventory policy. We use periodic-review policy for the raw materials inventory at the distribution center [3]. Under this policy, if the inventory is below the defined level, the order is placed to restore the inventory to level . Since the storage devices are reusable, the policy is modified to account for the returning devices after the delivery of the drug.

Given the demand distribution, the potential locations of distribution and manufacturing centers, costs and time for processing, inventory, and transportation, we determine the location of different centers, inventory levels at the distribution centers, response time, and cost of delivery for each sample, with objective function of minimizing the cost and response time. The problem is formulated as a single objective optimization where response time is multiplied by a factor and added to the cost function. A case study of China is considered, with 20 hospitals situated in different provinces and a total of 130 patients over a period of 1 month. Also, the location of 20 potential locations of the distribution centers and 20 manufacturing centers are given. We found out that the majority of the cost is incurred in setting up manufacturing and distribution centers. The cost accounting for the response time depends on the multiplicative factor we choose. In addition, we also explored the importance of batching the samples at the distribution centers and transporting to the manufacturing center subjected to the response time constraint. Decisions related to inventory of raw material are critical to have an efficient supply chain. Also, sensitivity analysis is conducted to study the effect of different parameters on the optimal network design and overall cost.

References:

1. Tang, J., et al., *The global landscape of cancer cell therapy.* Nat Rev Drug Discov, 2018. **17**(7): p. 465-466.
2. Bach, P.B., S.A. Giralt, and L.B. Saltz, *FDA approval of tisagenlecleucel: promise and complexities of a $475 000 cancer drug.* Jama, 2017. **318**(19): p. 1861-1862.
3. Brunaud, B., et al., *Inventory policies and safety stock optimization for supply chain planning.* AIChE Journal, 2019. **65**(1): p. 99-112.

600490 Two-Stage Control Using Model-Based Reinforcement Learning and Predictive Control for Fed-Batch Bioreactor

Abstract:

As the demand for biomass and pharmaceutical materials is increasing in the market, various genetically modified cells are tested in a batch bioreactor in order to increase production rate [1]. At the same time, a robust and efficient control strategy is required to successfully produce the product with uncertainty of cellular behaviors [2]. However, limitation of the information on the system dynamics makes it challenging problems to optimize the operating conditions of a bioreactor and to implement an automatic controller. In general, the biosystem is less reproducible and shows higher nonlinearity compared to other engineering systems. In particular, most of the parameters in the first-principle model are lumped type and these values would be varied not only in a single batch operation with changes of the states but in a batch-to-batch fashion. Therefore, implementing an automatic control to track the optimal control trajectories obtained off-line is not a robust approach to produce fine products for a large number of batch reactors.

In this work, we propose a two-stage control strategy to optimally control a fed-batch type bioreactor that produces penicillin. The proposed strategy utilizes model-based Reinforcement Learning (RL) off-line to optimize the operating trajectories of state and input. In on-line, the Moving Horizon Estimation (MHE) is applied to estimate the parameter values, and an adaptive Model Predictive Control (MPC) is utilized to simultaneously modify and track the trajectories obtained in off-line. First, the first-principle model of the system dynamics is represented in Stochastic Differential Equations (SDEs) form which could theoretically handle the white noise in a continuous system. In addition, the variance of the process noise could shift a mean value of drifting terms of the system so that the changes in parameter values could be presented by adjusting the variance of the process noise [3]. These SDEs are numerically integrated and served as a virtual plant. The next step is obtaining the operating trajectories of the reactor off-line. The optimal control method could be classified in three classes, direct, indirect and RL (dynamic programming-based method). The direct method is the one mostly applied in practice since the problem formulation and the procedure of obtaining the solution is intuitive. On the other hand, RL has several advantages over the direct method. RL provides a closed-loop solution of the optimal control problem so that the control policy could be obtained instead of just a control trajectory [4]. This is a more robust approach since RL evaluates the control cost for the case where the states have deviated from the optimal trajectories. In addition, value based RL provides not only the optimal trajectories but also the value of state which is an expected total cost of the control policy starting from the state [4]. Although this value is just a by-product to find the optimal control policy, it is useful information when the trajectories are modified on-line.

The trajectories suggested by RL could serve as a baseline of optimal control. However, due to the change in parameter values and disturbances, tracking those trajectories might not be feasible which requires a modification of trajectories. For this purpose, MHE and adaptive MPC could be utilized to estimate the parameters and modify the trajectories on-line. However, even though changes in parameter values are successfully estimated by MHE, the modification of trajectories by MPC for the entire horizon could not be feasible by the lack of computational time. This forces MPC to solve the optimal control problem with a limited length of prediction horizon where the terminal state of the bioreactor is not considered. This becomes a problem because most of the important cost is imposed only at the terminal state of the bioreactor. Therefore, providing a proper terminal cost of MPC becomes a crucial point to properly modify the trajectories. At this point, the value calculated from the RL could be used as the terminal cost of MPC. Although this value would not be exact by the changes of parameter values, the simulation results suggest that this approximation is valid for MPC to stably calculate the new trajectories.

The key to this combination of MPC and RL is selecting the proper length of the prediction horizon. Under this scheme, the length of the prediction horizon is not only determined by the computational limits, but it is related to the reliability of the parameter values. The long prediction horizon implies that MPC is more rely on the newly estimated parameter values than the one used for RL. The presence of the parameter changes is one support on using the longer prediction horizon. However, the reliability of the estimated parameter values is related to both the quality and quantity of measurement data. Therefore, the choice of prediction horizon length should be carefully determined by evaluating the reliability of the parameter values. We propose one method for evaluating such reliability and propose a method of adjusting the prediction horizon length.

[1] C. Larroche, Sanromán Maria Ángeles, G. Du, and A. Pandey, Current developments in biotechnology and bioengineering. Amsterdam: Elsevier, 2017.

[2] Shuler and Kargi, Bioprocess engineering. United States: Academic Internet Publishers, 2007.

[3] H.-H. Kuo, Introduction to stochastic integration. New York: Springer, 2006.

[4] R. S. Sutton, Reinforcement learning: an introduction. Cambridge, Mass: MIT Press, 2018.

608716 Sustainable API Development: Developing a Strategy for Measuring and Reducing Environmental Impact in the Pharmaceutical Industry

Abstract:

In January 2020, AstraZeneca announced its ambition to eliminate greenhouse gas emissions from its sites and fleet by 2025 and to become carbon negative across the entire value chain by 2030. The development and manufacture of active pharmaceutical ingredients (API) contributes significantly to AstraZeneca's carbon footprint and process engineers have a responsibility to play a leading role in achieving the zero-carbon target.

This presentation will share the approach taken to implement a strategy for measuring and reducing environmental impact during the development and manufacture of APIs. This approach comprises of influencing ways of working during API process development, implementing robust measurement of process environmental impact and identifying opportunities for innovation through science and technology to minimise AZ's carbon footprint.

This approach describes developing a procedure to appropriately consider sustainability alongside other key drivers for process design such as product quality, manufacturability and regulatory factors. Similarly, creating a culture to deliver continuous improvement against established metrics for sustainability in every aspect of API development and manufacturing.

This case also describes initiation of novel technology investigations to significantly reduce the carbon footprint of AstraZeneca's API manufacturing processes. This will be a leading factor in achieving a 50% reduction in carbon footprint for API development and the zero-carbon ambition of the company.

609303 An Integrated Process Automation System in API Manufacturing: Key Characteristics and Case Study

Abstract:

Safety is always first priority and a critical concern in the chemical industry. Jiangsu Provincial Emergency Management Administration (JSEMA) in China released <Basic Requirements for Diagnosis and Improvement of Instinct Safety> (Su Emer. NO.53) in June 2019, which specifies the requirements for the main aspects of chemical production. The concept of "integrated process automation" in Su Emer. No. 53 poses a great challenge to API manufacturing. It emphasizes mandatory safety and intensified automatic implementation. This paper illustrates the key characteristics in the integrated process automation of API manufacturing and gives some case studies.

Such automation system includes sophisticated solution and cutting-edge tools. According to the recipe and process of the product, the consultant optimizes the control strategic with expertise to enhance the safety and meet the regulation requirements. The process control modules or batch control program are deployed in DCS system for the purpose of multifunctional and flexible manufacturing. In case study it gives example of the mixing control strategic in the reactors which is critical and should consider all concerned factors. Usually the geared motor electricity is ignored to monitor, the absence may interfere the cooling system activation in time, its real time monitoring and interlock with cooling media are necessary; another example is the utility facilities configuration and interlock with automatic systems which should be tuned duly to eliminate the hidden trouble. The integrity of safety instrumented functions (SIFs) should be determined through risk assessment tools and rating tools, such as HAZOP (Hazard and Operability Analysis) and LOPA (Layer of Protection Analysis). The related case study reports sharing control components between SIS and DCS will increase the possibility of failure, which should be validated. The allocation of center control room and controller assignment should be concerned with HSE guideline. It also discusses the roadmap of information technology and MES (Manufacture Execution System) for API manufacturing. Our implementation shows that the integrated process automation system can help to reach instinct safety, which increase the productivity and improve the quality of products with regulatory compliance.

605455 A Quantitative Assay of Sodium Triacetoxyborohydride

Abstract:

Sodium triacetoxyborohydride (STAB) is a common reducing agent for reductive amination reactions. Due to sensitivities to water and air, the potency of this reagent tends to degrade over time. Traditional analytical assays for STAB fail to accurately assess the content of the active borohydride species, and often report assay purities of 97% or greater, when the true potency is lower. An incorrect or unknown potency of the STAB being used may lead to significant challenges when scaling up reductive aminations. An excess of STAB may be required and reactions may not achieve full conversion with the amount of STAB charged, requiring additional charges. The use of an excess of STAB comes with greater raw material costs and the thermal and H_2 evolution hazards associated with quenching the excess reagent.

This presentation outlines a simple assay, based on an aldehyde reduction, that was developed to determine the active borohydride content of STAB. The HPLC assay yield of a salicylaldehyde reduction has been shown to accurately determine this potency and has been validated against an H_2 evolution method, as well as yields obtained from a reductive amination.

607708 Delivery Strategies for Live Therapeutic Bacteria

Abstract:

The microbiome is a collection of bacteria, viruses and fungi that perform essential functions for human health. Dysbiosis, or the dysregulation of the microbiome, can contribute to chronic inflammation, metabolic disorders, neurological diseases, and cancer. Clinically, one of the only methods for altering the composition of the microbiome is with broad-spectrum antibiotics, which contribute to dysbiosis and create selective pressure for antibiotic resistant pathogens. As such, live therapeutic bacteria (LTB) have been identified as an attractive alternative to antibiotics, as they can modulate the composition of the microbiome without eliminating commensal species. However, LTBs are commonly delivered orally and experience significant challenges during gastrointestinal (GI) transit, including survival in the harsh conditions of the stomach, adherence to the intestinal lumen, and navigation of interpatient variability in the microbiome due to factors such as diet. Few engineered drug delivery platforms have been rationally developed to address these challenges. In this work, we describe two complementary platforms that separately adapt LTBs to improve viability during gastric transit and modify LTB surfaces to aid in adhesion along the GI tract. In the first system, we sought to understand the influence of factors such as diet on LTB adaptation and colonization. We delivered an LTB to female BALB/c mice on two distinct diets and found that diet significantly influenced both the appearance of a small colony variant of the LTB, as well as its distribution along the GI tract. As adaptation appeared to significantly affect LTB colonization, we developed an approach to control the adaptation of LTBs to specific environmental insults during GI transit. We found that growth in an *in vitro* adaptation media can prime LTBs for the acidic conditions of the stomach, significantly improving their viability during a challenge in simulated gastric fluid. Following transit through the stomach, LTBs must adhere to the intestinal lumen to persist and colonize in the GI tract. Therefore, for our second system, we developed a platform to chemically conjugate targeting ligands to the surface of LTBs to improve their adhesion to the GI mucosa. This platform significantly improves both the rate of colonization, as well as the concentration of LTBs in the GI tract of female BALB/c mice. Collectively, this work represents a simple approach to adapt LTBs to the intestinal environment during *in vitro* growth, while enabling subsequent modification of the LTB surface for improved intestinal adherence. Together, these delivery techniques for LTBs may help overcome challenges associated with GI transit, colonization and adaptation.

609030 In Vitro Biomimetic Tumor Model with Cancer-Specific Gene Regulators and Green Fluorescence Reporter for High Throughput Drug Screening

Abstract:

Although much scientific effort has been expended for cancer treatment, this disease is still one of the leading causes of death around the world. The limitations and shortcomings of current therapies highlight the urgent need for more effective therapeutic strategies. So far, most of the chemotherapy drugs act negatively on the cell division without selection, leading to serious adverse effect on normal cells. The current mainstream of anti-cancer drug development is to find the drugs that have specific targets (DNA, mRNA or proteins). Those targets are overexpressed in cancerous cells and are critical for carcinogenesis. In this study, we engineered breast and lung cancer cells to express the enhanced green fluorescent protein (EGFP) under cancer-specific gene promoters for three known carcinogenesis genes - survivin, Bcl-2 and hTERT, which enable online high-throughput screening of drugs that may up- or down-regulate these cancer genes as indicative of efficacy in inhibiting breast and lung cancers. Additionally, the mechanism of drugs and the crosstalk between cancer genes can be elucidated at the early stage of drug discovery. Another issue in the pharmaceutical industry is the high drug failure rate owing to poor predictability of early drug assays in a two-dimensional cell culture system. In such traditional assay, multiple cell-cell and cell-matrix interactions are deficient, leading to incomprehensive cellar responses to drug agents. To improve the drug assay predictability, our assay has focused on the construction of three-dimensional biomimetic cancer cell models with the recapitulation of multiple elements of tumor microenvironment. While developing the new drug screening assay, we also combined the rapid cell detection approach to achieve the high-throughput capability. Our optimized drug screening process can help to discover more anti-cancer therapies.

609939 Preparation and Use of Fine Grade Engineered Excipients for Direct Compression of Binary Blends of Cohesive Drug Powders

Abstract:

Novel, surface modified excipients are prepared using a fluid energy mill (FEM) to simultaneously mill and dry coat microcrystalline cellulose (MCC) with nano-silica to produce different sizes of fine, well flowing MCC excipient [1]. The purpose of these new class of excipients is to enable direct compression (DC). DC route is one of the most efficient and economic methods for tablet manufacturing since only blending and compression operations are required, greatly simplifies the process-train. However, the popularity of direct compression is limited by its stringent requirements, such as good flowability, high packing density, and excellent compaction for excipient-active pharmaceutical ingredient (API) blends. When the blend has high drug loading with a poorly flowing API, the flowability and tabletability of blends is generally decreased sharply. In a recent work, it has been shown that surprisingly, fine excipients of 20 mm size that have improved flow and bulk density after silica dry coating, can greatly improve direct compression performance [2]. Unfortunately, it could not work for very fine cohesive APIs such as 10 mm acetaminophen (APAP) at higher drug loading. In this work, binary blends of several fine sized dry coated MCC-based excipients are considered with three different fine, cohesive APIs at high drug loadings. Bulk density, FFC, and tablet compactability for all these blends are measured and compared to determine which excipient sizes provide the best overall performance. The results are analyzed for the relative sizes of the excipients and APIs to better understand the interplay between the components of the blends and assessment of the extent of silica transfer from coated excipient powders to API powders and its effect on blend bulk density and flowability. It is expected that using such novel surface engineered fine excipients in blends with fine APIs may also help in reduced segregation tendency during overall processing. Hence, these results are expected to provide industry relevant guidelines for the use of finer surface modified excipients in DC tableting.

596229 Applying Machine Learning to Predict Therapeutic Antibody Viscosity

Abstract:

A general machine learning feature selection protocol to predict antibody stabilities is proposed. These features only require information from sequences and molecular structures of antibodies. In addition, the physically relevant features are easily interpretable and help to design experiments to understand the underlying mechanisms of antibody stabilities. Two case studies are demonstrating using this protocol. First, we applied this framework to predict aggregation behaviors of 21 marketed monoclonal antibodies at high concentration (150 mg/mL), yielding an accuracy of 97% on validation tests with only three features. Models trained using structural features, rather than sequence features, correlated better with aggregation. Second, the concentration dependent viscosity behavior of 27 FDA approved antibody drugs was measured in concentrated solutions at high concentration (150 mg/mL). Combining molecular modeling and machine learning feature selection, we found that the net charge in the antibodies and the amino acid composition in the variable region are key features which govern the viscosity behavior. We presented two predictive models for aggregation and viscosity based on machine learning trained models, facilitating early stage design.

602107 Development of Multi-Adjuvant Systems for Vaccines Against Infectious Diseases

Abstract:

The ability to elicit potent humoral and cellular immune responses is a key element required for the development of efficacious prophylactic and therapeutic vaccine formulations. For vaccines against intracellular pathogens, the induction of a balanced, long-lived, and potent $CD4^+$ and $CD8^+$ T cell activation and proliferation is critical, combined with robust antibody responses. One strategy is to activate innate immunity and proinflammatory responses through Toll-like receptor-independent cytosolic pathways. Importantly, engagement of these mechanisms has demonstrated its ability to provide protection against bacterial challenge, and several small molecules have been evaluated in preclinical and clinical studies as vaccine adjuvants against infectious diseases.

Small molecules can bind to cytoplasmic receptors triggering signaling cascades that activate innate immunity. Despite their potential as adjuvants, small molecules have some limitations that need to be overcome. For example, there may be low exposure in target tissues due to their size and charge, and are prone to enzymatic biodegradation; with some off-target toxicities. In order to address these challenges, the use of drug delivery systems is being explored.

In this work, three different polymeric nanoparticle platforms were evaluated for small molecule delivery and for their capability to act synergistically as co-adjuvants and elicit robust immune responses. In these studies, nanoparticles based on natural and synthetic polymers were fabricated using water/oil/water and oil/water emulsions. Particle characterization was performed using dynamic light scattering, scanning electron microscopy, and zeta potential measurements that confirmed the size, morphology, and charge of the resulting nanoparticles. Small molecule compounds were loaded into each of the platforms, and their loading and release efficiency were determined using liquid chromatography. Finally, these systems were evaluated using *in vivo* murine models following parenteral administration to assess their ability to induce cytokine secretion, antibody and cell-mediated immunity. The results obtained demonstrate the potential of polymeric particulate systems as dual delivery and co-adjuvant platforms for the development of small molecule-adjuvanted prophylactic and therapeutic vaccines.

603577 Development of a Harvest Small Scale Model to Simulate Continuous Centrifugation

Abstract:

Disk stack centrifugation is one of the main clarification steps used in large scale mammalian cell culture processes. During process development, pilot scale disc stack centrifuge is normally used to develop and characterize the harvest process for new cell culture processes due to the lack of smaller scale model for disc stack centrifugation. In order to increase speed of development, a small scale model (SSM) of a disk stack centrifuge is of high demand to enable harvest development in an earlier stage of process development. A small scale model can also improve throughput and reduce cost. In this study, we have developed a SSM for disc stack centrifugation using a capillary shear device combined with benchtop centrifugation. Capillary tube size, length, and flow rate were optimized to generate energy dissipation rates comparable to large scale disk stack centrifuges. Additionally, shear force from multiple centrifuges at different scales were standardized and compared. Mathematical correlations were established for energy dissipation, viable cell density, cell viability, and lactate dehydrogenase (LDH) activity. Our SSM offers an approach to understand how the separation process, such as centrifuge designs, sizes, and configurations affect filtration performance and product quality. This work can speed up harvest development during process scale-up in biopharmaceutical mammalian cell culture processes.

599615 Optimizing and Reducing Cleaning Production Loss in Pharma, Agrochemical and Specialty Chemical Companies Using a New Expert System.

Abstract:

Pharmaceutical, Biopharmaceutical, Agrochemical and Specialty chemical companies make many different products in their plants. This is true both for the active ingredient stages as well as the secondary of formulation processing. Cleaning and set up between products is critical and reduce productivity time up to 50% in some cases. The presentation will show the critical areas important in the modelling development in this expert system and how it will aid companies in scheduling their productions to reduce cleaning time and to reduce issues with cleaning.

599656 Revolutionize of Pharmaceutical Product Changeovers By Enhanced Cleaning with Microbubbles

Abstract:

Pharmaceutical, Biopharmaceutical, Agrochemical and Specialty chemical companies make many different products in their plants. This is true both for the active ingredient stages as well as the secondary of formulation processing. Cleaning and set up between products is critical and reduce productivity time up to 50% in some cases. Microbubbles have been shown to enhance many unit operations e.g. distillation. The benefits of microbubbles in cleaning will be presented along with the impact this will enable in regulatory/ registration processes.

613297 Keynote: Cybersecurity for Critical Infrastructure Systems

Abstract:

Industrial control systems (ICS) are used throughout the chemical sector for activities such as auditing, quality control, chemical mixing and process control systems, and safety systems. Cyber-attacks on ICS at chemical manufacturing plants could cause not only a disruption to chemical supply, but equipment damage, environmental effects, and health and safety impacts. Cyber-attack consequences can include improper mixing, over-pressurization, overheating, or lack of monitoring on the system. Manual overrides can ensure system safety, but an override of control could lead to a plant operating in unsafe conditions. The solution for many cyber-attack vectors is system shutdown, which can cause missed production quotas and economic loss to operators.

Our cyber defense strategy heavily depends on the level of sophistication of the adversaries we are defending against. As part of a national laboratory, the Cyber and Infrastructure Resilience program at LLNL is focused on defending our nation's critical infrastructure against highly sophisticated nation state adversaries. This particular threat requires a layered defense approach, where each layer increases the difficulty for the attacker by impeding their ability to execute the attack successfully. Our research focus can be mapped to the NIST Cybersecurity framework. As part of this talk, we will present representative capabilities and research efforts focused on each step of the framework specifically designed for industrial control system cybersecurity including: mapping and modeling our Operational Technology (OT) networks, understanding communication pathways and potential attack vectors and entry points for an attacker, analyzing system level impacts of cyber attacks on ICS, advanced detection of vulnerabilities in the supply chain, real-time anomaly detection of cyber attacks, and enhancing the resilience of our systems allowing for operation through compromise. These capabilities are broadly applicable across infrastructure sectors and their deployment can significantly enhance the cybersecurity of the chemical sector against sophisticated cyber attacks.

613296 Keynote: Integrating Cybersecurity into PHA and PSM

Abstract:

As operational and information technologies converge to allow for remote and real-time access to plant operating data and control functions, the process industry could become increasingly susceptible to cyber-attacks. Traditional hazard and risk analysis methods appear inadequate to identify, prevent, and mitigate such attacks. As plant technologies evolve, such analysis methods should also be updated to address new and heightened risks of cyber-attacks on critical systems. Integrating cybersecurity into hazard and risk analyses as well as other elements of process safety management (PSM) could be done at an early design stage of a new plant, such as through Prevention through Design or Inherently Safe Design practices. It could also be applied throughout the life of a facility such as part of design/safety reviews to support operations/maintenance or for a capital expansion/ retrofit project. As anonymous adversaries could attack at any place or time, one must stay vigilant and implement a holistic approach to enhance awareness and capabilities to defend against and recover from cyber-attacks throughout the life cycle of a plant.

https://www.aiche.org/resources/publications/cep/2019/june/integrate-cybersecurity-process-hazard-analyses

https://www.sciencedirect.com/science/article/pii/S0950423019307995

610523 Machine Learning Approach for Tool Condition Monitoring in Manufacturing Systems

Abstract:

Condition monitoring is receiving great attention in modern (smart) manufacturing systems in an effort to increase productivity and efficiency by minimizing quality losses and scrap. Mourtzis et al. [1] showed that the production downtime in a machining process due to cutting tool wear makes up 75% of the total production downtime. In precision machining specifically, flank wear is the most common wear type and affects the durability and dimensional accuracy of the machining part. In addition, a gradual increase in wear during the process can lead to overall failure of the process and damage to the machine tools. Thus, monitoring of cutting tool conditions is important to improve the efficiency of overall machining process.

Modern tool condition monitoring systems leverage advances in machine learning (ML) for analysis of the state of cutting tool, classification of tool wear, or prognosis of its future state [2]. The application of ML in tool condition monitoring has been extensively studied over the past decade. Dominant methods involve ML algorithms that use techniques such as K-star algorithms [3] and Decision Trees [4]. Although these methods have shown good performance, the fundamental methodology and complete tool chain that is suitable to predict tool condition still need to be investigated. The purpose of this study is to propose an ML-based approach for predicting tool condition in CNC systems. We use convolutional neural networks (CNN) as the ML algorithm, which can provide insights into tool wear prediction in the machining process and assist with productivity improvement for the overall system.

The proposed ML-based framework in this study is presented in the Fig. 1. The method is divided into three process blocks: data acquisition, data pre-processing and CNN model development. In the data acquisition stage, we collect, tabulate and label vibration and acoustic data of the CNC process under varying cutting conditions. The vibration and acoustic data are types of the time waveforms collected at a specific time epoch for a condition. The signal pre-processing stage is necessary to extract information from the signal as features. In this step, we transform the data to the frequency domain, using Fast Fourier Transform (FFT). The data collected are time discrete and finite, thus we apply Discrete Fourier Transformation (DFT). Sometimes, it is necessary to reduce the number of features in the data, increase the information content and eliminate features of small variance to increase accuracy, reduce overfitting and computational effort. A feature selection step such as Principal Components Analysis (PCA) or Non-Linear PCA can be employed for this purpose. The selected features of audio and vibration signals are modeled and analyzed with a CNN model. The first input layer of the CNN performs operations on the features, while further layers include convolution and pooling blocks, followed by a single fully connected layer. The final layer reduces the features to map the number of tool condition classes for condition assessment. For the training of CNN, features from the signal processing stage are processed through the convolution and pooling layer for the best feature selection and then through fully connected layers to obtain prediction results. This process is repeated during the training step for given each batch size. After the final batch in training step, the model is tested with the testing dataset in testing step to generate final prediction results. This overall process is repeated for the given epoch number to optimize the model for better prediction results.

The framework of Fig. 1 was applied to two different precision machining processes. A Mazak Variaxis 630-5X II T Mill/Turn Machine 40 hp 5-axis CNC machine, representative of the type of machine in use across global manufacturing for precision machining of metals. The machine is instrumented to acquire in-process data for machine power, spindle power, spindle vibration, and to record audio/video and is MTConnect capable. The second application was on a Gerber Technology Paragon Cutter (used for fabric cutting) in which errors were injected to the real machine including a dull knife, loose knife, loose screw(s), loose swivel, no belt, worn Lanchester balancer, etc. Application of the framework of Fig. 1 for the purpose of condition monitoring and classification of the machine state will be presented. Overall classification accuracy up to 100% (counts of correct classification in the confusion matrix of the testing set) were achieved. Then, the use of the ML algorithm for the prediction of parameters of a physics-based model that accurately describes precision machining of metals will be discussed.

Acknowledgment: This material is based upon work supported by the U.S. Department of Energy's Office of Energy Efficiency and Renewable Energy (EERE) under the Advanced Manufacturing Office Award Number DE-EE0007613.

References

[1] D. Mourtzis, E. Vlachou, N. Milas, N. Xanthopoulos, A Cloud-based Approach for Maintenance of Machine Tools and Equipment Based on Shop-floor Monitoring, Procedia CIRP. 41 (2016) 655–660. doi:10.1016/j.procir.2015.12.069.

[2] P. Jahnke, Machine learning approaches for failure type detection and predictive maintenance, TU Darmstadt, 2015.

[3] S. Painuli, M. Elangovan, V. Sugumaran, Tool condition monitoring using K-star algorithm, Expert Syst. Appl. 41 (2014) 2638–2643. doi:10.1016/j.eswa.2013.11.005.

[4] C.K. Madhusudana, H. Kumar, S. Narendranath, Face milling tool condition monitoring using sound signal, Int. J. Syst. Assur. Eng. Manag. 8 (2017) 1643–1653. doi:10.1007/s13198-017-0637-1.

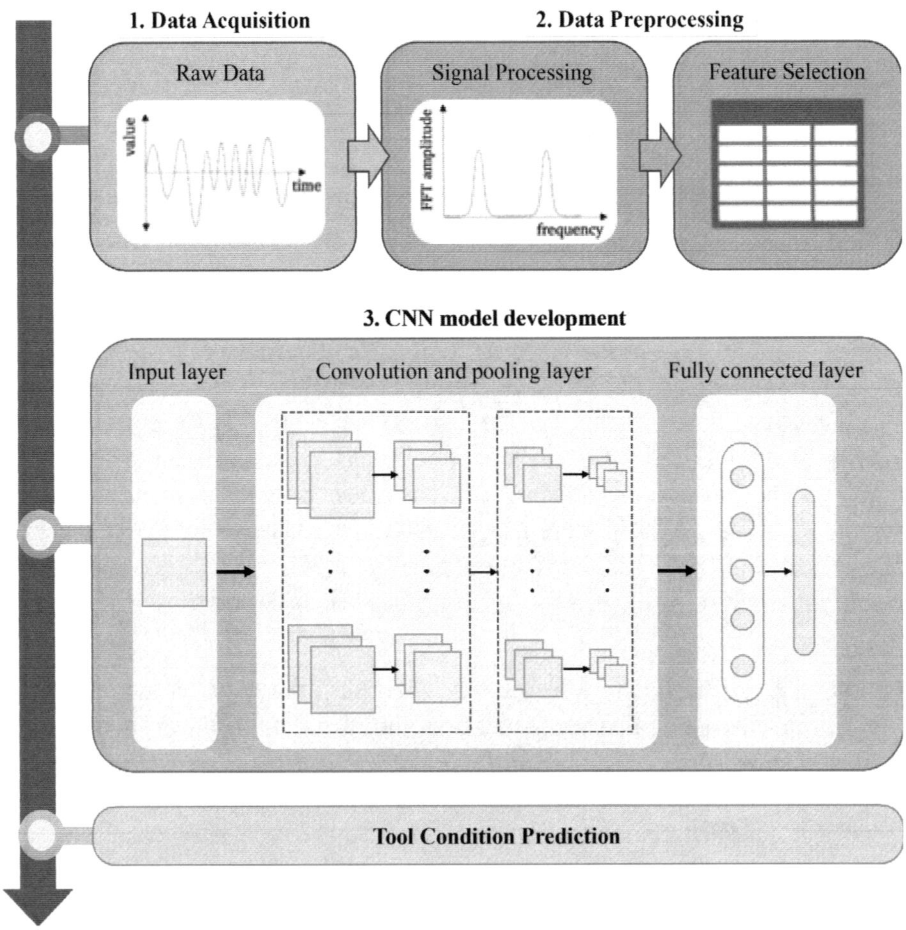

Figure 1. Schematic diagram of the machine learning-based framework for tool condition prediction.

602384 Fast-Convergence of Deep Reinforcement Learning Controller: Application to a Continuous Stirred Tank Reactor

Abstract:

Reinforcement Learning (RL) originated several decades ago in computer science and operations research to solve complex sequential decision-making problems but its application to process control has been recent and limited. RL involves an agent that learns the optimal policy by interacting with the environment in real-time [1]. Many approaches have been proposed in the past to solve the RL problem but recent advancements in deep learning have made it possible to combine deep neural networks (DNNs) with RL. This combination has already delivered tremendous success in video games like Atari games [2] etc. More recently, in the context of process control, a Deep RL (DRL) controller, a model-free off-policy actor-critic algorithm, was proposed based on temporal difference (TD) learning [1] and Deterministic Policy Gradient (DPG) algorithm [3] for controlling discrete-time nonlinear processes [4]. The DRL controller utilizes two DNNs to generalize the actor and the critic to the continuous state and action spaces, and two more as target networks for their learning. Ideas like replay memory and gradient clipping were used in DRL controller to make learning appropriate for process control applications. The DRL controller was able to solve set-point tracking problems for single-input-single-output (SISO), multi-input-multi-output (MIMO), and a nonlinear system with external disturbances [4].

Despite its success, DRL controller has few limitations including the requirement of a large amount of data and high computational loads, and careful selection and initialization of hyperparameters for fast convergence, etc. Additionally, one glaring limitation of DRL controller, as with many other RL methods, is the long training time before it can deliver satisfactory control performance [5]. In order to overcome this challenge, we propose to train the actor and the critic offline using historical process data before using it for online control. For the actor network which approximates the policy function, we use the information of past states and control actions to train it offline until convergence within the training region is achieved. To train the critic network offline, which approximates the action-value function, we use the information of reward gain, calculated based on a pre-defined reward function, to train it offline until convergence is achieved. Once trained offline, we use the learned actor-critic as the starting point in the DRL controller. This pre-trained DRL controller is implemented to track concentration and temperature set-points for a continuous stirred tank reactor (CSTR) process, and we successfully demonstrate its ability to adapt and learn to track set-points outside the training region faster compared to a DRL controller that was randomly initialized. We also compare the control performance of this pre-trained DRL controller against a model-predictive controller in tracking a set-point.

Literature cited:

[1] Sutton, R.S., Barto, A.G. Reinforcement Learning: An Introduction. *Cambridge: MIT Press*, 1998.

[2] Mnih, V., Kavukcuoglu, K., Silver, D., Graves, A., Antonoglou, I., Wierstra, D., Riedmiller, M. Playing Atari with Deep Reinforcement Learning. *arXiv Preprint, arXiv:13125602*, 2013.

[3] Silver, D., Lever, G., Heess, N., Degris, T., Wierstra, D., Riedmiller, M. Deterministic policy gradient algorithms. *In ICML*, 2014.

[4] Spielberg, S., Tulsyan, A., Lawrence, N.P., Loewen, P.D., Bhushan Gopaluni, R. Toward self-driving processes: A deep reinforcement learning approach to control. *AIChE J*, **65**(10), 2019.

[5] Shin, J., Badgwell T.A., Liu, KH., Lee, J.H. Reinforcement Learning – Overview of recent progress and implications for process control. *Computer Aided Chemical Engineering*, **44**:71-85, 2018.

594894 Smart Proxy Modeling for Cfd; Application of Artificial Intelligence & Machine Learning to Numerical Simulation

Abstract:

Computational Fluid Dynamics (CFD) is a computationally expensive numerical simulation modeling technology. In the past few decades traditional proxy models (statistical response surfaces, or reduced order models) have been developed in order to expand the utilization and practicality of CFD models for applications such as optimization and uncertainty quantification.

Smart Proxy is a new technology that is defined as the engineering application of Artificial Intelligence and Machine Learning. Smart proxy modeling replicates the results of complex CFD models with high accuracy without reducing the physics of the models or its resolution in space and time. Unlike response surface approach, Smart Proxy modeling replicates the physics of the CFD model simultaneously for all the cells and all the required output variables. Once developed, Smart Proxy modeling accomplishes this objective at an incredibly fast pace when it is deployed on a laptop or a desktop. Development of this technology requires only a handful of CFD simulation runs.

The CFD model that is the subject of this particular study, simulates the combustion of natural gas within more than 9.3 million cells in three segments: Inlet-Combustion-Exhaust as shown in the figure below. A single run of this CDF model using a commercial CFD software application (FLUENT) takes 24 hours on an HPC that includes 40 cores. The output of interest in this model was the detail and comprehensive distribution of Pressure and Temperature, as well as Nitrogen, Oxygen, and CO_2 volume fractions throughout the 9.3 million cells. The Smart Proxy model for this application was developed using only 8 CFD simulation runs.

Without compromising the complex physics that has been modeled or the original space-time resolution, the data-driven Smart Proxy replicates the CFD with high accuracy and at very high speed. Smart Proxy takes advantage of pattern recognition capabilities of Artificial Intelligence and Machine Learning to achieve its objectives.

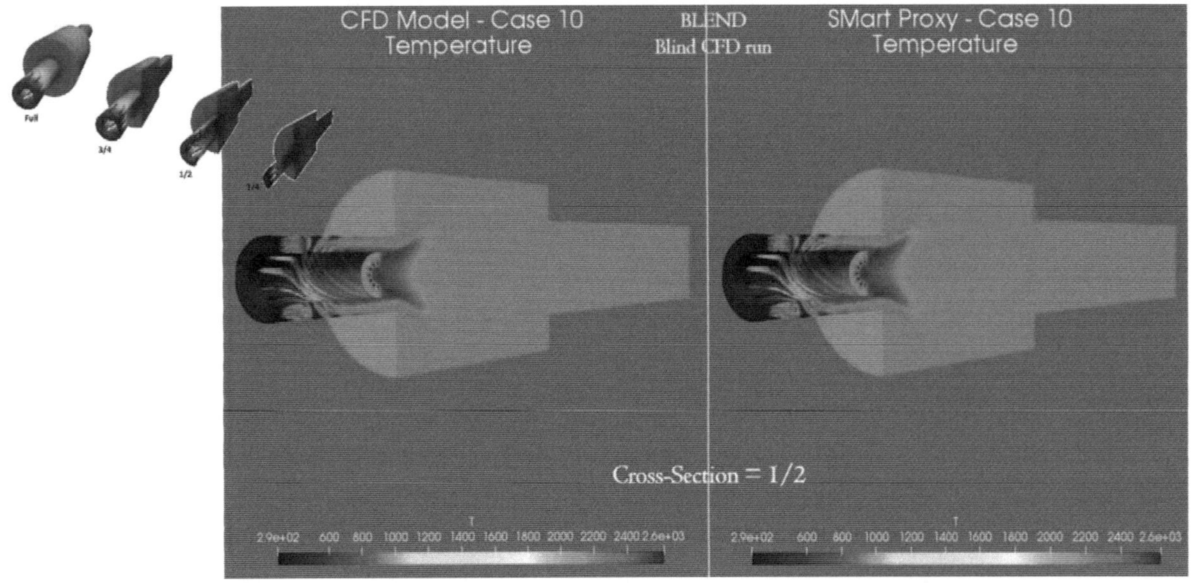

603320 Comparison of State-of-the-Art Dynamic Machine Learning Methods for MPC of Coal-Fired Utility Generator Performance

Abstract:

The identity of the world's electricity generating sources is evolving at an unprecedented rate. Massive additions of variable renewable energy sources (VREs) are fundamentally changing operating methods of existing fossil-fuel power plants, as they attempt to maintain reliability and meet demand [1], [2]. Many large coal-fired generators originally designed as baseloaded units are now performing load following in a manner never anticipated at ramp rates once thought to be impossible [3], [4]. This significant departure from design conditions can lead to substandard operating performance, decreased efficiency, and increased emission rates [5]. Efforts by many countries to significantly reduce their carbon footprints further escalates the need to reduce the heat rate of thermal utility generators in order to achieve these goals. As these practices are adopted by more generation facilities, advanced control strategies must be developed to maintain appropriate operation. Real-time artificial intelligence optimization systems have been shown to consistently reduce emission rates of coal-fired power plants in closed-loop control and are a preferred method for reducing heat rate. [6]–[8], but observed benefits are limited during transient operation. Model predictive control (MPC) is a proven method of optimizing dynamic systems during transient operation [9], [10]. To date, the application of MPC to combustion control has been precluded by the complexity of dynamically modeling the combustion process. Advancements in data-driven machine learning modeling methods are reducing the barrier to entry of model generation and application of MPC to combustion control. This work examines current state-of-the-art dynamic machine learning modeling methods and their suitability for use within an MPC framework to minimize heat rate and NOx emission rates at a coal-fired utility generator. Existing dynamic modeling methods, including forms of autoregressive models (VAR, ARX, NARX), recurrent neural networks (RNN, LSTM, bidirectional RNN and LSTM, CTRNN, DBN), and recurrent support vector machines, are compared based on prediction accuracy and other criteria over multiple time horizons on operational data from a 500 MW coal-fired utility generator. Advantages and disadvantages of each method are discussed. Appropriate modeling methods for use in future work of developing and operating an MPC system in closed-loop control at this utility generator are identified.

[1] W. Cole *et al.*, "Variable Renewable Energy in Long-Term Planning Models : A Multi-Model Perspective Variable Renewable Energy in Long-term Planning Models : A Multi-model Perspective," Golden, 2017.

[2] P. Denholm, G. Brinkman, and T. Mai, "How low can you go? The importance of quantifying minimum generation levels for renewable integration," *Energy Policy*, vol. 115, no. October 2017, pp. 249–257, 2018.

[3] E. Moiseeva, S. Wogrin, and M. R. Hesamzadeh, "Generation flexibility in ramp rates: Strategic behavior and lessons for electricity market design," *Eur. J. Oper. Res.*, vol. 261, no. 2, pp. 755–771, 2017.

[4] S. M. Safdarnejad, J. F. Tuttle, and K. M. Powell, "Dynamic modeling and optimization of a coal-fired utility boiler to forecast and minimize NOx and CO emissions simultaneously," *Comput. Chem. Eng.*, vol. 124, pp. 62–79, 2019.

[5] J. F. Tuttle and K. M. Powell, "Analysis of a thermal generator's participation in the Western Energy Imbalance Market and the resulting effects on overall performance and emissions," *Electr. J.*, vol. 32, no. 5, pp. 38–46, 2019.

[6] J. F. Tuttle, R. Vesel, S. Alagarsamy, L. D. Blackburn, and K. Powell, "Sustainable NOx emission reduction at a coal-fired power station through the use of online neural network modeling and particle swarm optimization," *Control Eng. Pract.*, vol. 93, no. September, 2019.

[7] R. Cass and B. Radl, "ADAPTIVE PROCESS OPTIMIZATION USING FUNCTIONAL-LINK NETWORKS AND EVOLUTIONARY OPTIMIZATION," *Control Eng. Pract.*, vol. 4, no. 11, pp. 1579–1584, 1996.

[8] D. Hasler and W. Rosenquist, "Coal-Fired Power Plant Heat Rate Reductions," 2009.

[9] J. D. Hedengren, R. A. Shishavan, K. M. Powell, and T. F. Edgar, "Nonlinear modeling, estimation and predictive control in APMonitor," *Comput. Chem. Eng.*, vol. 70, pp. 133–148, 2014.

[10] J. B. Rawlings, D. Q. Mayne, and M. M. Diehl, *Model Predictive Control: Theory, Computation, and Design*, 2nd ed. Madison: Nob Hill Publishing, 2017.

605753 Integrating Machine-Learning Methods for Adaptive Model Predictive Control Design

Abstract:

Automatic control systems which are widely used in the chemical process industries have faced many challenges such as nonlinearity, multi-variable interactions, and process constraints. To address these problems, advanced control strategies, in particular model predictive control (MPC), have been developed and successfully applied to the process industries (see e.g., [1]). The choice of an appropriate and reliable model is an essential task for successful development of any MPC design. Due to several challenges that exist to develop and maintain first-principles models and increasing availability of data, data-driven models and machine-learning based techniques have received significant attention in recent years. Among these approaches, artificial neural networks (ANN), especially recurrent neural networks (RNN), have been widely employed for modeling process dynamics and model-based control designs owing to their ability to capture process nonlinearity (see e.g., [2], [3], [4], [5], and [6]).

The first step for developing data-driven models is data generation. The model ability to predict the future behavior of the system can be strongly influenced by type of data which is generated under specific operation condition. Several methods consider perturbing process inputs using typically uncorrelated pseudo-random binary input signals (PRBS) (see e.g., [7]). However, applying standard uncorrelated input signals for process excitation is not sometimes possible due to the unstable nature of the process and economic and safety issues. To solve this problem, alternative methods have been proposed in the literature using closed-loop data (see e.g., [8], [9], and [10]). In the situations where data is generated under the closed-loop condition that may induce co-linearity in the input signals, the negative implications of the over-fitting problem would get exacerbated. To address this problem, the pre-processing step should be applied to training data in order to remove any co-linearity present in input signals. In this direction, principal component analysis (PCA) can be used to reject the redundancy.

Motivated by above considerations, this presentation addresses the problem of model identification in the presence of co-linearity in the input training data. A base case scenario is used for comparison where the training data is used to build an RNN model, and used directly in the MPC formulation. As expected, under MPC implementation, breaking of the correlation between the inputs leads the process to a region where the model is no longer valid, leading to poor performance. To solve this problem, two methods are proposed. In the first method, a combination of PCA and RNN is utilized to first model the process dynamics. The model is then employed in MPC implementation in a way that naturally respects the correlation observed in the training data. In the second approach, the base case RNN model is used in MPC, but a new constraint on squared prediction error (SPE) is included to ensure that the control moves remain in the same plane as the training data. Next, an approach is presented to fully exploit the performance enhancement capabilities of MPC, i.e., allow the MPC to break the input correlation seen in the training data, while not loosing model validity. To achieve this, the process is first run under MPC, albeit with slightly relaxed constraints on the SPE, thus generating richer data (while only slightly compromising on the performance). A model identified using this new data is next used in the MPC implementation to achieve superior performance.

[1] Qin, S. J., & Badgwell, T. A. (2003). A survey of industrial model predictive control technology. *Control engineering practice, 11*(7), 733-764.

[2] Nikravesh, M., Farell, A. E., & Stanford, T. G. (2000). Control of nonisothermal CSTR with time varying parameters via dynamic neural network control (DNNC). *Chemical Engineering Journal, 76*(1), 1-16.

[3] Kittisupakorn, P., Thitiyasook, P., Hussain, M. A., & Daosud, W. (2009). Neural network based model predictive control for a steel pickling process. *Journal of process control, 19*(4), 579-590.

[4] Sadeghassadi, M., Macnab, C. J., Gopaluni, B., & Westwick, D. (2018). Application of neural networks for optimal-setpoint design and MPC control in biological wastewater treatment. *Computers & Chemical Engineering, 115*, 150-160.

[5] Wu, Z., Tran, A., Rincon, D., & Christofides, P. D. (2019). Machine learning-based predictive control of nonlinear processes. Part I: Theory. *AIChE Journal, 65*(11), e16729.

[6] Wu, Z., Tran, A., Rincon, D., & Christofides, P. D. (2019). Machine-learning-based predictive control of nonlinear processes. Part II: Computational implementation. *AIChE Journal, 65(11)*, e16734.

[7] Shariff, H. M., Rahiman, M. H. F., & Tajjudin, M. (2013). Nonlinear system identification: Comparison between PRBS and Random Gaussian perturbation on steam distillation pilot plant. In *2013 IEEE 3rd International Conference on System Engineering and Technology*, 269-274.

[8] Forssell, U., & Ljung, L. (1999). Closed-loop identification revisited. *Automatica, 35*(7), 1215-1241.

[9] Qin, S. J., & Ljung, L. (2003). Closed-loop subspace identification with innovation estimation. *IFAC Proceedings Volumes 36*(16), 861-866.

[10] Shardt, Y. A., & Huang, B. (2011). Closed-loop identification with routine operating data: Effect of time delay and sampling time. *Journal of Process Control, 21*(7), 997-1010.

606512 An Efficient Reinforcement Algorithm Approach to Optimal Control and Environmental Sustainability

Abstract:

In this paper, we present a new and efficient Reinforcement Learning approach to solving an optimal control problem, based on the Batch Q-learning algorithm. To improve on the convergence of the RL algorithm, we explore the use of k-dimensional uniformity of advanced sampling procedures, namely employing the Halton and Hamersley sequences (HSS). These sequences are used to randomly sample the discrete controls from the action space for the RL optimal control problem. Specifically, the Neural-fitted Q-iterative algorithm is applied to solve an optimal control problem for a first-order state dynamical system. We found that the HSS-based RL improves the convergence significantly as compared to the current RL practice. We implement the schemes in MATLAB and present the simulation results, comparing the optimal policy resulting from using the Pontryagin Maximum Principle with that from using the Neural fitted Q iteration approach. Our results indicate that the RL approach performs reasonably well, especially considering that it is model-free when compared to the objective function value for the deterministic case and error from the optimal trajectory for the stochastic case.

In an age of climate consciousness due to growing environmental concerns, biodiesel has shown enormous promise in replacing traditional fossil fuels like gasoline and diesel. Using optimal control, we aim to optimize the production of biodiesel and improve efficiency to increase its viability as a source of alternative energy. As such, in this paper, we also present a real-world case study involving optimal control problems for biodiesel production, employing the new HSS-RL algorithm.

613618 Performance of Sweeping Gas Membrane Distillation for Treating Produced Water: Modeling and Experiments

Abstract:

Membrane distillation (MD) has shown promise for desalination of PW, but there are very few studies reported on the performance of MD with real PW and the characteristics of the fouling and scaling in MD. Among different MD configurations, sweeping gas MD (SGMD) is the least studied configuration. Here we report the application of SGMD for treating high-salinity PW (120 g L^{-1}). The SGMD operating conditions were first carefully optimized through pure water MD experiments by varying different operating conditions. The flux evaluation was accompanied by a detailed Aspen Custom Modeler simulation based on the electrolyte-NRTL thermodynamic model to better understand the heat and mass transfer in the feed and permeate channel. Supported by experiments, modeling results confirmed the partial condensation of the vapor in the permeate channel. Additionally, it was concluded that the saturation of the sweeping gas is controlling the overall mass transfer through the membrane. Selected membranes and optimized conditions were then used to conduct short-term and long-term performance tests with PW. In long-term MD tests, the concentration of the PW increased by a factor of two (50% water recovery), while the membrane exhibited small flux decline. Elemental analysis showed that strontium and sodium were the major scalant compounds.

602825 Generalized Modular/Collocation Framework for Representation and Synthesis of Intensified Multi-Scale Reaction Systems

Abstract:

Reaction engineering, the heart of many chemical production processes, can drive radical revolutions in chemical and energy industry by designing novel catalysts and employing intensified modular reactor designs, such as micro-reaction systems, to boost volumetric or energic efficiency [1,2]. The conceptual design of chemical reactors, especially micro-reactor, normally depends on first-principle modeling with detailed descriptions on mass/heat balances and mass/heat transport with respect to specific reactor types and catalysts. However, a holistic approach to rapidly screen and systematically generate intensified and optimal reaction process alternatives by simultaneously taking into consideration multi-scale reactor options (e.g., meso- and micro- reactors) is still lacking.

A promising means to address this challenge is via the phenomena-based synthesis-intensification strategy [3-6], which represents a chemical process with fundamental building blocks without requiring any pre-postulation of plausible equipment and/or flowsheet configurations. However, key open questions remain on: (i) how to capture micro-reactors via such synthesis representation, (ii) how to capture meso- and micro- reaction systems under a unified synthesis approach, and (iii) what type of minimum process input information or data is needed to reveal the differentiation of multi-scale reactors.

In this work, we propose a synthesis representation approach for reaction systems based on the Generalized Modular Representation Framework (GMF) [7,8]. Herein, the reaction systems are represented as aggregated multifunctional mass/heat exchange modules and pure heat exchange modules to intensify the fundamental mass and/or heat transfer phenomena. Driving force constraints, derived from total Gibbs free energy change, are employed to characterize the mass transfer feasibility from a general thermodynamic perspective. Diffusion and transport rates are also incorporated into the driving force constraints to particularly account for the miniaturization effects in micro-reactors. These rate terms can be activated or de-activated respectively based on the corresponding values of Damkohler number and Thiele modulus which dictate the choice of micro- or meso- reactor size. Spatial distribution information within GMF modules is extracted via orthogonal collocation [9] in a physically compact and computationally efficient manner. A comparative case study on micro-reactor and conventional reactor for methane steam reforming [10,11] is presented to showcase the proposed approach. Extensions of catalyst impact on reactor design will also be discussed.

References

1. Stankiewicz, A. I., & Moulijn, J. A. (2000). Process intensification: transforming chemical engineering. *Chemical engineering progress*, *96*(1), 22-34.
2. Tian, Y., Demirel, S. E., Hasan, M. M. F., & Pistikopoulos, E. N (2018). An overview of process systems engineering approaches for process intensification: State of the art. *Chemical Engineering and Processing: Process Intensification*, *133*, 160-210.
3. Babi, D. K., Lutze, P., Woodley, J. M., & Gani, R. (2014). A process synthesis-intensification framework for the development of sustainable membrane-based operations. *Chemical Engineering and Processing: Process Intensification*, *86*, 173-195.

4. da Cruz, F. E., & Manousiouthakis, V. I. (2016). Process intensification of reactive separator networks through the IDEAS conceptual framework. *Computers & Chemical Engineering, 105*, 39-55.
5. Demirel, S. E., Li, J., & Hasan, M. F. (2017). Systematic process intensification using building blocks. *Computers & Chemical Engineering, 105*, 2-38.
6. Kaiser, N. M., Flassig, R. J., & Sundmacher, K. (2018). Reactor-network synthesis via flux profile analysis. *Chemical Engineering Journal*, *335*, 1018-1030.
7. Papalexandri, K. P., & Pistikopoulos, E. N. (1996). Generalized modular representation framework for process synthesis. *AIChE Journal*, *42*(4), 1010-1032.
8. Tian, Y., Pappas, I., Burnak, B., Katz, J., & Pistikopoulos, E. N. (2020). A systematic framework for the synthesis of operable process intensification systems – Reactive separation systems. *Computers & Chemical Engineering*, *134*, 106675.
9. Proios, P., & Pistikopoulos, E. N. (2006). Hybrid generalized modular/collocation framework for distillation column synthesis. *AIChE journal*, *52*(3), 1038-1056.
10. Stefanidis, G. D., Vlachos, D. G., Kaisare, N. S., & Maestri, M. (2009). Methane steam reforming at microscales: Operation strategies for variable power output at millisecond contact times. *AICHE journal*, *55*(1), 180-191.
11. Maestri, M., Vlachos, D. G., Beretta, A., Groppi, G., & Tronconi, E. (2008). Steam and dry reforming of methane on Rh: Microkinetic analysis and hierarchy of kinetic models. *Journal of Catalysis*, *259*(2), 211-222.

608050 Dynamic Data Reconciliation, Parameter Estimation, and Plant-Wide Modeling of a Microwave (MW)-Assisted Direct Non-Oxidative Methane Dehydroaromatization (DHA) Process

Abstract:

The microwave (MW) technology has been commercially used for many years with applications in wood drying, pharmaceuticals, and food industry for cooking, thawing, pasteurization and preservation of food materials. Over the past several years, it is being increasingly investigated for assisting chemical reactions both for the homogenous and heterogeneous reactive systems [1]. Enhancement of reaction rate and improvement of product yields in MW-assisted reactors have been the two main reasons for the increased interest in application of MW for reaction engineering.

Modeling of MW-assisted catalytic fixed bed reactors and theoretical analysis of MW-assisted endothermic reactions has been pursued by a few authors. Gerasev [2] has developed a 1-D heterogeneous model of a packed bed catalytic reactor for an irreversible endothermic reaction. The gas in the reactor is assumed to absorb the MW energy changing its composition. Their study used Lambert's law for MW power propagation. A study comparing the conventional and microwave heating efficiency of a packed column with dielectric catalyst has been performed by Cherbanski [3]. They used Lambert's law for microwave power propagation inside the reactor. Bhattacharya et al. [4] performed a comprehensive theoretical analysis of the effect of MW heating for a first order endothermic reaction in reactor packed with dielectric material. The current literature lacks kinetic and reactor models for a MW-assisted methane dehydroaromatization (DHA) process. In this process deactivation due to coke formation takes place making it difficult to develop an accurate kinetic model and estimate its parameters due to the transient operation and complex interaction between the MW and changing reactant composition and coke formation in the reactor.

A kinetic model including a model of coke formation is proposed for an in-house catalyst for the direct non-oxidative methane DHA reactor. Formation of coke is a dynamic process affecting also the dynamics of conversion and yield from the reactor. However with the current measurement technology, it is not feasible to measure the changing mass and composition of coke inside the reactor. To satisfy the carbon and hydrogen balance in the experimental data of the product flowrates from the reactor, a dynamic data reconciliation problem is solved. Subsequently a dynamic parameter estimation problem is solved using the reconciled data. This rate model is then used to develop a multi-scale reactor model where mass transfer, heat transfer, reaction kinetics and MW propagation in the catalyst particles are modeled and integrated with the bulk reactor model where MW propagation is modeled based on the configuration and design of the MW generator and the wave guide. The MW reactor model is then used to develop and optimize a plant-wide model. Economics of the MW-assisted process are compared with a commercial process for conversion of natural gas to aromatics via methanol synthesis.

References

[1] Lidström P, Tierney J, Wathey B, Westman J. Microwave assisted organic synthesis - A review. Tetrahedron 2001. https://doi.org/10.1016/S0040-4020(01)00906-1.

[2] Gerasev AP. Emergence of traveling wave endothermic reaction in a catalytic fixed bed under microwave heating. Energy 2017. https://doi.org/10.1016/j.energy.2016.11.042.

[3] Cherbański R. Numerical simulations of heat transfer in a packed column: comparison of microwave and convective heating. Heat Mass Transf Und Stoffuebertragung 2015. https://doi.org/10.1007/s00231-014-1447-5.

[4] Bhattacharya M, Basak T, Senagala R. A comprehensive theoretical analysis for the effect of microwave heating on the progress of a first order endothermic reaction. Chem Eng Sci 2011. https://doi.org/10.1016/j.ces.2011.08.003.

611591 Multiphase Microchannel Separator: A Microfluidic Platform for the Intensification and Modularization of Solute Extraction and Phase Separation

Abstract:

The Multiphase Microchannel Separator (MMS) is a microscale-based, multiphase separation technology for chemical processing. This microfluidic platform relies on the inherent attributes of micro-scale operations, such as enhanced mass transport, refined fluidic control, reduced size and improved energy efficiency, to intensify solute extraction and phase separation processes.

In the first stage of MMS operation, an accelerated solute extraction is achieved via the generation of uniform solvent micro-droplets dispersed in a solute-rich feed. The arising benefits of this system relative to conventional extraction, including smaller characteristic diffusion time, higher interfacial surface area and lower interfacial film thickness, lead to the intensification of solute transfer from the aqueous feed to the organic solvent.
In the second stage, a continuous and gravity-independent phase separation is realized via a careful utilization of the platform's flow geometry and surface properties. The pillar of this microfluidic manipulation consists of passing the two-phase flow through a series of micro-posts with gradually increasing gap spacing. In turn, this micro-architecture design creates a capillary force gradient 'field' opposite to the direction of the flow, which guides the non-wetting phase (solvent) to exit through the outlet with the least resistive path (the organic outlet).

In this work, we present the experimental findings of solute extraction and phase separation in MMS 2.0. The investigated system of Isopropanol (IPA solute) - Water (feed) - Ethyl Hexanoate (organic solvent) is of current industrial interest. Three parameters were used to assess the extent of two-phase separation; the aqueous and organic composition % in the aqueous and organic effluents, respectively. The third parameter is the 'two-phase separation %' describing the distribution of the organic phase among the two outlets (100% corresponding to a total exit of the solvent from its appropriate organic outlet). The level of solute extraction was determined by measuring the concentration of IPA in the effluent solvent.

IPA extraction and phase separation in MMS 2.0 was investigated under a constant total flow rate of 2 ml/min and varying inlet organic/aqueous (O/A) flow ratios (0.33:1 to 1:1). The studied IPA contents in the aqueous feed were 5%, 10% and 20% % w/w.

The results showed that the highest phase separation performance was achieved when lower IPA amounts were introduced in the aqueous feed. At 5% and 10% w/w IPA in the feed, the two-phase separation % were consistently measured at near 100% across all the investigated inlet O/A flow ratios. These values dropped to a minimum of 50% and a maximum of 85% under 20% IPA w/w in the feed. This outcome was expected given that increasing the IPA amount in the water reduces the aqueous-solvent interfacial surface tension, which in turn diminishes the capillary pressures between the micro-posts needed to separate the phases.
Additionally, it was observed that for all the investigated IPA amounts (5%, 10% and 20% w/w), increasing the inlet O/A flow ratio resulted in a higher proportion of organic phase leaving the organic outlet. It was determined that increasing the inlet O/A flow ratio will lead the now more abundant organic phase to displace the aqueous phase exiting the organic outlet, thus improving the separation

performance.
As for solute extraction in MMS 2.0, it was noticed that the IPA extraction was enhanced under higher inlet O/A flow ratios. This effect was attributed to the increase in fresh solvent availability for IPA extraction. Moreover, it was observed that for any given inlet O/A flow ratio, the IPA extraction increased under higher IPA amounts in the feed. This outcome can be explained by the improved mass transfer of IPA from the aqueous phase to the organic one when more IPA existed in the water.

On the other hand, it was shown that varying the flow ratio of the effluent streams can alter the proportion of phases exiting from a given outlet, leading to major effects on the separation performance. More importantly, this parameter can serve as a simple yet effective tool to maximize the separation performance under a given set of process conditions. Operating under a constant total flow rate of 2 ml/min, inlet O/A flow ratio of 1:1 and in the absence of IPA, it was shown that changing the outlet O/A flow ratio from 1:1 to 1.6:1 can increase the resulting two-phase separation from 75.6% to 94.8%.

Finally, as part of our efforts to develop the next-generation high-throughput integrative MMS 3.0 platform, we have investigated the effects of increasing the channel depth, and thus the plate processing capacity, on the resulting separation performance. Comparing the 0.3 mm deep plate of the MMS 2.0 to novel plates with channel depths of 0.3, 2, 4 and 6 mm, we have determined that the separation performance was significantly improved under all the investigated deeper channels. This enhancement was mainly attributed to a possible contribution of the buoyant forces in deeper channels to phase separation. These results strongly suggest that the development of a high-throughput modular platform for the intensification of solute extraction and phase separation processes is within technical reach.

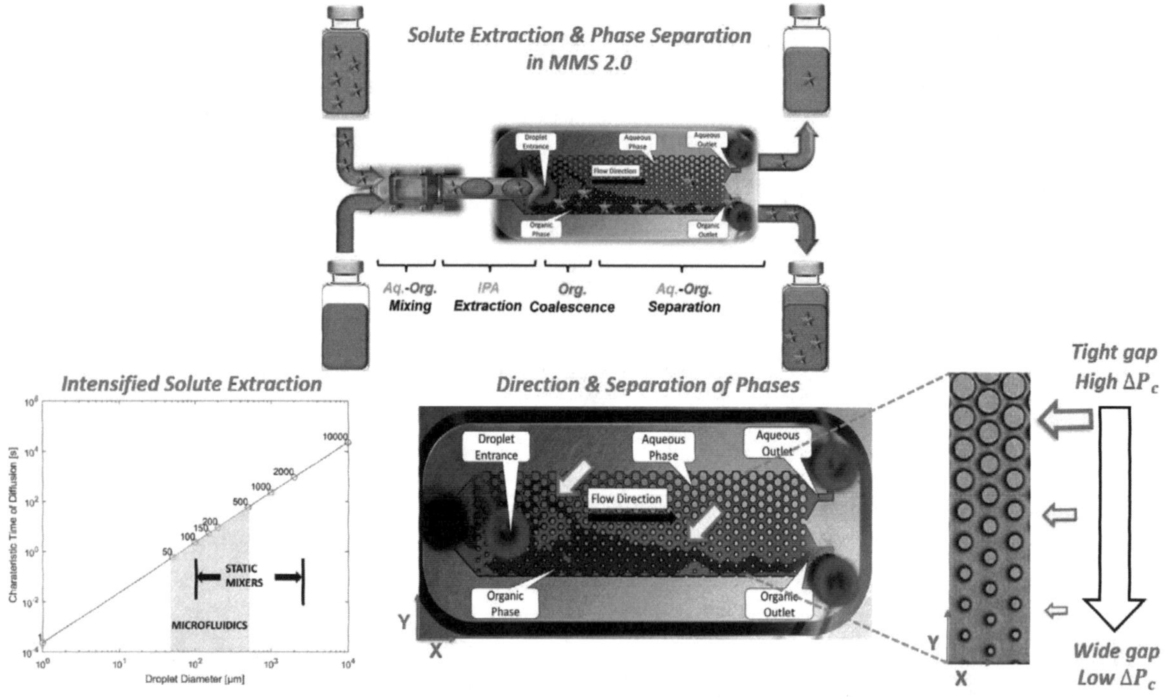

612838 Data-Efficient Probabilistic Model Learning with Embedded High-Fidelity Knowledge for Biomanufacturing in Deep Space Manned Missions

Abstract:

The NASA Center for the Utilization of Biological Engineering in Space (CUBES) leverages systems biology and systems engineering to realize life support for long, deep space manned missions on Mars. The aim of the center is to advance the practicality of multi-function, multi-organism biomanufacturing for production of pharmaceuticals, cell-based treatments/therapeutics, and biopolymers for on-demand 3D printing using *in situ* Martian resources [1-2]. Robust predictive modeling and control capabilities for the biomanufacturing system are crucial for decision-support and mission optimization under strict safety-critical requirements for life support on Mars [3].

In this talk, we will discuss how the abundancy of data generated from high-fidelity simulations, offline experiments, and real-time process operation can be systematically leveraged for learning probabilistic data-driven models for biomanufacturing systems. In particular, we will present a Bayesian learning framework that embeds high-fidelity knowledge and uncertainty quantification for data-efficient probabilistic model learning for a Mars life support biomanufacturing system. The proposed data-driven model learning framework is comprised of deep neural network surrogates for first-principles bioprocess models, trained based on simulations, combined with a Bayesian neural network with variational layers that acts as a real-time "corrector" to model hard-to-model phenomena that cannot be sufficiently captured by the first-principles models. This probabilistic nature of the learning-based model, expressed through the uncertainty of the variational layers, allows for quantifying the probabilistic uncertainty of model predictions, making the models especially suitable for robust decision-making and optimization. Furthermore, embedding the first-principles knowledge in the probabilistic learning-based models enhances data efficiency in model learning, while the ability to adapt the models using real-time data improves their predictive capability by accounting for unmodeled system behaviors.

We will demonstrate the proposed model learning framework on an integrated system of bioprocesses for Martian biomanufacturing. The process consists of production of biological feedstocks through microbial processing of *in situ* resources to fix carbon and nitrogen from the atmosphere, followed by downstream production of food, pharmaceuticals, and biopolymers suitable for 3D printing. The integrated bioprocesses are modeled using an advanced first-principles modeling tool, echusOverlook, developed in CUBES. We will discuss how the probabilistic learning-based models with embedded high-fidelity knowledge can enable evaluating different Mars mission scenarios and life support technologies in terms of the equivalent system mass metric, as well as reliability and cost-benefit considerations, under uncertainties in models and data.

[1] Menezes A.A., Montague M.G. , Cumbers J., Hogan J.A. , Arkin A.P., "Grand challenges in space synthetic biology," J. R. Soc. Interface, vol. 12, no. 113, p. 20150803, 2015.

[2] Menezes A.A., Montague M.G. , Cumbers J., Hogan J.A. , Arkin A.P., "Towards synthetic biological approaches to resource utilization on space missions," J. R. Soc. Interface, vol. 12, no. 102, p. 20140715 2015.

[3] Makrygiorgos, G., Gupta, S. S., Menezes, A., & Mesbah, A. "Fast Probabilistic Uncertainty Quantification and Sensitivity Analysis of a Mars Life Support System Model," IFAC-PapersOnLin In Press, 2020.

[4] Liang, F. "Bayesian neural networks for nonlinear time series forecasting," Statistics and Computing, vol. 15,13-29, 2005.

612712 Recognizing and Avoiding Big Data Analytics Traps in Applications

Abstract:

Advances in sensors and low-cost storage are leading to large increases in the collection of manufacturing data for use in process modeling, monitoring, and control. At the same time, the number of readily available software tools has grown exponentially, which has led to a large increase in applications of big data analytics and machine learning. While big data analytics can be effective using applied to the right problem in the right way, many of these traps are embedded in widely used software produced by major big data companies.

This talk will discuss some common mistakes made when applying big data analytics, ranging from the simple to the more subtle. The traps include (1) using leave-one-out cross-validation, (2) drawing inferences based on matching statistics, (3) selecting the best model from many models/methods, (4) finding false correlations by analyzing many two-variable combinations, and (5) not paying close enough attention to the intended model use when selecting methods. Specific examples of the traps are provided from a broad spectrum of applications from polymers to health care.

610621 Data-Driven and Hybrid Modeling of Integrated Paper Production Systems

Abstract:

The paper industry is instrumented with sensing systems that generate large streams of data. These rich data streams provide distinctive opportunities for improved optimization and control of these energy and water intensive processes. Recently, researchers have been trying to utilize these data streams to achieve better system performance [1], predict system variables [2–4], or to find the root cause of the problem [5]. At the same time, paper production is comprised of a set of process steps for mixing and drying streams containing fibers, for which partial first-principle knowledge is available. However, due to the complex nature of data and models, previous approaches are limited to single process-level analysis, which leads to suboptimal operation.

To overcome these challenges, we propose a system-level analytical framework which combines first-principle knowledge and low-dimensional learning from high-dimensional streaming data. We take a hybrid approach that combines the domain knowledge with data-driven modeling, through model calibration techniques utilizing Bayesian analysis [6]. To achieve more accurate predictions and improve interpretability, we present a scheme for the integration of physics-based equations for the prediction of pulp fiber orientation and final paper properties with Gaussian process model training. Due to the high-dimensionality of the data, dimensionality reduction techniques are required to compress the input space into the important reduced-space features. The system studied includes main downstream process steps for paper manufacturing, including the headbox, wire forming, wet pressing, and drying sections. Through this approach, we are able to develop a tractable integrated process model and quantitatively link process-design-material variables to final paper quality attributes for the first time. We will also discuss how the techniques presented for the paper production system can be generally applied to a variety of different manufacturing systems.

[1] P. Zhou, H. Wang, M. Li, Z. Zhao, and T. Chai, "Data-driven ALS-SVR-ARMA2K modelling with AMPSO parameter optimisation for a high consistency refining system in papermaking," *IET Control Theory Appl.*, vol. 10, no. 14, pp. 1620–1629, 2016, doi: 10.1049/iet-cta.2015.0850.

[2] K. Ciesielski and K. Olejnik, "Application of neural networks for estimation of paper properties based on refined pulp properties," *Fibres Text. East. Eur.*, vol. 22, no. 5, pp. 126–132, 2014.

[3] A. Alonso, C. Negro, A. Blanco, and I. San Pío, "Application of advanced data treatment to predict paper properties," *Math. Comput. Model. Dyn. Syst.*, vol. 15, no. 5, pp. 453–462, 2009, doi: 10.1080/13873950903375445.

[4] O. T. Okan, I. Deniz, and S. Tiryaki, "Application of artificial neural networks for predicting tensile index and brightness in bleaching pulp," *Maderas Cienc. y Tecnol.*, vol. 17, no. 3, pp. 571–584, 2015, doi: 10.4067/S0718-221X2015005000051.

[5] J. Fu and P. W. Hart, "Leveraging mill-wide big data sets for process and quality improvement in paperboard production," *Tappi J.*, vol. 15, no. 5, pp. 309–319, 2016, doi: 10.32964/tj15.5.309.

[6] M. C. Kennedy and A. O'Hagan, "Bayesian calibration of computer models," *J. R. Stat. Soc. Ser. B (Statistical Methodol.*, vol. 63, no. 3, pp. 425–464, 2001, doi: 10.1111/1467-9868.00294.

596529 Optimal Operation of Plasma Enhanced Atomic Layer Deposition Via Machine Learning

Abstract:

Over the past decade, plasma enhanced atomic layer deposition (PEALD) has been extensively studied in experimental labs and gradually become a popular deposition method in the semiconductor industry. Its distinguished ability to deposit ultra-thin high-k dielectric films makes it the perfect candidate in the production of contemporary high aspect ratio designs like fin field-effect transistor (FinFET) [1]. With the introduced high-energy plasma species, PEALD ideally requires much less thermal energy input, while enabling high throughput with temperature-sensitive precursors. Despite the advantages of PEALD, the complicated combination of remote plasma chamber, the showerhead reactor geometry and the plasma induced surface reaction makes the optimal operation hard to locate [2]. In addition, the process operation still remains costly, which discourages the extensive experimental exploration [3].

In light of this, a comprehensive multiscale computational fluid dynamics (CFD) model has been proposed to capture the integrated dynamic profile of a remote plasma PEALD process with Tetrakis-dimethylamino-Hafnium (TDMAHf) and oxygen plasma as precursors. Based on the simulation model, in this work, the operation of a PEALD reactor under different operating conditions is discussed, and the optimal operating regime of this reactor is analyzed. Specifically, an operational database is constructed using the multiscale CFD model to map a variety of reactor operation input combinations to their resulting film qualities and deposition profile. This database is then processed through machine learning analysis to explore the feasible operating domain, and within which, the optimal operating decision can be identified according to the desired production throughput and economic demand.

[1] Ishikawa, K., Karahashi, K., Ichiki, T., Chang, J.P., George, S.M., Kessels, W., Lee, H.J., Tinck,S., Um, J.H., Kinoshita, K., 2017. Progress and prospects in nanoscale dry processes: How can we control atomic layer reactions? Japanese Journal of Applied Physics 56, 06HA02.

[2] Profijt, H., Potts, S., Van de Sanden, M., Kessels, W., 2011. Plasma-assisted atomic layer deposition:basics, opportunities, and challenges. Journal of Vacuum Science & Technology A: Vacuum,Surfaces, and Films 29, 050801.

[3] Joo, J., Rossnagel, S.M., 2009. Plasma modeling of a PEALD system for the deposition of TiO2and HfO2. Journal of Korean Physical Society 54, 1048.

602726 System Visualization Using Real-Time Data-Driven Models Derived from High-Resolution Sensor Profiling

Abstract:

Control systems play a critical role in wastewater/water treatment plants (WWTPs) for guaranteeing effluent quality, energy conservation, and economics [1, 2, 3, 4]. Recently, model predictive control (MPC) has been proposed to maintain the effluent water quality in WWTPs [5, 6, 7]. Development of an effective model that can accurately predict the effluent water state as feedback to a controller is necessary and significant for MPC system design and implementation. The data-driven approach could allow engineers to develop predictive models for use by the control system using data from the actual process under realistic operating conditions. Compared with traditional mechanistic modeling approaches, data-driven models are flexible and labor-saving, and many researchers have made it useful in the processes of WWTP [8, 9, 10, 11, 12]. However, most existing data-driven models have been established upon big data without knowing physical properties of a given system. Since all unit operations in WWTPs are governed by the physical laws, it is pragmatic to establish models revealing the "black box" accounting for some fundamental principles, such as mass conservation. In turn, we achieve highly-predictive, yet extremely simple models which lend themselves for use within MPC.

In this paper, we discuss the development of novel physics-informed data-driven models to simulate dynamic conductivity, pH, and temperature profiles in a continuously-stirred tank. We establish reduced-order multi-compartment models for profiling conductivity and pH in the tank. Specifically, the tank is partitioned into four nominal zones: high, middle, and low zones represent the compartments corresponding with the location of the sensors deployed in the experimental apparatus and a "mixing zone" denotes the bottom region where the stirrer resides. Mixing models are established based on an assumed mass transfer phenomenon between each zone and the conservation of mass. For heat transfer modeling, we invoke the dynamic heat equation with forced convection for homogeneous media.

We conduct a series of experiments to simulate the mass and heat transfer phenomena in the continuous-stirred tank. Specifically, three types of at milli-electrode array (MEA) sensors [13, 14] were deployed lengthwise along the tank's inner surface for profiling conductivity [15], pH [16], and temperature at the high, middle, and low zones. For conductivity and pH experiments, we injected a very small volume of concentrated solution (sodium chloride for conductivity shock and potassium hydroxide for pH shock) and recorded the signals measured by each sensor to profile heterogeneous mixing in the tank. For the temperature profiling experiment, we heated the tank from the bottom and measured the temperature in the three zones.

Deterministic global optimization methods were applied to determine the optimal transport parameter values for model validation. For the conductivity and pH models, we reformulated the dynamic optimization problems into large-scale nonlinear programs using the explicit Euler discretization of the ODEs. We solved these parameter estimation problems to global optimality using the ANTIGONE v1.0 solver [17] in GAMS v24.7.4 [18]. For the heat transfer model, we derived the closed-form analytical solution from the energy conservation equation and solved the global optimization problem using the EAGO v0.2.1 solver (EAGO.jl) [19] in the Julia programming language [20]. We conclude that the physics-informed data-driven modeling methodology developed in this study has great potential to predict physicochemical state and heterogeneity within WWTPs. Such models enable the next-generation WWTPs with energy-optimizing control systems and improved robustness when implemented within MPC .

References

[1] K. V. Gernaey, M. C. van Loosdrecht, M. Henze, M. Lind, and S. B. Jørgensen, "Activated sludge wastewater treatment plant modelling and simulation: state of the art," *Environmental modelling & software*, vol. 19, no. 9, pp. 763-783, 2004.

[2] S. R. Qasim, *Wastewater treatment plants: planning, design, and operation*. Routledge, 2017.

[3] M. Benosman, "Model-based vs data-driven adaptive control: An overview," *International Journal of Adaptive Control and Signal Processing*, vol. 32, no. 5, pp. 753-776, 2018.

[4] E. Aydın, Y. Arkun, G. Is, M. Mutlu, and M. Dikbas, "Plant-wide optimization and control of an industrial diesel hydro-processing plant," *Computers & Chemical Engineering*, vol. 87, pp. 234-245, 2016.

[5] W. Shen, X. Chen, M. Pons, and J. Corriou, "Model predictive control for wastewater treatment process with feedforward compensation," *Chemical Engineering Journal*, vol. 155, no. 1-2, pp. 161-174, 2009.

[6] N. Wahab, R. Katebi, J. Balderud, and M. Rahmat, "Data-driven adaptive model-based predictive control with application in wastewater systems," *IET Control Theory & Applications*, vol. 5, no. 6, pp. 803-812, 2011.

[7] O. Bello, Y. Hamam, and K. Djouani, "Coagulation process control in water treatment plants using multiple model predictive control," *Alexandria Engineering Journal*, vol. 53, no. 4, pp. 939-948, 2014.

[8] L. Corominas, X. Flores-Alsina, L. Snip, and P. A. Vanrolleghem, "Comparison of different modeling approaches to better evaluate greenhouse gas emissions from whole wastewater treatment plants," *Biotechnology and Bioengineering*, vol. 109, no. 11, pp. 2854-2863, 2012.

[9] C. Rosén and J. Lennox, "Multivariate and multiscale monitoring of wastewater treatment operation," *Water research*, vol. 35, no. 14, pp. 3402-3410, 2001.

[10] L. Belanche, J. J. Valdés, J. Comas, I. Roda, and M. Poch, "Prediction of the bulking phenomenon in wastewater treatment plants," *Artificial Intelligence in Engineering*, vol. 14, no. 4, pp. 307-317, 2000.

[11] Z. Ge, "Review on data-driven modeling and monitoring for plant-wide industrial processes," *Chemometrics and Intelligent Laboratory Systems*, vol. 171, pp. 16-25, 2017.

[12] W. Yan, D. Tang, and Y. Lin, "A data-driven soft sensor modeling method based on deep learning and its application," *IEEE Transactions on Industrial Electronics*, vol. 64, no. 5, pp. 4237-4245, 2016.

[13] Z. Xu, Q. Dong, B. Otieno, Y. Liu, I. Williams, D. Cai, Y. Li, Y. Lei, and B. Li, "Real-time in situ sensing of multiple water quality related parameters using micro-electrode array (MEA) fabricated by inkjet-printing technology (IPT)," *Sensors and Actuators B: Chemical*, vol. 237, pp. 1108-1119, 2016.

[14] Z. Xu, F. M. Dehkordy, Y. Li, Y. Fan, T. Wang, Y. Huang, W. Zhou, Q. Dong, Y. Lei, M. D. Stuber, et al., "High-fidelity profiling and modeling of heterogeneity in wastewater systems using milli-electrode array (MEA): Toward high-efficiency and energy-saving operation," *Water research*, vol. 165, p. 114971, 2019.

[15] E. Levlin, "Conductivity measurements for controlling municipal waste-water treatment," in *Proceedings of a Polish-Swedish-Ukrainian Seminar, Utron*, 2010.

[16] I. A. AI-Ghusain, J. Huang, O. J. Hao, and B. S. Lim, "Using pH as a real-time control parameter for wastewater treatment and sludge digestion processes," *Water Science and Technology*, vol. 30, no. 4, p. 159, 1994.

[17] R. Misener and C. A. Floudas, "Antigone: algorithms for continuous/integer global optimization of nonlinear equations," *Journal of Global Optimization*, vol. 59, no. 2-3, pp. 503-526, 2014.

[18] GAMS Development Corporation, "General algebraic modeling system (GAMS) release 24.7.4," September 2016.

[19] M. Wilhelm and M. D. Stuber, "EAGO: Easy advanced global optimization Julia package," 2018. https://github.com/PSORLab/EAGO.jl.

[20] J. Bezanson, A. Edelman, S. Karpinski, and V. B. Shah, "Julia: A fresh approach to numerical computing," *SIAM review*, vol. 59, no. 1, pp. 65-98, 2017.

595988 Towards the Use of Multi-Relational Data Mining (MRDM) in Bioprocess Models

Abstract:

In the era of digitalization, the complexity of bioprocesses stands in the way of the biomanufacturing industry to fully adopt and benefit from the Industry 4.0[1] standards. Even though several aspects of next-generation manufacturing can be adopted from other industries as "of the shelf", such as the management of data and AI algorithms, making use of the underlying process knowledge and models is key to achieve smart biomanufacturing.

So far, a significant investment has been made in developing mechanistic models for bioprocesses, both by academia and industry. However, little thought has been put into how to store and keep on developing and exploiting these models. Hence, even if they contain vast and comprehensive knowledge about the process, this information might be misplaced or even lost.

A potential solution is the use of relational databases. They store available information by applying logical connections and consequently allow the creation of rationally interconnected repositories. Furthermore, this structure enables the use of multi-relational data mining (MRDM) as a tool to find inter-table patterns[2]. MRDM has been successfully applied to other disciplines, such as in bioinformatics[3], once it provides a basis for finding new premises and relationships within the information discriminated and disseminated across the multiple tables. Therefore, in order to get a step closer to smart biomanufacturing, this study has a two-fold objective: (i) to design and implement a relational database which aims at collecting mechanistic models for bioprocesses, and related process information; as well as (ii) to explore the potential behind MRDM with regards to biomanufacturing.

The relational database, developed in this work, collects knowledge in several interconnected tables. In these tables, the following data is recorded: (I) process conditions (e.g. pH, temperature, mode of operation, etc.); (II) model details (type and characteristics); (III) model parameters; and, (IV) uncertainty ranges associated with the parameters (when available)[4]. Additionally, the database contains a table of functional problems, based on the operational conditions (e.g. the cooling system is not working), and a table with the corresponding feasible solutions (e.g. adjust the cooling water flow). Subsequently, an inductive logic programming and relational data mining algorithm[5] are implemented and applied to the database. This allows, for example, finding connections between new models and the library of operational problems simply based on the implicit characteristics of the model. In order to highlight the applicability and usefulness of the proposed strategy, it is demonstrated on an industrial anaerobic microbial conversion case study.

(1) Narayanan, H.; Luna, M. F.; von Stosch, M.; Cruz Bournazou, M. N.; Polotti, G.; Morbidelli, M.; Butté, A.; Sokolov, M. Bioprocessing in the Digital Age: The Role of Process Models. *Biotechnol. J.* **2019**, *1900172*, 1–10. https://doi.org/10.1002/biot.201900172.

(2) Ferreira, C. A.; João, G.; Victor, S. C. Sequential Pattern Mining in Multi-Relational Datasets. In *Current Topics in Artificial Intelligence*; Meseguer, P., Mandow, L., M. Gasca, R., Eds.; **2009**; pp 121–130.

(3) Page, D.; Craven, M. Biological Applications of Multi-Relational Data Mining. *ACM SIGKDD Explor. Newsl.* **2003**, *5* (1), 69. https://doi.org/10.1145/959242.959250.

(4) Caño de las Heras, S.; Krühne, U.; Mansouri, S. S. Relational Database for the Description of Fermentation inside a Simulation Software. In *ECAB 5 The 5th EUROPEAN CONGRESS OF APPLIED BIOTECHNOLOGY*; Florence, **2019**; pp 1328–1329.

(5) Srinivasan, A. Aleph Manual http://www.cs.ox.ac.uk/activities/programinduction/Aleph/aleph.html#SEC4 (accessed Mar 6, 2020).

600691 The Algorithm to Discriminate Factors of Defects and Determine the Countermeasures from Process Data and Application to Highly-Functional Film Process

Abstract:

The highly-functional film produced by coating various varnishes on the base film depending on the purpose has been used in a wide range of industries such as the electronics field and the pharmaceutical field.In general, the highly-functional film manufacturing process is composite (e.g. coating and drying) and continuous process. Thus, **when defects are generated, production line operators take countermeasures through trial and error because of difficulty of discriminating factors of defects and it takes long production downtime for countermeasure.**

To address this problem, recently, the development of the factor analysis system by using process data from sensors installed in the production line has been promoted[1]. In the present system, features of sensor data are statistically analyzed and the degrees of influence of each feature to the product quality are scored. Therefore, the sensors which mostly affect product quality can be extracted. However, on the present system, the operators need to estimate the factors of defects and determine the countermeasure based on the extracted sensors because the present system does not automatically select factors of defects and countermeasures. To solve this issue, **we have proposed the algorithm that discriminates factors of defects and determine the appropriate countermeasure by using process data and past countermeasures record**.

In the proposed algorithm, firstly, countermeasures implemented in production line are linked with factors of defects extracted by Fault Tree Analysis of manufacturing process in advance. Secondly, various sensors related to the factors are selected. **Thirdly, for each factor, the regression model of the factor occurrence is constructed by Gaussian Process Regressor**. Here, the normal state data where no defects are generated are used as training data. Finally, the anomaly score of the factors are calculated during manufacturing, then the factor with the highest anomaly score and appropriate countermeasure can be determined.

We applied the proposed algorithm to the highly-functional film production line. **As a result of verifying the algorithm by using actual process data, it was confirmed that the accuracy of the predicted countermeasure was higher than 90%**. By automating the determination of countermeasure by using the algorithm, downtime for countermeasure is reduced and productivity is improved.

Reference

[1]Haifeng Chen, Mizoguchi Takehiko, Yan Tan, Kai Zhang, Geoff Jiang, proceedings of the 2015 45th Annual IEEE/IFIP International Conference on Dependable Systems and Networks, 529–536, 2015

610913 Maintenance Testing in Precision Machining

Abstract:

Manufacturing processes like face milling that involve controlled removal of material are energy intensive [1]. Here, the digital twin of a precision machining process is developed that can be used to minimize the cost and power consumption of machining operations and for model-based maintenance. For the latter, a methodology for simultaneous selection of sensors and test settings is developed for the digital twin to determine faults in the machining operations. Robust maintenance algorithms and standard work target energy consumption reduction through early detection of faults in the system that prevent losses from downtime. The method is applied on a simple case study of face milling of metals. Face milling is a precision machining process that involves material removal, scrap generation from metal to metal interaction and cutting fluid utilization for lubrication and heat removal [2]. The digital twin of face milling was formulated in this work as a physics-based model based on [3] to compute power consumption and scrap generation. The model of a Mazak Variaxis 630-5X II T Mill/Turn Machine 40 hp 5-axis CNC machine was validated against experimental data, shown in Fig.1, of the power consumption for one pass dry face milling. The model was formulated as a system of algebraic equations with inputs, outputs, parameters and faults as detailed in [5]. The experimental facility is instrumented to acquire process data for machine power, spindle power, spindle vibration, and to record audio/video and is MTConnect capable. Moreover, the machine can be controlled to perform milling at various speeds, widths of cut, depths of cut and feed rates, which provides a wide space for maintenance testing. In this work, we explore the issue of maintenance testing as a non-linear program that selects maintenance test conditions and sensors on the basis of information criteria. Common uncertainty in the milling process include the material hardness of the workpiece, friction factor, depth of cut and the width of cut. The most common fault (target of maintenance testing) is the tool wear [4]. Optimal selection of test settings and sensors for the purpose of tool wear prediction is performed by applying the model-based method proposed in [5] for fault detection and isolation (FDI). The method formulates the selection of test settings and sensors as discrete and continuous variables, respectively; and the selection is done based on the contribution of each to the information gain with respect to the estimation of tool wear. The problem is formulated as a constrained mixed integer non-linear problem and solved for all possible combinations of available sensors. The objective of the optimization problem is to maximize a measure of the normalized (for the number of sensors used) Fisher Information Matrix (FIM), which is calculated from the sensitivities of the outputs with respect to system faults and uncertainties. The isolation capacity of the test settings is determined from the Kullback-Leibler divergence of healthy and faulty system measurements and the confusion matrix of tool wear classification. Classification of the tool wear is performed using the k-nearest neighbor algorithm for the optimal maintenance test with the most informative set of sensors. It is shown that classification accuracy is variable within the operating space of the machine, and that noisy sensors can further deteriorate tool wear condition assessment. Therefore, a maintenance test for the digital twin of the precision machining equipment that optimizes test conditions and sensor network is imperative.

Acknowledgment: This material is based upon work supported by the U.S. Department of Energy's Office of Energy Efficiency and Renewable Energy (EERE) under the Advanced Manufacturing Office Award Number DE-EE0007613.

for the accuracy, completeness, or usefulness of any information, apparatus, product, or process disclosed, or represents that its use would not infringe privately owned rights. Reference herein to any specific commercial product, process, or service by trade name, trademark, manufacturer, or otherwise does not necessarily constitute or imply its endorsement, recommendation, or favoring by the United States Government or any agency thereof. The views and opinions of authors expressed herein do not necessarily state or reflect those of the United States Government or any agency thereof."

References

1. EIA, 2011. Annual Energy Review. http://www.eia.gov/totalenergy/data/annual/index.cfm.
2. https://www.sandvik.coromant.com/en-us/knowledge/milling/pages/face-milling.aspx
3. Shao, H., Wang, H. L., & Zhao, X. M. (2004) A cutting power model for tool wear monitoring in milling. *International Journal of Machine Tools and Manufacture*, *44*(14), 1503–1509. https://doi.org/10.1016/j.ijmachtools.2004.05.003.
4. Altintas, Y., Yellowley, I., & Tlusty, J. (1988) The detection of tool breakage in milling operations. *Journal of Manufacturing Science and Engineering, Transactions of the ASME*, *110*(3), 271–277. https://doi.org/10.1115/1.3187881
5. U. Awasthi, K.A. Palmer, G.M. Bollas (2020) Optimal test and Sensor Selection for Active fault diagnosis using Integer programming, Journal of Process Control, (under review).
6. U. Awasthi and G.M. Bollas (2020) Sensor network design for smart manufacturing { Application on precision machining, 21st IFAC World Congress, Berlin, Germany.

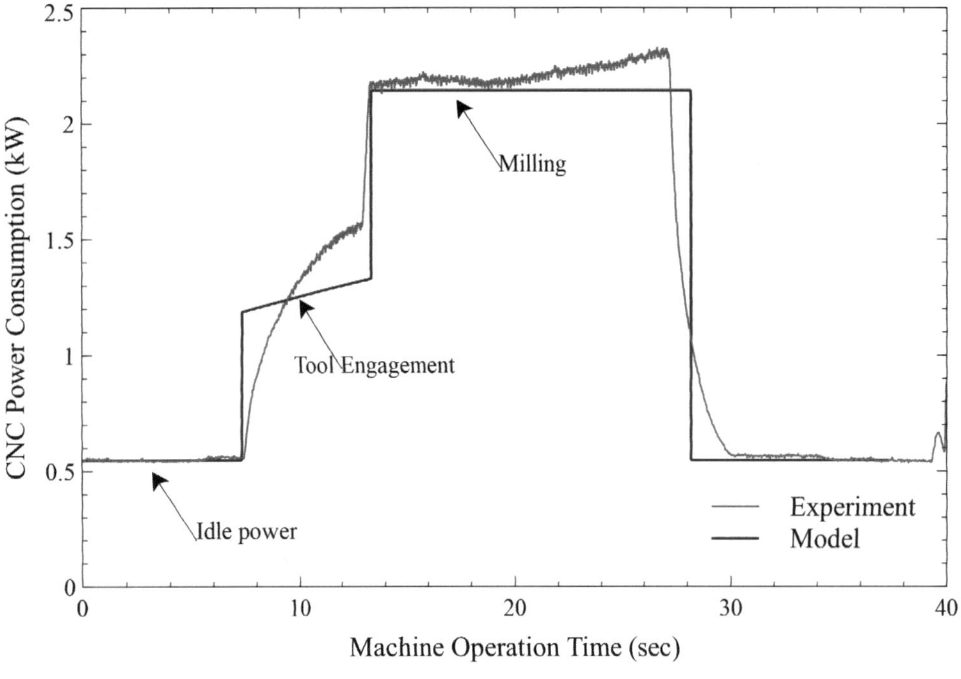

Figure 1: Model vs. Experimental data for face milling operation.

606609 Convolutional Neural Network for Predicting Morphology, Flow and Transport Properties of Complex Materials

Abstract:

Physical properties of porous media and materials, such as their permeability, electrical conductivity, and elastic moduli are controlled by their morphology, i.e., their pore-size distribution, pore shapes, and connectivity. Experimental studies provide much insight, but they often are costly. Computational studies that involve model of such materials and media involve assumptions, approximation, and simplifications. On the other hand, advancements in imaging techniques have made it possible to use two- and three-dimensional (3D) images of materials in direct computations without any assumptions or simplifications. With advances in the development of convolutional neural networks (CNNs), it is now possible to combine the CNNs with images of porous materials and media in order to make accurate predictions for their physical properties. But one must first address the issue of lack of a large dataset, as well as the required resolution in the available images.

Using a recently developed stochastic reconstruction technique, we have generated big data for use in a CNN in order to enhance the resolution of porous materials' images, with the enhancement evaluated quantitively. We have also developed a hybrid and physics-guided deep learning approach for estimating the permeability of porous materials, which enables us to establish a direct link between the morphology and to the flow properties. Since several stochastic reconstruction algorithms have been developed recently, we have also quantified their using a deep-learning method.

References

Kamrava, S., Tahmasebi, P., & Sahimi, M. (2019). Enhancing images of shale formations by a hybrid stochastic and deep learning algorithm. Neural Networks, 118, 310-320.

Kamrava, S., Tahmasebi, P., & Sahimi, M. (2020). Linking morphology of porous media to their macroscopic permeability by deep learning. Transport in Porous Media, 131(2), 427-448.

Kamrava, S., Sahimi, M., & Tahmasebi, P. (2020). Quantifying accuracy of stochastic methods of reconstructing complex materials by deep learning. Physical Review E, 101(4), 043301.

602234 Developing a Closed-Loop Optimum Pumping Sequence with Shear Thinning Fluid for Enhanced Hydraulic Fracturing

Abstract:

In the United States, 30% of the total energy comes from natural gas, and shale gas continues to make a major contribution to the total Natural Gas production of the nation[1]. Hydraulic fracturing and horizontal drilling are the two major technologies that have made the commercial extraction of hydrocarbons from unconventional reservoirs economically feasible. Specifically, hydraulic fracturing is carried out with slick water, which leads to huge water leak off into the formation and produces (artificial) hydraulic fractures[2].

In recent years, hydraulic fracturing with high-viscosity gels has been one of the most explored topics both in theory and practice. The use of high viscosity gels has resulted in fractures with higher fracture conductivity and longer propped length[3-5]. The most commonly used gels in hydraulic fracturing are shear thinning fluids which sufficiently reduce pressure drop at the wellbore and fractures. One of the major problems associated with using these gels is that they leak off into the reservoir and eventually cause formation damage by chocking pore throats. These gels also lead to sufficient damage inside the fractures by mechanisms like proppant crushing and cake deposition. The formation and fracture damage phenomenon are sometimes very severe and can have a significant negative impact on the oil and gas production[6-8]. However, there is a dearth of study that considers all these factors and proposes an optimal operation sequence for practice. Consequently, the current challenge for the operators is the choice of a gel viscosity that would lead to minimum formation and fracture damage along with determining the proper pumping schedule to achieve optimal fracture geometry.

The work presented here attempts to address this challenge by proposing a novel methodology that can be used for deriving the optimal operating parameters for hydraulic fracturing with gels. Specifically, a novel model describing fracture propagation and multiphase reservoir behavior is developed. This model precisely predicts the formation damage phenomenon due to the leak-off of high-viscosity gel and also predicts the fracture damage due to filter cake deposition. Once the formation and fracture damage are modeled, they are used as the initial conditions for predicting the oil and gas production from a reservoir by utilizing a three-dimensional in-house reservoir simulator. The sensitivity analysis of hydrocarbon production is performed by adjusting three parameters, which are the gel viscosity, average fracture conductivity and propped surface area. The sensitivity analysis gives us a value of gel viscosity at which the maximum oil and gas production is achieved. The optimal value of gel viscosity is then used in developing a pumping schedule to attain the optimal values of average fracture conductivity and propped surface area. Since pumping operation in practice is a multi-input and multi-output system, a closed-loop model predictive controller (MPC) is designed to get an optimum pumping schedule. Specifically, the flow rate of gel and concentration of proppant are the two manipulated inputs, and average fracture conductivity and propped surface area are the two outputs. Additionally, several practical considerations are incorporated through constraints like the maximum amounts of fluid and proppant to be injected. This closed-loop MPC scheme is developed by utilizing a computationally less intensive reduced-order model for the process followed by the development of a Kalman filter to estimate the unmeasurable states using feedback measurements from the process. The MPC problem is formulated including the constraints and the reduced order model, and since, the operation time is fixed it becomes a shrinking horizon closed-loop optimization problem which is solved to obtain the optimal input for each time interval. This closed-loop structure of the proposed methodology extends its application to real-time operations where a plant-model mismatch is plausible. In summary, this work shows that for a given reservoir, we can calculate the viscosity of fracture gel using which an optimum pumping schedule can be obtained for the maximum oil and gas production from that reservoir.

Literature Cited:

[1] Wang, Q., et al., Natural gas from shale formation–the evolution, evidences and challenges of shale gas revolution in United States. Renewable and Sustainable Energy Reviews, 2014. 30: p. 1-28.

[2] Jr., Alfred Jennings. "Fracturing Fluids - Then and Now." *Journal of Petroleum Technology*, vol. 48, no. 7, 1996.

[3] Clark, J.b. "A Hydraulic Process for Increasing the Productivity of Wells." *Journal of Petroleum Technology*, vol. 1, no. 01, Jan. 1949.

[4] Seright, R.s. "Use of Preformed Gels for Conformance Control in Fractured Systems." *SPE Production & Facilities*, vol. 12, no. 01, Jan. 1997.

[5] Al-Muntasheri, Ghaithan A. "A Critical Review of Hydraulic Fracturing Fluids over the Last Decade." *SPE Western North American and Rocky Mountain Joint Meeting*, 2014.

[6] Wang, Yilin, et al. "Simulation of Gel Damage on Fracture-Fluid Cleanup and Long-Term Recovery in Tight-Gas Reservoirs." *SPE Eastern Regional/AAPG Eastern Section Joint Meeting*, 2008.

[7] Maloney, Daniel R., et al. "Non-Darcy Gas Flow Through Propped Fractures: Effects of Partial Saturation, Gel Damage, and Stress." *SPE Production Engineering*, vol. 4, no. 04, 1989, pp. 417–422.

[8] Yi, X. "Model for Displacement of Herschel-Buckley Non-Newtonian Fluid by Newtonian Fluid in Porous Media and Its Application in Fracturing Fluid Cleanup." *SPE International Symposium and Exhibition on Formation Damage Control*, 2004.

602200 Developing a Hybrid Model of a Biochemical Fermentation Process

Abstract:

With ever-increasing energy demands, rising pollution, and over-reliance on the crude oil industry, there is a definite need for a cheap, environment-friendly energy source. Among various renewable sources, ethanol has been perceived as one of the most promising candidates, one which can meet the growing global energy demands. At present, primary methods of ethanol production are the chemical conversion of food sources like corn with a high starch content [1], and thermochemical conversion of lignocellulosic biomass to ethanol [2]. However, both have drawbacks like sustainability in both food and fuel market or high energy demands, respectively. Hence, the recent focus is shifting towards biochemical production of ethanol, which has the potential to utilize many different cellulosic biomass substrates and pretreatment technology options without the use of food resources, which solves the issues of the previous methods [3].

It is important to develop a mathematical model to understand and describe the fermentation process, predict optimal parameters, and help in future automation. However, representing such a process is quite complicated as it has coupled nonlinear differential equations that are difficult to solve and often have uncertain parameters or states. It is difficult to incorporate everything occurring inside a plant, in a model which may result in a plant-model mismatch between optimal and actual ethanol production rate. This mismatch is mainly due to assumptions of constant parameters, a lack of complete process information, and homogeneity of properties for making the process computationally less intensive. Hence, it is crucial to identify the uncertain parameters resulting in this plant-model mismatch. To this end, we perform a global sensitivity analysis by varying the nominal parameters in a stipulated range to obtain the set of the most important model parameters, which profoundly affects the model outputs.

Based on the global sensitivity analysis results, we built a neural network model with temperature as input and ethanol rate as the output maximizing the productivity and keeping the ethanol concentration constant. We then combined it with the first principle model to develop a hybrid model that gives us better prediction accuracy, extrapolation capabilities, and generalization enabling us to reduce the existing plant-model mismatch [4]. Once we had this hybrid model, we needed to estimate the state variables in the process to maintain them at optimal operating conditions adequately. For this, we designed a state observer that estimated the internal states of the biochemical system like the ethanol production and substrate concentrations. We then validated our hybrid model and the designed observer against an entirely new set of process data, and results showed that the model works well in describing the process. Then, we determined optimal conditions to achieve a set of target values of the ethanol production rate while ensuring the safety of the operation and separation of the product from biological residues. It is essential to develop a relationship between the estimated optimal production rate of ethanol and the variation of temperature needed to attain it. In this regard, we developed a Model Predictive Control (MPC) algorithm to maintain the temperature at suitable levels considering all practical plant constraints. Finally, we combined these algorithmic models and implement them on a real industry-scale ethanol fermentation plant and demonstrated the performance of the proposed MPC algorithm with real plant data and monitoring.

1. Thompson, P., The Agricultural Ethics of Biofuels: The Food vs. Fuel Debate. Agriculture, 2012.

2: p. 339-358.

2. Niziolek, A.M., et al., Coal and Biomass to Liquid Transportation Fuels: Process Synthesis and Global Optimization Strategies. Industrial & Engineering Chemistry Research, 2014. 53: p. 17002-17025.

3. Raftery, Jonathan Patrick(2017). Continuous Biochemical Processing: Investigating Novel Strategies to Produce Sustainable Fuels and Pharmaceuticals. Doctoral dissertation, Texas A&M University.

4. Bangi, Mohammed Saad Faizan, and Joseph Sang-Il Kwon. "Deep Hybrid Modeling of Chemical Process: Application to Hydraulic Fracturing." *Computers & Chemical Engineering*, vol. 134, 2020, p. 5. *Crossref*, doi:10.1016/j.compchemeng.2019.106696.

607784 A Combined Computational-Fluid-Dynamics Model and Advanced Control Strategies for Direct Perfusion Bioreactor Systems

Abstract:

Tissue engineering (TE) involves growing cells within supporting scaffolds to obtain structures for in vivo implantation with adequate functionality. For the cultivation of cells, bioreactor systems are mandatory to receive tissue-engineered grafts with uniform cell distribution, growth, and viability in a reproducible way. The application of bioreactor systems gives rise to improved tissue quality with respect to static cultivation by providing proper cultivation conditions to mimic an in vivo environment [1]. To allow for the investigation of proper cultivation conditions and the reproducible generation of tissue engineered grafts, a bioreactor system, which comprises the control of crucial cultivation parameters, i.e. flow rate and nutrient concentration, in bioreactors, is beneficial. Furthermore, the use of a bioreactor as an automated cell seeding tool enables even cell distributions on stable scaffolds. The use of a perfused bioreactor allows for an optimal supply with nutrients, while toxic metabolites are effectively removed from the cell culture.[2] One of the major challenges bioreactor technology is facing at present is to translate research-scale production models into clinically applicable manufacturing designs that are reproducible, clinically effective and economically acceptable, while complying with Good Manufacturing Practice (GMP) requirements [3].

The development of mathematically and computationally orientated research has failed to catch up with the recent developments in biology. This can be attributed to a lack of true integration between engineering and biological disciplines [4]. However, it is becoming more apparent that computational methods can be a powerful and cost-effective tool for the design, optimization and control of bioreactor systems. With all the new recent advances, TE is now at the stage where it can shift from research-based technology into large-scale and commercially successful products. Nevertheless, we are ultimately faced with the fact that even the most clinically successful products will need to demonstrate cost-effectiveness and cost-benefits over existing therapies, absolute safety for patients, manufacturers, and the environment, and compliance to the current regulations.

This work sets the foundations for the design of several control algorithms to facilitate manufacturing for any type of cell culture using a continuous perfusion bioreactor. The algorithms are also capable of dealing with any disturbances in the process. Different types of control strategies are designed, implemented and tested starting from classic PID controller as well as advance model predictive control strategies. These strategies are designed to be able to work with different manipulated and controlled variables depending on the needs of the process. As Computational Fluid Dynamics (CFD) has the capability of describing the interplay between the flow field in a perfusion bioreactor and the cell growth kinetics the latter will be used to develop a comprehensive high fidelity mathematical model. Usually the mathematical model for such process is too complex to be used directly for control studies and therefore, a simplified version is approximated using discrete time models in state space form via model order reduction techniques or system identification techniques. The reduced model is then used to facilitate the implementation of advanced control strategies. Finally, the performance of the control strategies is validated against the original high-fidelity CFD model.

To test the performance and limitations of the developed control strategy, the controllers are tested for varying operating targets, process disturbances as well as changes in the parameters and for different cell cultures. This work illustrates how using model-based control approaches greatly improves the time and resource utilization during bioreactor operation. It will reduce or even eliminate the need for Design of Experiments (DoE) to design new processes. Moreover, the developed mathematical model can be further used to gain in depth understanding of the process and as a testing platform for the designed controllers. By automating and standardizing tissue manufacture controlled closed systems, bioreactors could reduce production costs and time, thus facilitating a wider use of engineered tissues. Moreover it can assure consistency of product quality and of the time spent producing the product which will bring great benefits from a scheduling point of view.

1. Schmid, J., et al., *A Perfusion Bioreactor System for Cell Seeding and Oxygen-Controlled Cultivation of Three-Dimensional Cell Cultures.* Tissue Eng Part C Methods, 2018. **24**(10): p. 585-595.
2. Coletti, F., S. Macchietto, and N. Nlvassore, *Mathematical modelling of three-dimensional cell cultures in perfusion bioreactors. Part II*, in *Computer Aided Chemical Engineering*, W. Marquardt and C. Pantelides, Editors. 2006, Elsevier. p. 1699-1704.
3. Martin, I., D. Wendt, and M. Heberer, *The role of bioreactors in tissue engineering.* Trends in Biotechnology, 2004. **22**(2): p. 80-86.
4. Kiparissides, A., et al., *'Closing the loop' in biological systems modeling — From the in silico to the in vitro.* Automatica, 2011. **47**(6): p. 1147-1155.

600639 Modeling a Direct- Fired Industrial Tunnel Furnace Using an Implicit Multiple 1D Approach

Abstract:

Today's expectations regarding CFD simulations for industrial furnaces are ambitious since they represent a viable option to reduce the need of conducting usually expensive and time-consuming experiments. Furthermore, CFD simulations provide the possibility to estimate certain process parameters, as for example local concentrations or temperature profiles, where conventional experiments and measurements fail. Although offering a broad spectrum of advantages, most simulations require immense computational power because of the high number of cells and algorithms, which must be iteratively solved. Due to these disadvantages the time required to deliver results is high. Therefore, both experiments and CFD simulations, are incapable of providing essential data needed for process related decisions, raising the need of reduced furnace models that can generate results faster.

This paper proposes a method that can decrease computational cost significantly, by separating the simulation domain into several coupled one-dimensional regions. Thus, hours of process time and associated furnace parameters can be predicted within minutes of calculation. The advantage by introducing reduced furnace models is emphasized by several authors who are using similar mathematical concepts and discretization strategies for computation. The fundamental difference between reduced furnace models and CFD simulations is the significant difference in cell resolution and connectivity. In consequence of the interconnected one- dimensional discretization the total number of cells is reduced drastically, hence reducing the cell resolution and improving the simulation speed. Additionally, the connectivity of the respective cell regions is only towards one dimension which further improves the calculation performance because of the reduced number of equations to solve. Based on this approach the generated results represent global process parameters like the average gas temperature or concentration within a specific furnace compartment. Those results can be utilized afterwards to determine the energy efficiency of a specific furnace setup or the effect on the heat-treated products in regard of quality control.

Although, if using reduced furnace models as a data generating method in addition to CFD simulations and experiments seem to be promising, caution must be exercised when using them. More specifically, a detailed knowledge of the heat-treated products, as well as the required dew temperatures are of uppermost importance in order to implement optimal solutions. This work concentrates on a direct- fired industrial tunnel furnace that treats products at temperatures above 1400°C, making radiation to the major mode of heat transfer. The modeling of the direct- firing process bases on an equilibrium approach utilizing calculated gas concentrations to determine the released energy supply in the burning zone. Furthermore, the additional modes of heat transfer, conduction and convection, are added to give a complete physical description. The product feed is done on a continuous basis depending on the pulling time of the respective products, which is deposited in a production plan provided to the simulation. Subsequently to the model implementation, an existing furnace of industry partners is modeled and afterwards validated by measurement data. Ultimately, the generated temperature profiles show a high potential to predict the influence of the applied heat on the processed products properties along the furnace axis.

610393 Particle-Resolved Packed Bed Reactor Simulation: Advances in Contact Point Treatment

Abstract:

Packed bed reactors are one of the working horses in the chemical and process industry. They are used, amongst others, for highly exothermic or endothermic catalytic reactions like (partial-)oxidation or (dry, steam) reformation of hydrocarbons. These types of reactor are used for several decades, but there is still a lack of understanding and room for improvement of the performance of these reactors. Recent advances in modeling particle-resolved packed beds [1,2,3] allows a detailed insight in the flow, species and temperature distribution as well as into the reactions in the beds. One challenge from the simulation point of view is the treatment of the contact points. The local flattening approach by [1] is an efficient and accurate way to handle this, but in the past it was mainly used for spherical particles.

In this contribution we show how this methodology can be modified to enable the handling of non-spherical particles like cylinders or Raschig-rings, which are more commonly used in industry. The modification lead to a well defined gap and do not affect the process regarding runtime nor accuracy. This will be shown for several industrial relevant particle shapes as well as for a highly complex concave-convex shape.

With this methodology new and more efficient particle shapes can be virtually designed and developed.

1. Eppinger, et al., "DEM-CFD simulations of fixed bed reactors with small tube to particle diameter ratios", *Chem. Eng. J.* 166 (1), pp 324-331 (2011)
2. G. D. Wehinger et al., "Detailed numerical simulations of catalytic fixed-bed reactors: Heterogeneous dry reforming of methane", *Chem. Eng. Sci.* 122, pp 197-209 (2015)
3. N. Jurtz et al., "Advances in fixed-bed reactor modeling using particle-resolved computational fluid dynamics (CFD)", *Rev. Chem. Eng.* 35(2), pp 139-190 (2019)

605936 Continuous Mixing Technology: The Impact of Blend Properties and Process Parameters on Mixing Performance

Abstract:

Continuous powder mixing technology application during continuous direct compression process has emerged as a leading technology used in the development and manufacture of solid oral dosage forms [1, 2]. The critical quality attributes (CQAs) of the final product is heavily depending on the performance of the mixing step as the quality of mixing will directly influences the content uniformity of the tablet.

Therefore, the aim of this study was to investigate the impact of blend properties (bulk density, API sizes) and process parameters (process throughput, hold up mass and impeller speed) on the mixing performance. Mixing performance was characterized using the residence time distribution (RTD) of the process, which has been broadly used to characterize unit operations in pharmaceutical processes. In this work, the RTDs for a vertical continuous mixing device termed Continuous Mixing Technology (CMT) were obtained in order to evaluate mixing performance over a defined operating space. The RTDs were determined using a spike injection of tracer material into the mixer and measuring the concentration of the tracer in the outgoing material. This was achieved using a PAT interface situated at the exit of the CMT that presents the material to a near-infrared (NIR) probe. The theoretical residence time of the mixer is given by the hold-up mass divided by the throughput [1], and for each experiment, NIR spectra of the exiting material was measured for five theoretical residence times after tracer injection to obtain the full RTD.

As the CMT contains an upper de-lumping screen and a lower mixing chamber, the residence time distributions of this system can be described as two CSTRs in series with different residence times [1] where the parameter ‘r’ describes the ratio between these residence times. The best value of ‘r’ for all experiments was determined through simultaneous optimization of all measured RTDs. The results showed that the CMT operates close to a single CSTR over the whole operating space, has good ability to dampen the fluctuations from the gravimetric feeders, and that the mixing performance of the CMT is not significantly impacted by blend properties (i.e. bulk density and API particle size).

References:

[1] Blackwood, D.O., Bonnassieux, A. and Cogoni, G. Continuous Direct Compression Using Portable Continuous Miniature Modular & Manufacturing (PCM&M). In *Chemical Engineering in the Pharmaceutical Industry: Drug Product Design, Development, and Modeling*, 2nd Edition, am Ende, T., am Ende, D. J., Wiley, 2019, pp 547-560.

[2] Markarian, J. (2 April 2018). Modernizing Pharma Manufacturing. www.pharmtech.com

600731 Investigation of Critical Process Parameters of Screw Elements of a Twin Screw Extruder By an Improved Simulation Method

Abstract:

INTRODUCTION

A commonly used technique for the production of pharmaceutical products is hot melt extrusion on a co-rotating intermeshing twin screw extruder [1]. An active pharmaceutical ingredient is embedded in a molten polymeric carrier and the product is conveyed to the die [2]. The extruder screws have a modular design with, for example, conveying, kneading or mixing elements. Each element has specific properties in order to combine different unit operations in one process unit. The extrusion process is mainly influenced by the chosen screw configuration and the geometry of each screw element. [3]

Conveying elements are, depending on their pitch, responsible for the pressure build-up and the conveying of the material to the die. Kneading elements are mainly used in the melting zone because they introduce shear forces into the material according to the staggering angle of the kneading discs. Mixing elements are responsible for distributive and dispersive mixing. [4] The different types of elements can be seen in Figure 1.

Each specific screw element can be characterized by dimensionless numbers regarding the pressure and power characteristic (Figure 2). These parameters are linearly related to the dimensionless volume flow. The intersection points with the axes result in the dimensionless numbers A_1 and A_2 for the pressure build-up and B_1 and B_2 for the power input into the material. [6]

All parameters mentioned affect the residence time distribution, which indicates how long the material is exposed to the process conditions. For this reason, especially this parameter is of crucial importance for the manufacturing of pharmaceutical products with a constant product quality. Furthermore, it requires process understanding and the prediction of important process parameters. [7] Based on these challenges, the modelling has been gaining substantial interest for pharmaceutical applications within the last decade [8]. Especially the 3D simulation has moved into the focus of development due to its resolution in all special directions and its closeness to reality. Currently, it is not possible to calculate complete screw configurations of a production extruder in a reasonable time because of the high computational effort. The high amount of time is caused by the use of dynamic meshes. [9]

The aim of this work is the implementation of a 3D simulation in Ansys Fluent with the possibility to calculate a complete extruder screw configuration in a reasonable time. Furthermore, the impact of abrasion of the screw elements, which is unavoidable by the daily use in the production process, on important parameters for the extrusion process should be investigated on the basis of the explained dimensionless numbers. In addition to these correlations, the residence time distribution will be used as an important process parameter to characterize the screw elements. For this purpose, a dimensionless relationship analogous to the pressure and power characteristics will be derived.

METHODS AND MATERIALS

The simulation was performed in Ansys Fluent based on the Navier Stokes equations. A stationary, laminar and isothermal flow with an incompressible Newtonian fluid was assumed. Normally, a dynamic mesh is used for an exact representation of the screw rotation. In this case, however, an apparent movement using the method of pulse transmission was implemented. In a first step conveying elements with different pitches (20, 30, 40 mm) were constructed in Solid Works and then imported to Ansys. For the determination of the dimensionless pressure and power characteristics, the mass flow, the pressure difference across the screw element and the torque, which can be converted into a power, were evaluated. In order to investigate the effect of abrasion, screw elements currently used in the laboratory were compared with new elements and those specified by Leistritz as the lower limit. The abrasion of the barrel was not considered.

The simulation was validated by extrusion experiments performed on a twin screw extruder (ZSE27 MAXX, Leistritz Extrusionstechnik, Nuremberg, Germany) with a screw diameter of 27 mm and a customized length of five barrel elements. Only the last two elements (8D) before the die were part of the extrusion process to avoid disturbances. A silicon oil with a viscosity of 85 Pa s (Bluesil FLD 47V100000, Brenntag GmbH, Duisburg, Germany) was extruded. The pressure difference was measured with two pressure sensors (S-11, WIKA Alexander Wiegand SE & Co. KG, Klingenberg, Germany) with a distance of 30 mm. The mass flow was measured with a scale (MSE2203S-100-DR, Satorius, Göttingen, Germany) and the torque was automatically recorded by the extruder. The screw speed was set at 50 rpm and the mass flow was varied by using different die diameters (2, 4, 6 mm).

For the determination of the residence time distribution a passive scalar tracer was placed on the stationary solution of the fluid field simulation. The discretization was performed with the second order upwind method.

Experiments to validate the simulation were carried out in analogy to the measurements for the dimensionless numbers. A tracer (Eosin, Merck, Darmstadt, Germany) was added at the material inlet and the resulted signal was detected at the die with an inline UV/Vis-Spectrometer (InspectroX, ColVisTec AG, Berlin, Germany).

From the obtained residence time distribution, the Bodenstein number was calculated, which indicates the ratio of convective flows to dispersive flows and is thus a measure of the remixing within the extruder. The computations were performed with an internal python script, which fits the residence time density functions based on the axial dispersion model. This number was plotted over the screw speed for different die diameters, in order to clearly determine the resulting residence time distribution for each screw element. By a convolution of the individual residence time distributions, the resulting residence time distribution for a desired screw configuration can be determined.

RESULTS

The 3D simulation was successfully implemented in Ansys Fluent with the pulse transmission method. The dimensionless numbers regarding the pressure and power characteristics of the screw elements could be validated by extrusion experiments. Furthermore, the impact of the abrasion on the process parameters was investigated. The pressure and power characteristics were determined for conveying elements with different inner (D_{in}) and outer (D_{out}) screw diameters. Table 1 shows the results for the dimensionless pressure numbers. New, unused screw elements (new), screw elements at the lower limit (old) and actually used screw elements (used) were compared.

It can be seen that the gap between the screws and the barrel and the two screws have a great impact on the pressure build up during the extrusion process. The greater the abrasion of the elements, the less pressure build-up is possible. Next to the pressure build up, the gap width affects other process parameters, like the viscosity or the residence time distribution.

For this reason, it should be also possible to predict the residence time for individual screw elements. The residence time density function determined with Ansys Fluent was used to determine the Bodenstein number and the mean residence time and plotted against screw speed for different die diameters (Figure 3). It is obviously that the Bodenstein number increases with increasing volume flow while the mean residence time decreases.

CONCLUSION

In this work a 3D simulation in Ansys Fluent was implemented with the aim of a shorter calculation time for the representation of the extrusion process. Dimensionless numbers were used to validate the simulation and to show the impact of abrasion of the behavior of screw elements during the process. The deviations may also be significant in relation to the application of models to predict important process parameters. One of the most important parameter is the residence time, which is therefore used as a further parameter for the characterization of screw elements. By determining the Bodenstein number and the mean residence time, it is possible to predict the residence time distribution of the entire process for the specific extrusion process.

REFERENCES

1. Repka, M., et al.; Applications of hot-melt extrusion for drug delivery, Expert Opin Drug Deliv, 5 (12), 2008
2. Lang, B., McGinity, J.W., Williams, R.O.; Hot-melt extrusion - basic principles and pharmaceutical applications, Drug Dev Ind Pharm, 40 (9), 2014
3. Crowley, M. M., et al.; Pharmaceutical applications of hot-melt extrusion. Drug Dev Ind Pharm, 33(9), 2007
4. Gogos, C.G., Huiju, L., Wang, P.; Laminar Dispersive and Distributive Mixing with Dissolution and Applications to Hot-melt Extrusion, in: Douroumis, D. (Ed.), Hot-melt Extrusion: Pharmaceutical Applications, 2012
5. Sarhangi Fard, A., et al.; Tools to Simulate Distributive Mixing in Twin-Screw Extruders, Macromol Theor Simul, 21 (4), 2012
6. Eitzlmayr, A., et al.; Mechanistic modeling of modular co-rotating twin-screw extruders, Int J Pharm, 474 (1-2), 2014
7. Wesholowski, J., Berghaus, A., Thommes, M.; Investigations Concerning the Residence Time Distribution of Twin-Screw-Extrusion Processes as Indicator for Inherent Mixing, Pharmaceutics, 10 (4), 2018
8. Bochmann, E. S.; Gryczke, A.; Wagner, K. G.; Validation of Model-Based Melt Viscosity in Hot-Melt Extrusion Numerical Simulation, Pharmaceutics, 10 (3), 2018
9. Bravo, V. L., Hrymak, A. N., Wright, J. D.; Numerical simulation of pressure and velocity profiles in kneading elements of a co-rotating twin screw extruder, Polym Eng Sci, 40 (2), 2004

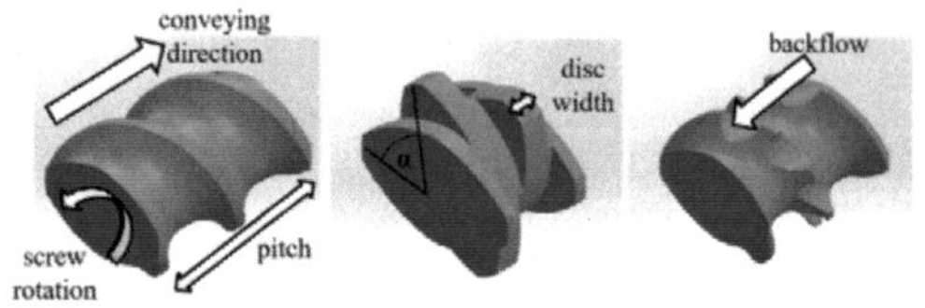

Figure 1: Conveying, kneading and mixing element, screw rotation and conveying direction are equal, according to [5]

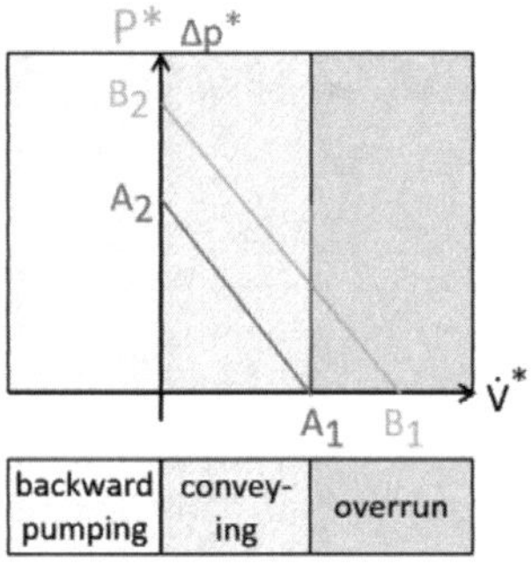

Figure 2: Dimensionless pressure and power characteristic for one specific screw element, according to [6]

Table 1: Comparison of dimensionless pressure numbers (A_1 left, A_2 right) for different geometries of GFA-2-20-30

	D_{in_used}		D_{in_new}		D_{in_old}	
D_{out_used}	0.187	3422				
D_{out_new}			0.183	3665	0.196	3372
D_{out_old}			0.198	2075	0.215	1957

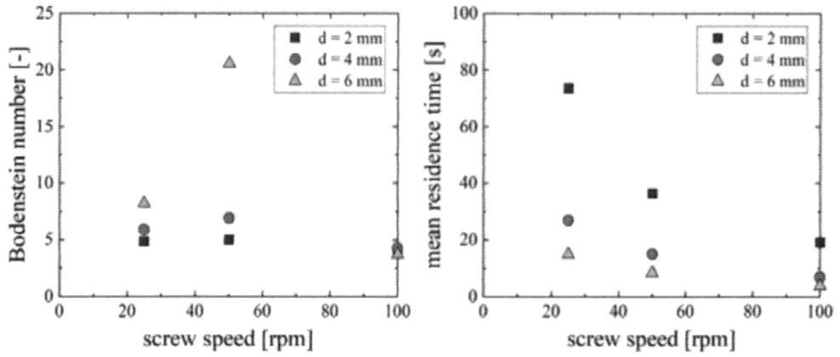

Figure 3: Bodenstein number and mean residence time plotted over the screw speed for different die diameters (new unused screw element GFA-2-20-30)

601134 Drug Product Continuous Powder Feeding for Low-Percentage Formulation Components

Abstract:

Drug product formulations typically contain low-percentage components, such as lubricants, glidants, disintegrants and sometimes drug substance. For continuous manufacturing these materials require relatively low mass feed rates. Low-end limitations of twin-screw feeders commonly used for cDC led to the investigation of a novel feeding device that uses brief, frequent pulses of compressed gas to temporarily fluidize and propel powder through a narrow, adjustable gap. For cDC relevance, the device was outfitted with an appropriately sized hopper, and additional modifications were made including vibration and agitation to disrupt powder bridging. A series of experiments were conducted with common low-percentage pharmaceutical excipients, and additionally with cohesive drug substance. An initial target of 60 grams per hour was identified which most materials could achieve, and some materials were able to be fed below 10 grams per hour. Preliminary results are promising; however, additional development of this technology is needed prior to integrating into a continuous manufacturing process train.

598164 Stochastic Analysis and Modeling of Pharmaceutical Screw Feeder Mass Flow Rates

Abstract:

Screw feeders serve as the critical first unit operation in a continuous manufacturing of drug product (CMDP) processes, influencing the mass flow rate of active pharmaceutical ingredients and excipients downstream. There is great industrial interest modeling the behavior of screw feeders accurately. Existing flow sheet models focus on deterministically simulating the powder mass flow rate [2]–[4]; however, these models only capture the average mass flow rate, and lack the stochastic behavior of the mass flow. The use of custom Discrete Element Method (DEM) models provide highly realistic simulations of particles' residence time and by extension mass flow, but prohibitively require long computation times to simulate only minutes of operation [5]. In the pursuit of designing better processes and improved controllers, there is a need for a computationally quick-to-solve flow sheet model that can simulate the stochastic nature of a real screw feeder. This is precisely the gap addressed in the present paper.

This work presents the novel characterization and modeling of the stochastic nature of a screw feeder's mass flow rate using statistical time series analysis and a deterministic flowsheet model. First, a battery of experiments was performed using different powders and a variety of operating speeds on a Coperion K-tron KT20 screw feeder. Then, for each experiment, the parameters of a hybrid mechanistic-empirical screw feeder model, based upon the work of [4], were estimated. After, the stochastic element of the mass flow is isolated by subtracting the flowsheet model's deterministic mass flow rate from the feeder-reported instantaneous mass flow rate. Next, each experiment had an autoregressive moving average model (ARMA) [6] fit to its stochastic remainder, characterizing its mass flow variation. Finally, the set of ARMA models was used to develop a predictive model, relating the flow rate stochasticity to powder properties and operating conditions. This predictive model was integrated into the current deterministic feeder model, yielding a novel hybrid mechanistic-empirical-stochastic flow sheet model that simulates a realistic, high variance mass flow rate and is suitable for the development of CMDP processes and controllers.

[1] S. García-Muñoz, A. Butterbaugh, I. Leavesley, L. F. Manley, D. Slade, and S. Bermingham, "A flowsheet model for the development of a continuous process for pharmaceutical tablets: An industrial perspective," *AIChE Journal*, vol. 64, pp. 511–525, 2017.

[2] Y. Yu, "Theoretical modelling and experimental investigation of the performance of screw feeders," PhD thesis, 1997.

[3] F. Boukouvala, V. Niotis, R. Ramachandran, F. J. Muzzio, and M. G. Ierapetritou, "An integrated approach for dynamic flowsheet modeling and sensitivity analysis of a continuous tablet manufacturing process," *Computers & Chemical Engineering*, vol. 42, pp. 30–47, 2012.

[4] D. Bascone, F. Galvanin, N. Shah, and S. Garcìa-Muñoz, "A hybrid mechanistic-empirical approach to the modelling of twin screw feeders for continuous tablet manufacturing," *Industrial & Engineering Chemistry Research*, 2020.

[5] P. Toson and J. G. Khinast, "Particle-level residence time data in a twin-screw feeder," *Data in brief*, vol. 27, p. 104672, 2019.

[6] G. E. P. Box, G. M. Jenkins, G. C. Reinsel, and G. M. Ljung, *Time series analysis: Forecasting and control*. John Wiley & Sons, 2015.

595586 Tablet Pan Coating Performance during Scale up and between Batch and Continuous Processes

Abstract:

Abstract

Dimensionless analytical models for cylindrical pans under fixed bed conditions were developed that provided mechanistic relationships between material, equipment, and process parameters; and coating uniformity for development- and production-scale processes in batch and continuous operation. The batch coating model was verified with the exact analytical solution derived using the binomial distribution and was found to agree well with literature data. The models were then used to provide insights into formulation changes that resulted in higher coating uniformity for batch and continuous coating processes. A stochastic model was used to check the validity of the non-cylindrical pans and dynamic bed conditions. Both analytical and practical solutions for bridging coating performances between batch and continuous processes, and different formulations are presented.

Extended Abstract

The main goal of a tablet coating process is to arrive at processing conditions that result in high-quality, evenly coated tablets. It follows that the control strategy for coating processes should be the coating uniformity endpoint but, apart from the drug-layering process, this is rarely done in the industry largely due to a lack of knowledge and/or the high cost of coating uniformity characterization. This work was undertaken to provide an analytical comparison for easier visualization of performance differences between batch and continuous coating processes, as well as between development-scale and production-scale processes for seamless technology transfers. This is achieved through derivations of dimensionless mechanistic models by applying simplification to a few process/equipment parameters provided in a coating uniformity simulation patent (Choi, 2009).

Model Derivation

Statistical and analytical coating uniformity models were derived based on the coating schematics shown in Figures 1 and 2 for the batch and continuous mode of operation, respectively. The models assume the same quantity of coating is received per pass through the coating zone and the probability of entering the coating zone per cycle is x_C. This assumption proved to be a good approximation regardless of the shape of the coating distribution since all approach normal distribution with an increasing number of cycles according to the central limit theorem.

The exact analytical models were derived by summing all probabilities of tablets passing through the coating zone. The resulting binomial distribution models for the batch and continuous (see Choi and Meisen, 1997) coating processes, respectively, were:

Equation (see attachment) (1)

And Equation (see attachment) (2)

The statistical model for the batch coating schematic shown in Figure 1 was derived using the central limit theorem as follows:

Equation (see attachment) (3)

Substituting the geometric relationship for a cylindrical pan shown in Figure 3; assuming a static bed; and using the tablet wall velocity relationship given by Chen et al. (2010); Equation 3 could be expressed as

Equation (see attachment) (4)

where Equation (see attachment) (5)

and Equation (see attachment) (6)

Results

The exact analytical model (Equation 1) results were indistinguishable from those of the statistical model (Equation 3); therefore, Equations 1 and 3 are considered the same. The statistical model (Equation 4) agreed well with the Chen et al. data: the bed loading had the index of 0.58 compared with 0.5 in both Pandey et al. (2006) and Chen et al. (2010). The model predicted 3.85% compared to the experimental result of 3.8%[ii] when the pan speed was changed. The effect of tablet diameter could be matched exactly to the experimental data (Chen et al., 2010). The statistical model, by using a formulation coefficient for x_c, also predicted the higher coating uniformity results seen in a high-solids-load coating formulation (Karan et al., 2019) for batch and continuous coating processes.

To test the validity of the assumptions using static bed and cylindrical pan, stochastic simulations for a non-cylindrical pan and dynamic bed were conducted. Stochastic simulations of a 60" pan with 6" cone extensions on either side pan showed no discernable difference in coating uniformity from that of a cylindrical pan with the same diameter. The stochastic simulations showed 5% increase in coating uniformity for a 22% weight gain while the statistical model gave a 4% difference between the initial and the final bed size with 22% weight gain.

Despite the simplicity of assumptions used, the models derived in this work agreed well with the literature data and stochastic simulations.

Implications

The implications of this work is that the analytical models can be used to select and monitor coating uniformity, typically expressed in relative standard deviation or Acceptance Value, as the critical quality attribute to ensure the desired performance in a manufacturing control strategy for both functional and non-functional tablet coating processes.

For process bridging, the models can be used to calculate the relationship between different scales of equipment as well as between batch and continuous coating processes. Figure 5 shows an example of how the coating time needs to be adjusted with increasing pan diameter for the same relative bed loading and spray coverage. Figure 6 shows the characteristics of continuous coating design that is needed to match the coating uniformity of batch coating process.

Symbols

d_p tablet diameter, Sauter or surface-volume diameter used

D pan diameter

F_B coating distribution for the batch process

F_C coating distribution for the continuous process

h bed height

i denotes the position in the number of cycles or the number of passes through the spray zone

k coefficient

n number of cycles

P_i probability of i

RSD relative standard deviation

x_c probability of tablet going through the spray zone

x_e exit probability of tablets in continuous processes

Δ depth of the spray zone

φ pan rotational speed

σ population standard deviation

θ angle shown in Figure 1

θ_Δ angle accounting for the thickness of the coating zone depth

τ total coating time

References

Chen, W., Chang, S-Y, Kiang, S., Marchut, A., Lyngberg, O., Wang, J., Rao, V., Desai, D., Stamato, H., and Early, W., "Modeling of Pan Coating Processes: Prediction of Tablet Content Uniformity and Determination of Critical Process Parameters", J Pharm. Sci., 99 (7), July, 2010.

Choi, M., "Pan Coating Simulation for Determining Tablet Coating Uniformity", *US Patent #10/981,812*, December, 2009.

Choi, M. and Meisen, A., “Sulfur Coating of Urea in Shallow Spouted Beds”, Chemical Engineering Science, v52, p1073 (1997).

Karan, K., Hach, R., Lenick, L., During, T., Marjeram, J., Stevenson, G., Kuczynski, C., and Botnick, M. “Evaluation of Tablets Coated in a High Throughput Continuous Coater Using an Ultra-high Solids Film Coating System”, Poster T1330-08-56, presented at the Annual AAPS Meeting & Convention in San Antonio (2019).

Pandey, P., Katakdaunde, M., and Turton, R., “Modeling Weight Variability in a Pan Coating Process Using Monte Carlo Simulations”, *AAPS PharmSciTech*, 7 (4), 2006.

Equations

$$F_B = \begin{cases} P_0 = (1-x_c)^n \\ P_1 = \binom{n}{1} x_c \cdot (1-x_c)^{n-1} \\ \vdots \\ P_i = \binom{n}{i} x_c^i \cdot (1-x_c)^{n-i} \\ \vdots \\ P_n = x_c^n \end{cases} \quad (1)$$

$$F_C = \begin{cases} P_0 = \frac{x_e \cdot (1-x_c)}{1-(1-x_c)(1-x_e)} \\ P_1 = \frac{x_e \cdot x_c}{1-((1-x_c)(1-x_e))^2} \\ \vdots \\ P_i = \frac{x_e \cdot x_c^i \cdot (1-x_e)^{i-1}}{1-((1-x_c)(1-x_e))^{i+1}} \\ \vdots \end{cases} \quad (2)$$

$$RSD = \sqrt{\frac{1}{n} \cdot \left(\frac{1}{x_c} - 1\right)} \quad (3)$$

$$RSD = k \sqrt{\frac{\theta + \sin\theta}{\varphi^{2/3} \cdot d_p^{1/6} \cdot \tau} \cdot \frac{2\theta_\Delta - \sin(2\theta_\Delta)}{(2\theta - \sin(2\theta)) - (2\theta_\Delta - \sin(2\theta_\Delta))}} \quad (4)$$

where $\theta = \cos^{-1}\left(1 - \frac{h}{D/2}\right)$ (5)

and $\theta_\Delta = \cos^{-1}\left(1 - \frac{h - k_\Delta d_p}{D/2}\right)$ (6)

Figures

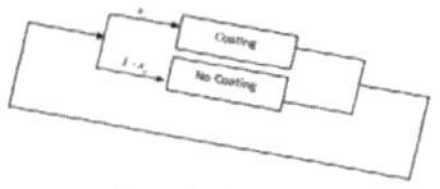

Figure 1: Schematic for the batch coating model.

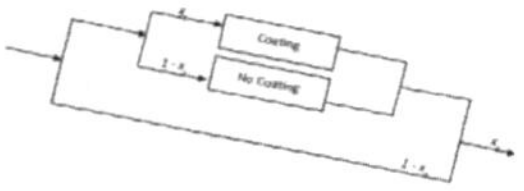

Figure 2. Schematic for the continuous coating model.

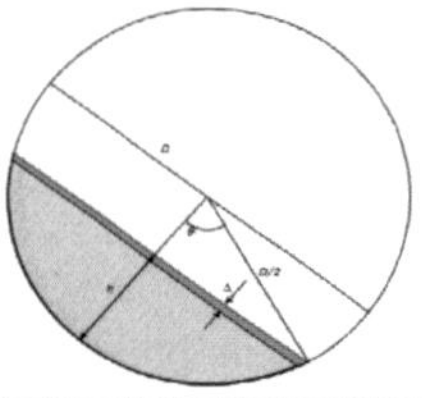

Figure 3. Geometry of pan and tablet bed (coating zone is depicted by the dark green shade)

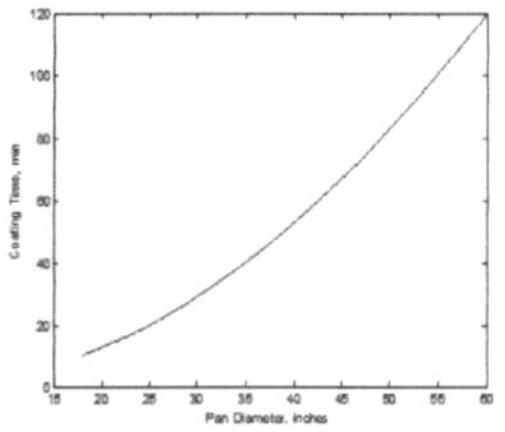

Figure 4: Coating times needed to achieve the same RSD with increasing pan size.

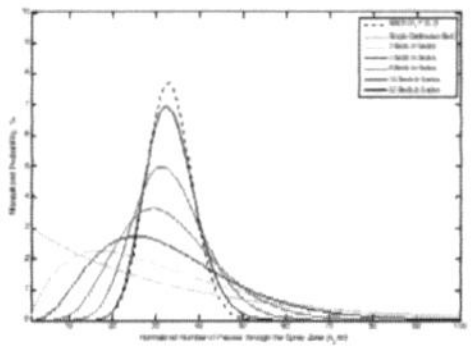

Figure 5: Comparison of coating distribution between continuous and batch coating processes.

608174 Design Method for Injectable Manufacturing Processes Considering Continuous and Batch Compounding

Abstract:

We present a process design method considering continuous and batch modes in the compounding unit for injectable manufacturing. The method can economically optimize the design variables such as the compounding mode, the lot size, and the number of filling needles. The method integrates two types of models. One is "process-oriented models" which captures the continuous nature of the process by residence time distribution (RTD), among others. The other is "discrete-oriented models" which describes the aspects such as failure events, operating shifts, and inventory by discrete and stochastic modeling. Know-how at the manufacturing shopfloors are fully used to estimate important model parameters such as the trend of failure events happening in the process. Net present value (NPV) is used as the objective function, considering pharma-specific costs such as change control and quality control. This economic evaluation model is an extension of our previously presented model for lot-sizing in the injectable manufacturing [1].

We applied the this method in an industrial case study. Using the parameters obtained in an actual injectable manufacturing process, the design variable, especially the choice between continuous and batch compounding, could be optimized. Furthermore, a generalized map was also created by varying the specified parameters in the possible ranges of products, process, and the market characteristics. The map provided quantitative insights on the cases where continuos compounding could be economically advantageous than batch compounding. The result could be useful for both retrofitting and grassroot design for injectable manufacturing processes.

[1] M. Yamada, S. Badr, S. Fukuda, M. Nakaya, Y. Yoshioka, H. Sugiyama, Journal of Pharmaceutical Innovation, accepted

600815 Efficient Production of Multi-Component Amorphous Solid Dispersions By Spray Drying

Abstract:

One of the major challenges in drug development nowadays is low water solubility (BCS classes II and IV). A widely used approach that had been proven successful in improving drug solubility is the formulation of amorphous solid dispersions (ASDs). Typically an ASD is composed of the poorly soluble drug and the hydrophilic polymer. If the combination is correct, the polymer propels the concentration of the poorly soluble drug into a supersaturated state during dissolution (spring effect). However, often the drug precipitates back into the solution, causing the concentration and thus bioavailability to drop. This issue can be circumvented by using a second polymer, which acts as a precipitation inhibitor (parachute effect). Yet, finding out the correct ratios in this tri-component system (drug: polymer 1: polymer 2) can often prove to be difficult.

In this work, we produced multi-component solid dispersions by spray drying in an efficient manner. As a widely recognized model system, we chose cyan, magenta, and yellow (CMY) printer dyes and lactose as a carrier. We can conclude that the ratio between the components in certain collected batch (experimental values) can be different from the ratio between the components in the inlet feed (theoretical values) as the residence time of particles in the system can be greater than the collection period. We evaluated the cross-contamination of samples by UV-Vis spectroscopy. Across all 15 samples, which formed a ternary diagram, the average deviation from the theoretical compositions was approximately 1.7 %. The first part of the figure below shows all collected samples arranged in the ternary diagram, forming visually confirmable color gradients. In the second part, the results from UV-Vis spectroscopy are plotted into the ternary diagram. These results show that we can produce multicomponent solid dispersions efficiently without significant cross-contamination between samples.

Acknowledgements:

Authors acknowledge financial support by the Technology Agency of the Czech Republic, project number TJ02000095 (Zeta Program).

a) Samples produced

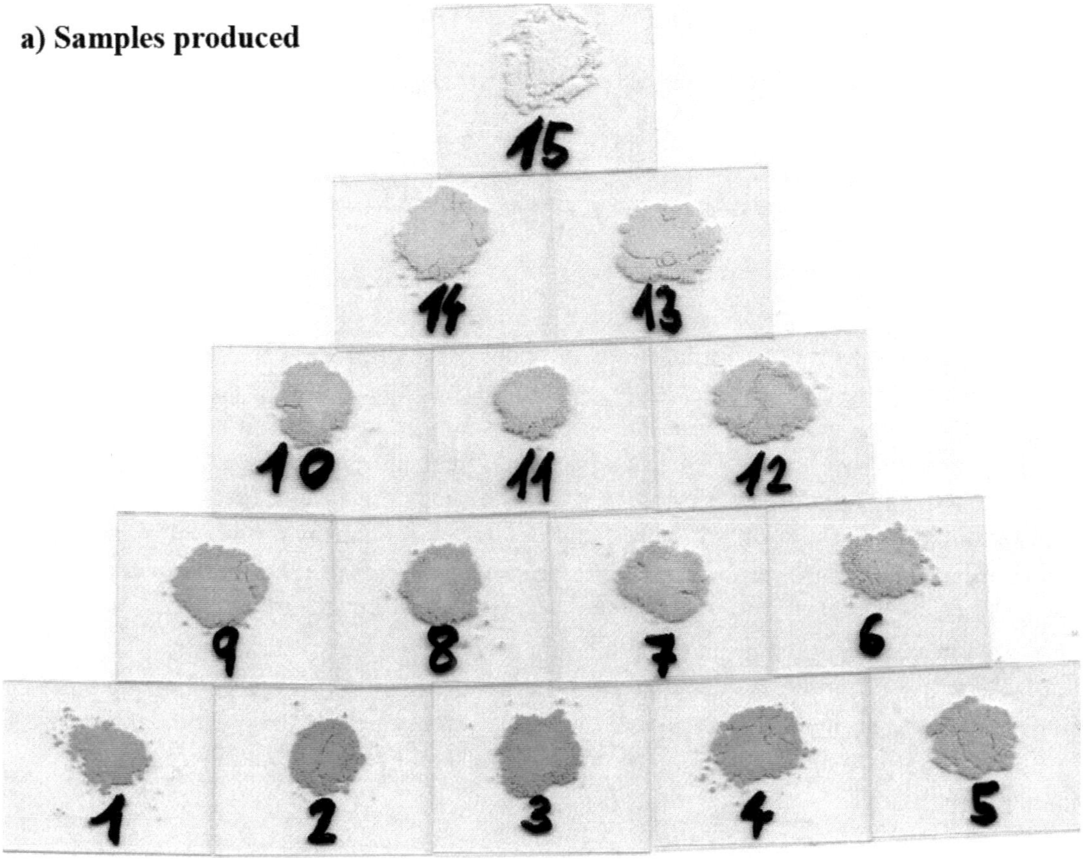

b) Results in ternary diagram

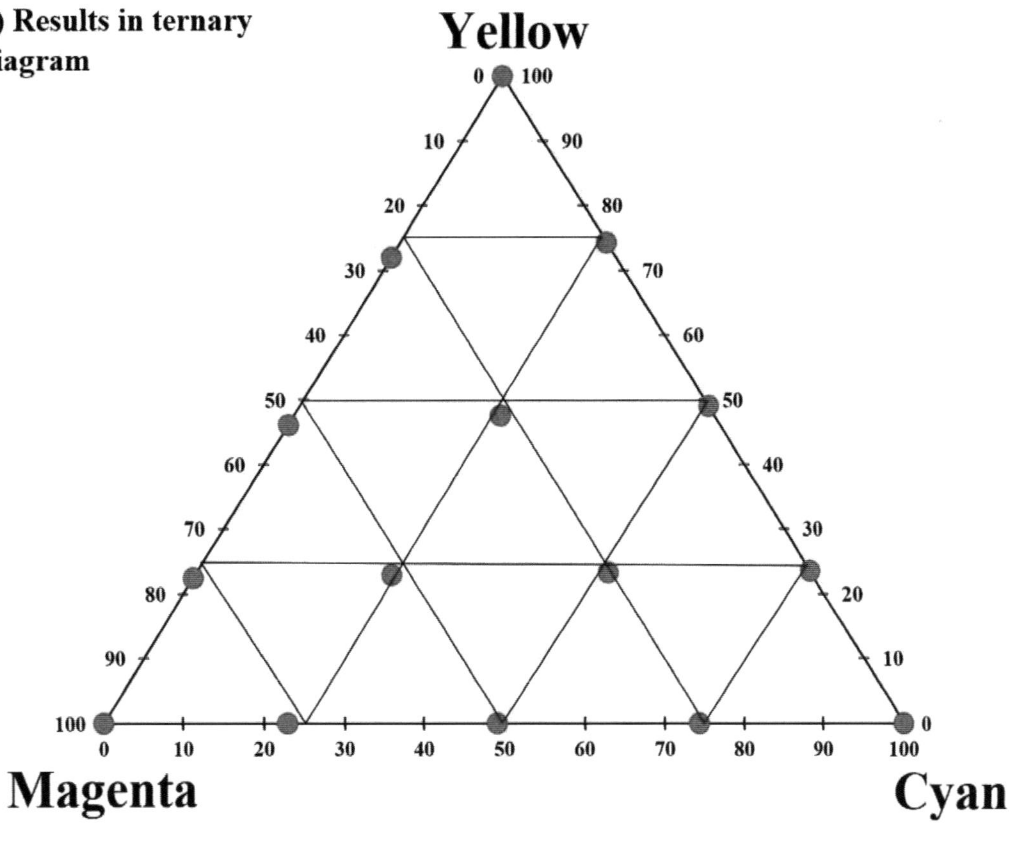

614901 Metal-Oxide Nanoarray based Catalysts for Water Electrolysis and CO_2 conversion at High Efficiency and Low Cost

Abstract:

Three-dimensional (3-D) assemblies of nanomaterials into applicable platforms or devices demand to meet the practical requirements in cost-effectiveness, structure sophistication, multi-function enabling, green and robust operations. Such integration process and product enabling involve dissimilar nano-building-blocks such as particles, wires, and films made of metals, ceramics, or polymers. In this talk, I will review our decade-long research effort in the assemblies and integrations of various ceramics and metals in the forms of nanostructure arrays (nanoarrays) onto different macroscale surfaces as function-carrying reactors and devices. Departing from automotive catalytic converters, sensors, water electrolysis, to CO_2-methanol conversion, a unique green process and product engineering pathway is being laid out with nanomaterial design, function-enabling, manufacturing capability, system integratability, and economic affordability as the underlying bases. Modular, multifunctional, and scalable nanoarray devices and systems are being designed and enabled for conversion, monitoring, production, and redistribution of fuels and chemicals such as carbon dioxide, hydrogen, and methanol.

References:

1. Zhang, J.Y. Sun, U. Salahudin, P.X. Gao, "Hierarchical and Scalable Integration of Nanostructures for Energy and Environmental Applications: A Review of Processing, Devices, and Economic Analyses," *Nano Futures*, **2020**, 4, 012002.
2. B. Wang, Z. Ren, Y.B. Guo, P.X. Gao, "Nano-array integrated monolithic devices: toward rational materials design and functional performance by scalable self-assembly of nanostructures", *CrystEngComm* **2016**, 18, 2980 - 2993.
3. Hoang, and P.X. Gao, "Nanowire Array Structures for Photocatalytic Energy Conversion and Utilization: A Review of Design concepts, Assembly and Integration, and Function Enabling," *Adv. Energy Mater.* **2016**, 6, 1600683.
4. Hoang, Y. Guo, A.J. Binder, W. Tang, S. Wang, H. Tran, J.J. Liu, X. Lu, Y. Wang, Y. Ding, E.V. Kyriakidou, J. Yang, T.J. Toops, T.R. Pauly, R. Ramprasad, and P.X. Gao, "Activating Low-temperature Diesel Oxidation by Single-Atom Pt on TiO_2 Nanowire Array" *Nature Communications*, **2020**, 11, 1062.
5. Y. Wen, R. Xu, C. Cui, W. Tang, Y. Mi, X.X. Lu, Z. Zeng, S.L. Suib, P.X. Gao, and Y. Lei, "Template-guided programmable Janus hetero-nanostructure arrays for efficient plasmonic photocatalysis" *Nano Letters,* **2018**, 18(8), 4914-4921.
6. S. Du, W. Tang, X. Lu, S. Wang, Y. Guo, P.X. Gao, "Cu-supported ZnO Nanoarray Integrated Structured Catalysts for Low-pressure CO_2 Hydrogenation to Methanol," *Adv. Mater. Int.* **2018**, 5, 1700730.

614902 Microwave Catalytic Reactor for Process Intensified Conversion of Stranded Energy Resources

Abstract:

This presentation introduces the applications of microwave catalysis for the conversion of stranded energy resources (shale gas and renewable power) into liquid transportable chemicals such as aromatics and ammonia. Microwave irradiation has shown a profound impact on catalyzed and uncatalyzed gas-solid reactions. In particular, it has been demonstrated that microwave-specific effects can manifest themselves through the enhancement of reaction rates, changes in the position of equilibria and the distribution of products.

Microwave irradiation provides a means for precise and site-selective activating of heterogeneous catalysts for organic reactions or materials processing. Microwave energy can be delivered directly to the reacting species and/or adsorbing catalyst where it influences localized electronic properties of the materials and reaction intermediates at interface [1-3]. Microwave-enhanced thermocatalytic reactions offer potential benefits to the process industry by providing: 1) precise and selective heating, applying heat directly to the catalyst for better energy efficiency and improved selectivity 2) much more rapid heating than conventional means, and 3) reduced process footprint, enabling process intensification. This paper presents thermal and non-thermal effects of microwave catalysis on two reactions: (1) simultaneous conversion of CH_4/N_2 to form ammonia and ethylene under ambient pressure, and (2) process intensified natural gas dehydroaromatization.

This novel process integrates system elements of electromagnetic sensitive catalysts and microwave reactor design to convert methane and/or nitrogen to value added products. The results indicate that stable molecules such as CH_4 and N_2 can be activated by microwave irradiation at appropriate reaction conditions to produce NH_3, aromatics and C_2 olefins. Microwave allows to achieve the selectivity and yield that cannot be obtained from conventional thermally heated reactor.

References

- Bai, S. Tiwari, B. Robinson, C. Killmer, L. Li, J. Hu, Catalysis Science & Technology, 8, 6302-6305,2018
- Bai, B. Robinson, C. Killmer, Y. Wang, L Li, J. Hu, Fuel, 243, 485-492, 2019
- Hu, US Patent Application :16/355,501, PCT/US19/22586, March 2019

614904 Multi-Scale Systems Analytics for Sustainable Process Engineering

Abstract:

As evidenced by the recent Paris Agreement on climate change and the public's increasing interests in carbon footprints, sustainability issues are becoming more prevalent in the thoughts and actions of many people around the world. Systems analysis and computational optimization methods from a life cycle perspective play an important role in the quest for sustainability. This talk will review recent advances in the area of life cycle optimization for sustainability analytics of energy systems. Applications to perovskite photovoltaics, shale gas and biofuels will be presented. The presentation will conclude with a brief review of multi-scale applications of these systems technologies and methods that range from molecular and materials systems, to unit operations and manufacturing processes, and to value chains and global ecosystems.

614903 Composite Ionic liquid alkylation(CILA) process produces high-quality gasoline

Abstract:

Isobutane alkylation is a common refinery process used to produce high-quality gasoline. The most desirable components in alkylate gasoline are trimethylpentanes (TMPs), which have research octane numbers (RONs) greater than 100. Conventional alkylation processes use either H2SO4 or anhydrous HF acid as catalysts. This has significant safety and environmental concerns due to the handling of large quantities of spent H2SO4 or hazardous HF. Ionic liquids（ILs） are ionic, salt-like materials that are liquid at less than 100° C. ILs have been historically used as solvents and homogeneous catalysts6 due to their negligible vapor pressure, good solubility to a wide range of organic and inorganic compounds, and ability to be recycled for reuse. The most important for alkylation process is its tunable super-acidity. A new composite ionic liquid (IL) was developed in the laboratory. It can be used in the isobutane alkylation process to produce high quality gasoline which has no sulfur, no olefin, no aromatic hydrocarbon, but has high RON. At this presentation, the development of composite ionic liquid, the deactivation and regeneration of catalyst, and the 100000 ton/year commercial process and its operation results will be discussed. Now several industrial plants are running,such as one 150,000 ton/year plant in Haerbin petrochemical company of CNPC, two 300,000 ton/year plant in Jiujiang and Wuhan petrochemical company of SINOPEC.

602824 Process Design and Intensification of Dividing Wall Column for an Industrial Methyl Methacrylate Separation Process

Abstract:

In pursuit of more energy efficient distillation technologies, dividing wall column (DWC) designs have received increasing attention from the chemical industry for multi-component separation [1,2]. Comparing to conventional distillation sequences, DWC features a compact and intensified design structure with less investment costs by integrating the whole process into a single shell column. Significant energy savings (approximately 30%) have also been reported due to reduced re-mixing effects occurring in the column and thus higher thermodynamic efficiency [3]. One DWC application of industrial importance is for purification of the reaction product mixture from preparation of methyl methacrylate, which also contains methanol, water, and oligomers of methyl methacrylate [4,5]. A patented design alternative for this heterogeneous azeotropic separation problem consists of a dividing wall column integrated with a decanter, showing 7.6% higher product recovery rate than a conventional two-column design flowsheet given the same energy consumption rate [4].

In this work, we revisit this industrial methyl methacrylate separation process based on our recently proposed framework for design, synthesis, and operational analysis of process intensification systems [6]. Specifically, a phenomena-based synthesis representation is developed using the Generalized Modular Representation Framework (GMF) to capture DWC systems with fundamental chemical building blocks (i.e., mass/heat exchange module and pure heat exchange module) [7-9]. To describe the non-ideal vapor-liquid-liquid behavior, rigorous thermodynamic models (e.g. UNQUAC) is explicitly incorporated in the synthesis model. The base case DWC design presented in [4] is then simulated with a pre-fixed building block structure to validate the accuracy and efficiency of GMF for this specific case study. Thereafter, a superstructure optimization model is formulated to systematically generate the optimal and intensified process alternative(s) for improved cost performance without any pre-postulation of plausible process unit or flowsheets. The resulting phenomenological process alternatives are translated to unit operation-based flowsheet for rigorous design and simulation using Aspen Plus. Given the inherent highly nonlinear dynamics with multiple steady-states in DWC systems, explicit/multi-parametric model predictive controllers designed via the PAROC framework [9] are applied to the above derived DWC-based flowsheets to ensure feasible operation under disturbance and uncertainty. Two alternative case studies are also presented for comparison of design strategies as well as optimality of the process solutions: (i) unit operation-based optimization of the base case design, and (ii) unit operation-based optimization of the conventional two-column flowsheet design with heat integration.

References

1. Yildirim, Ö., Kiss, A. A., & Kenig, E. Y. (2011). Dividing wall columns in chemical process industry: a review on current activities. *Separation and Purification Technology*, *80*(3), 403-417.
2. Tian, Y., Demirel, S. E., Hasan, M. M. F., & Pistikopoulos, E. N (2018). An overview of process systems engineering approaches for process intensification: State of the art. *Chemical Engineering and Processing: Process Intensification*, 133, 160-210.
3. Triantafyllou, C., & Smith, R. (1993). The design and optimization of dividing wall distillation columns. In *Energy Efficiency in Process Technology*(pp. 351-360). Springer, Dordrecht.

4. Jewell, D. W., Pendergast, J. G., & Worley, W. G. (2019). *U.S. Patent No. 10,392,337.* Washington, DC: U.S. Patent and Trademark Office.
5. Chang, W. L., & Chien, I. L. (2016). Energy-saving design and control of a methyl methacrylate separation process. *Industrial & Engineering Chemistry Research*, *55*(11), 3064-3074.
6. Tian, Y., Pappas, I., Burnak, B., Katz, J., & Pistikopoulos, E. N. (2020). A systematic framework for the synthesis of operable process intensification systems – Reactive separation systems. *Computers & Chemical Engineering*, *134*, 106675.
7. Papalexandri, K. P., & Pistikopoulos, E. N. (1996). Generalized modular representation framework for process synthesis. *AIChE Journal*, *42*(4), 1010-1032.
8. Proios, P., & Pistikopoulos, E. N. (2006). Hybrid generalized modular/collocation framework for distillation column synthesis. *AIChE journal*, *52*(3), 1038-1056.
9. Tian, Y., & Pistikopoulos, E. N. (2019). Generalized modular representation framework for the synthesis of extractive separation systems. In *Computer Aided Chemical Engineering*(Vol. 47, pp. 475-480). Elsevier.
10. Pistikopoulos, E. N., Diangelakis, N. A., Oberdieck, R., Papathanasiou, M. M., Nascu, I., & Sun, M. (2015). PAROC – An integrated framework and software platform for the optimisation and advanced model-based control of process systems. *Chemical Engineering Science*, *136*, 115-138.

601477 Thermal Analysis of Structured Reactors for Microwave Reaction Engineering

Abstract:

The application of microwaves to enable sustainable manufacturing is rapidly gaining importance as the pathway to process intensification. Under this manufacturing paradigm, sustainably generated photons in the MHz frequency range interact with carefully designed reactors and beds where the electromagnetic energy is converted to thermal energy enabling reaction chemistries. 3D printing is a novel strategy that enables the creation of these structured reactors with a wide array of materials. Thus, to develop this sustainable technology further, a fundamental understanding of thermal transport in these 3D printed structures is of paramount importance.

In this talk, we present insights from our computational investigation into the heating of structured reactors, such as foams, monoliths, and 3D printed structures, with microwaves and compare to our experimental data for the same systems. We will address the role of different length scales – the wavelength of microwaves (~100 mm) to the feature sizes of structured beds (10 mm) and the size of 3D printed pores (~1 mm). Additionally, we will compare structured beds with traditional random packed beds for microwave applications. We will discuss a criterion for materials selection and reactor design stemming analysis of electrical and thermal properties. Finally, we will share our perspective on leveraging these 3D printed reactors for endothermic chemistries.

607778 Cost-Effective Technology Identification & Uncertainty Quantification of Stranded Sour Gas Desulfurization Processes

Abstract:

The maximum utilization of natural gas resources motivates us to have efficient and cost-effective desulfurization strategies. There have been significant advancements in the development of new technologies, but the screening of these technologies is expensive and time-consuming. Computational models can help accelerate the screening of these novel desulfurization technologies. Optimization of detailed equipment models can help find the best process conditions and risks. Moreover, a better understanding of the models and their uncertainties can help transfer technologies from lab to commercial scale.

In this work, different natural gas desulfurization processes, e.g., SourCat® [1], iron-chelate redox systems LO-CAT® [2], Claus [3], and Triazine based scavenger [4] are simulated at different scales (1 to 1,000 MMSCFD) and inlet H_2S concentrations (500 to 2,500 ppm). For each combination of operating conditions, the desulfurization cost, which is defined as the net present value during the life period of the desulphurization system, is calculated for each process. The process with the lowest desulfurization cost is identified for each operating condition. This information is used to generate a map that envelopes the operating conditions of the cost-competitive processes. From our initial analysis, we found that the newly developed oxidative sulfur removal (OSR) SourCat® process is a viable and cost-competitive option for stranded gas sweetening in the wide range of 3,000 MSCFD to 75 MMSCFD at 500 to 2,000 ppm H_2S concentration, among the processes that were compared. Triazine based scavenger process becomes cost-effective at a lower flow rate (3,000 MSCFD) and low H_2S concentrations. In contrast, the Claus process is efficient at a much higher flowrate (45,000 MMSCFD) or higher H_2S concentrations.

The desulfurization cost depends on solvents, catalysts, equipment costs, electricity costs, and fuel costs. Of the parameters that impact desulfurization cost, solvent cost, catalyst cost, capital investment cost and fuel cost are most uncertain. A systematic sensitivity analysis was performed to identify the significant parameters to understand the impact of these uncertainties on the desulfurization cost and the resulting map of the operational envelope, specifically focusing on the boundaries. Then, the uncertainties of the significant parameters were propagated to estimate the uncertainty of desulfurization cost for each process. These estimates allowed us to update the map to include transition boundaries where there may be more than one process that is cost-competitive.

References:

1. "Sour Gas Has a Sweeter Future," Chemical Engineering Progress, January 2017 https://www.aiche.org/sites/default/files/cep/20170112.pdf
2. "Removing H2S from Gas Streams", https://www.merichem.com/technology/sulfur-recovery-with-lo-cat/?doing_wp_cron=1588006246.7422919273376464843750
3. "Steady State Simulation and Optimization of an Integrated Gasification Combined Cycle (IGCC) Plant with CO Capture," Bhattacharyya, D., R. Turton, and S. E. Zitney, Ind. Eng. Chem. Res. 50 (2011): 1674–1690.
4. "Hydrolysis of 1,3,5-Tris(2-hydroxyethyl)hexahydro- s -triazine and Its Reaction with H 2 S". Bakke, J.M.; Buhaug, J.; Riha, J. Ind. Eng. Chem. Res. 2002, 40, 6051–6054.

610128 Process Intensification Insights through Computational Catalysis Studies: Selective Ethylene Oxide Formation on Ag Catalysts

Abstract:

Process intensification (PI) is a design approach that leads to significantly cleaner, safer, and more energy-efficient process technology. Catalytic partial oxidation processes are responsible for an extremely large-scale industrial production, and also relatively expensive to perform. These processes typically operate at high thermodynamic inefficiencies because these catalytic reactions are very thermo-sensitive and also non-selective when even minor temperature excursions occur within the reactor. Thus, this industrial process is an excellent subject for PI studies. Herein, our studies focused on surface catalytic mechanisms of the ethylene oxide (EO) formation process. Periodic plane-wave Density Functional Theory (DFT) methods were used to analyze detailed reaction mechanisms comprised of elementary reactions on the Ag(111) surface facet with low coverage. Energetic changes of related species and pathways were calculated and will be discussed. DFT analysis is discussed in the context of microkinetic modeling techniques which provide detailed information on surface coverage and gas-phase species concentrations as a function of reactor process conditions and feed ratios. Key surface species have been identified to suggest the key factors for the selectivity during EO formation. Our results are consistent with previous kinetic modeling efforts in the literature which did not employ DFT analysis.[1] Lastly, our study demonstrates how fundamental theoretical investigations and multi-scale modeling techniques are currently impacting the advancement of rational catalyst design in the light hydrocarbon processing industry.

1. Stegelmann, C. et al, J. Catalysis 2004, 221 (2), 630-649.

610559 Simulation of Pressure Swing Adsorbers Using the Moving Bed Model

Abstract:

Simulation of pressure swing adsorption (PSA) units remains a challenge for practicing engineers, which hinders the process intensification potential of the technology. The Moving Bed model, a steady-state model which represents any adsorption process as a series of interconnected continuous adsorption units, has the potential to alleviate some of the difficulties in modeling PSA. [1, 2, 3] In the Moving Bed model, the cycle is divided into multiple constant-pressure operating states bearing similarity to the individual steps of the cycle which allows the properties of the state to be evaluated from equilibrium considerations alone. The operating steps are arranged from highest-to-lowest pressure to represent the driving force gradient exploited during the cycle and then connected by a virtual moving bed. The virtual moving bed does not physically exist in the system but represents the adsorbate and void space hold-up retained in the adsorption column at the end of the operating state. However, modeling a cyclic batch process as a continuous, equilibrium process neglects the rate-based transport phenomena involved. Using the idea of adsorption efficiencies [4] to account for non-idealities like mass transfer resistance and dispersion, similar to the tray efficiencies used in rate-based distillation modeling, the retentate flow rates can be adjusted to match the actual performance of the unit. [2, 3] To showcase the capabilities and current limitations of the model, a case study will be presented which investigates a lab-scale air separation unit. Guidelines for ensuring convergence and robustness will be given and the effect of important parameters will be discussed.

1. Sees, M.D. (2018, October). *Novel Steady-State Process Modeling Methodology for Pressure Swing Adsorption*. Paper presented at the 2018 Annual Meeting of the American Institute of Chemical Engineers, Pittsburgh, PA.
2. Kirkes, T. (2018, October). *Steady-State Modeling of Air Separation by Pressure Swing Adsorption Using a Bed-State Efficiency Model.* Poster presented at the 2018 Annual Meeting of the American Institute of Chemical Engineers, Pittsburgh, PA.
3. Sees, M.D. (2019, April). *Modeling of Gas Separations by Pressure Swing Adsorption Using a Novel Steady-State Methodology.* Paper presented at the 2019 Spring Meeting of the American Institute of Chemical Engineers, New Orleans, LA.
4. Smith, O.J. and A.W. Westerberg. *Eng. Sci.*, 1991, 48(12), 2967 – 2976.

602395 Dynamic Data Reconciliation, Parameter Estimation, and Validation for the Dynamic Model of a Supercritical Pulverized Coal Power Plant

Abstract:

The need for process systems engineering tools to improve the performance of existing fossil-fueled power plants in the presence of renewables in the energy market is well documented [1]. A powerful, low risk, and cost-effective tool for the improvement of these existing systems is the development of highly predictive process models. In the past, many such models for coal-fired power plants have been developed [2], [3]. However, many of these models focus on the steady-state design and operation of traditional subcritical pulverized coal systems, with less consideration given to dynamic models of supercritical pulverized coal (SCPC) power plant. A few papers are focused on the impact of carbon capture on the SCPC plant [4]–[6] where simple models of the SCPC plant were used. One fairly comprehensive study has been done using the commercial software Apros, but the equipment models used lack some of the fidelity needed to capture the dynamics of the system [7]. For a general SCPC plant, a past contribution established a plantwide model with higher fidelity for some systems in the plant [8]. A particular gap in the literature is the lack of industrial validation for the models that have been developed.

In this presentation, a first-principles comprehensive dynamic model of a SCPC plant is described comprising detailed models of the following sections - boiler, turbine, feed water heaters, and the balance of the plant. Distributed models for each component of the boiler including the economizer, superheaters, water wall, and platen superheater are discussed. Each section of the boiler model also contains a detailed heat transfer model with due consideration of conductive/convective/radiative heat transfer as appropriate. The boiler model takes into account the changes in the heat transfer coefficients due to the hydrodynamics of the system and thermal properties of the steam that can vary rapidly during load-following especially under sliding pressure operation. The model of the steam turbine can capture the change in the performance characteristics of the leading, non-condensing, and condensing stages. In addition, the steam turbine model can detect and account for moisture, if present, in the stages that may be non-condensing under nominal condition.

The SCPC dynamic model is validated with the full-load and transient data from an industrial partner plant (IPP). The data does not satisfy material and energy balances motivating the solution of a dynamic data reconciliation problem to ensure closure of the balances [9], [10]. One difficulty in the dynamic data reconciliation for this system is the poor estimate of the holdup - especially thermal holdup in the boiler components - and the poor initial guess of the initial state of the system. It is also impractical to mimic the complex feed-forward-augmented feedback controllers and logic-based approaches used in the coordinated control system in the IPP. Furthermore, there are large number of model parameters, especially equipment specific parameters that are not available and must be estimated. Due to the large number of partial differential algebraic equations in the plant-wide model, the optimization problem for dynamic data reconciliation and parameter estimation is large. A two-stage optimization problem is solved for reconciling the dynamic data in the first stage followed by optimal estimation of the parameters. Uncertainties in the initial estimates of the material and thermal holdup are incorporated in the form of sensitivity information in the proposed approach for reducing the computational expense. This work shows that steady-state and transient performance of the IPP can be very well predicted by the model. The validated SCPC dynamic model is used for developing advanced model-based controllers for load-following operation. The controllers maximize the plant efficiency during transient operation while satisfying temperature constraints at various locations. Results are also shown for important calculated performance indices, such as the plant heat rate.

[1] A. Subramanian, T. Gundersen, and T. A. Adams II, "Modeling and Simulation of Energy Systems: A Review," *Processes*, vol. 6, p. 238, 2018.

[2] E. Oko and M. Wang, "Dynamic modelling, validation and analysis of coal-fired subcritical power plant," *Fuel*, vol. 135, pp. 292–300, Nov. 2014, doi: 10.1016/j.fuel.2014.06.055.

[3] C. Chen, Z. Zhou, and G. M. Bollas, "Dynamic modeling, simulation and optimization of a subcritical steam power plant. Part I: Plant model and regulatory control," *Energy Convers. Manag.*, vol. 145, pp. 324–334, Aug. 2017, doi: 10.1016/j.enconman.2017.04.078.

[4] A. K. Olaleye, E. Oko, M. Wang, and G. Kelsall, "Dynamic Modelling and Analysis of Supercritical Coal-Fired Power Plant Integrated with Post-combustion CO2 Capture," in *Clean Coal Technology and Sustainable Development*, Singapore, 2016, pp. 359–363.

[5] K. Wellner, T. Marx-Schubach, and G. Schmitz, "Dynamic Behavior of Coal-Fired Power Plants with Postcombustion CO2 Capture," *Ind. Eng. Chem. Res.*, vol. 55, no. 46, pp. 12038–12045, Nov. 2016, doi: 10.1021/acs.iecr.6b02752.

[6] S. Ó. Garđarsdóttir, R. M. Montañés, F. Normann, L. O. Nord, and F. Johnsson, "Effects of CO2-Absorption Control Strategies on the Dynamic Performance of a Supercritical Pulverized-Coal-Fired Power Plant," *Ind. Eng. Chem. Res.*, vol. 56, no. 15, pp. 4415–4430, Apr. 2017, doi: 10.1021/acs.iecr.6b04928.

[7] J. Hentschel, H. Zindler, and H. Spliethoff, "Modelling and transient simulation of a supercritical coal-fired power plant: Dynamic response to extended secondary control power output," *Energy*, vol. 137, pp. 927–940, Oct. 2017, doi: 10.1016/j.energy.2017.02.165.

[8] P. Sarda, E. Hedrick, K. Reynolds, D. Bhattacharyya, E. S. Zitney, and B. Omell, "Development of a Dynamic Model and Control System for Load-Following Studies of Supercritical Pulverized Coal Power Plants," *Processes*, vol. 6, no. 11, 2018, doi: 10.3390/pr6110226.

[9] A. S. Chinen, J. C. Morgan, B. Omell, D. Bhattacharyya, and D. C. Miller, "Dynamic Data Reconciliation and Validation of a Dynamic Model for Solvent-Based CO2 Capture Using Pilot-Plant Data," *Ind. Eng. Chem. Res.*, vol. 58, no. 5, pp. 1978–1993, Feb. 2019, doi: 10.1021/acs.iecr.8b04489.

[10] P. Mobed, J. Maddala, R. Rengaswamy, D. Bhattacharyya, and R. Turton, "Data Reconciliation and Dynamic Modeling of a Sour Water Gas Shift Reactor," *Ind. Eng. Chem. Res.*, vol. 53, no. 51, pp. 19855–19869, Dec. 2014, doi: 10.1021/ie500739h.

605798 Forecasting Prices of Energy Feedstocks and Commodities Using Advanced Statistical and Machine Learning Methods

Abstract:

The prices of energy products and commodities are volatile [1] and sensitive to shifts in demand and supply mechanisms, technological breakthroughs, changes in monetary policy, geopolitical events or other major global challenges [2, 3]. Considering the fact that energy affects every single individual and entity in the world, it is crucial to quantify "the price of energy" accurately, and study how it evolves over time. To this respect, we have developed a novel forecasting framework to calculate the average as well as the market (spot) price of energy in the United States by introducing two indices: The Energy Price Index (EPIC) and the Energy Spot Price Index (ESPIC) respectively [4].

The complex energy landscape is thoroughly analyzed [5-9] to accurately determine the two key factors of this framework: the total demand of the energy products directed to the end-use sectors, and the corresponding price of each product in the form of either a monthly or a spot price [10-14]. We have already presented a rolling horizon forecasting methodology for the demands of energy in the future [4], and here we demonstrate a novel technique to forecast the future prices of energy feedstocks and commodities up to 12 months. The historical prices of each energy product and commodity are studied individually, so as to identify patterns, trends, cycles, outliers, relationships among different variables etc. Then, different forecasting models are fitted in training data sets i.e. Exponential Smoothing, ARIMA, Dynamic Regression, Neural Networks etc. The best forecasting model for each energy product and commodity is selected based on the validation results from testing data sets, and is then utilized to produce the 12 months forecasts. Being able to forecast the future prices of energy has tremendous potential for applications in economics, finance, engineering, law and policy.

References:

1. Regnier, E., 2007. Oil and energy price volatility. *Energy economics*, *29*(3), pp.405-427
2. I.W. Parry, M. D. Heine, E. Lis, and S. Li, Getting energy prices right: From principle to practice. International Monetary Fund, 2014.
3. Barnett, J.; Barron, J.; Oil market volatility is at all-time high. Energy Information Administration (EIA), Today in Energy, March 2020.
4. Baratsas, S.; Niziolek, A.; Onel, O.; Matthews, L.; Floudas, C. A.; Hallermann, D.; Sorescu, S.; Pistikopoulos, E. N. What is the price of energy to the end-users? A novel predictive framework and one of its applications to the energy and monetary policies. Nature Communications 2020 (Under Review).
5. Elia, J. A.; Baliban, R. C.; Xiao, X.; Floudas, C. A. Optimal energy supply network determination and life cycle analysis for hybrid coal, biomass and natural gas to liquid (CBGTL) plants using carbon-based hydrogen production. Computers and Chemical Engineering 2011, 35, 1399-1430.
6. Elia, J. A.; Baliban, R. C.; Floudas, C. A. Nationwide energy supply chain analysis for hybrid feedstock processes with significant CO2 emissions reduction. AIChE Journal 2012, 58 (7), 2142-2154.
7. Elia, J. A.; Floudas, C. A. Energy Supply Chain Optimization of Hybrid Feedstock Processes: A Review. Annual Review of Chemical and Biomolecular Engineering 2014, 5 (1), 147-179.

8. Onel, O.; Niziolek, A. M.; Floudas, C. A. Integrated biomass and fossil fuel systems towards the production of fuels and chemicals: state of the art approaches and future challenges. Current Opinion in Chemical Engineering 2015, 9, 66-74.
9. Floudas, C.A.; Niziolek, A. M.; Onel, O.; Matthews, L.R. Multi-Scale Systems Engineering for Energy and the Environment: Challenges and Opportunities. AIChE Journal 2016, 62 (3), 602-623.
10. Energy Information Administration (EIA), Annual Energy Outlook, 2019
11. Energy Information Administration (EIA), Monthly Energy Review, 2019
12. Lazard, Levelized Cost of Energy Analysis Version 13.0, 2019
13. Bureau of Labor Statistics (BLS), Producers Price Index, 2019
14. S. Department of Energy (DOE), Alternative Fuels Data Center, 2019

602323 Implementation of Hybrid Models to Perform System Analyses with Model Maintenance in Continuous Pharmaceutical Manufacturing

Abstract:

Recent development of Industry 4.0 technologies, including the advancements of cyber-physical systems (CPS) and Internet of Things (IoT), have increased the amount of data available from manufacturing shop floors and analytical laboratories[1-5]. The abundance of data contributes to the emergence and improvement of data analytic methods and cloud computing platforms, aiming to extract more information and knowledge from data to support predictive modeling and decision making[6, 7]. Traditionally, process industries, including continuous pharmaceutical manufacturing (CPM), mainly use white-box (WB) models for predictive modeling[8]. These WB models, including mechanistic, phenomenological, and first-principle models, rely heavily or solely on physical understandings of the system. Though the WB models provide a high degree of transparency on addressing the physical significance of variables and model structures, the development of the models is costly, and the prediction performance is largely dependent on the mastery of full process knowledge. To address these challenges utilizing the vast availability of data, recent efforts have focused on combining data with process knowledge to form hybrid models (HMs). HMs consist of at least one white-box and one black-box (BB) sub-model, where the BB sub-models take in historical process data and adopt different machine learning (ML) algorithms like artificial neural network (ANN) and support vector machine (SVM)[9]. The development of HMs of various structures provides greater flexibility and better predictability in complex engineering systems, making it applicable in modeling CPM processes.

Current research has been dedicated to developing proof-in-concept HMs for individual unit operations of CPM processes, but the capability of HMs to perform system analyses like sensitivity and feasibility analysis remains unexplored. These analyses can identify critical process parameters and design space of CPM systems, which is crucial for process design and operations[10]. Another challenge of implementing HMs in CPM systems is related to model maintenance. Parts of HMs are data-driven BB sub-models, and the offline training of BB sub-models with defined sets of historical data leads to static BB models[11]. These static BB sub-models only reflect the system at the timepoint that the models are developed, and they will not change unless the models are re-trained with new data and algorithms[12]. As time progresses, assuming the process stays the same, new datasets collected from plants should have the same or similar underlying distribution as the data used to train the original model. Using these new datasets to re-train should yield models with comparable performance to their predecessors. However, real datasets from manufacturing plants often contain normal operational data and abnormal data (e.g. highly noisy data, faulty data, missing data points, testing and maintenance data), and the quality of these new data can affect the re-training result considerably. On the other hand, like in the majority of manufacturing processes, a shift in the distributions of manufacturing data from CPM plants can happen over time due to changes such as environmental conditions, equipment performance, and control strategy. Because these changes are not captured in the model training process, deteriorating performance of HMs can be observed, which is a classic scenario known as model drift[11]. Model maintenance strategies therefore need to be in place to capture this shift and to ensure the accuracy of model predictions.

In this work, algorithms are developed to perform system analyses using HMs and to address model maintenance issues in CPM. For system analyses, HMs developed for individual unit operations of the continuous direct compaction (DC) line are connected in a flowsheet model, and they are then used to perform sensitivity and feasibility analysis[10, 13, 14]. The results from HMs are compared to those produced by WB models to compare the model performance. Next, two tasks are carried out to explore model maintenance schemes for HMs. Different types and amounts of new datasets are first fed into the re-training process to study the impact of the quality of additional data to HM performance. This analysis can yield practical guidance in data handling and maintenance frequency based on characteristics of newly available datasets. Then, to address the issue of model drift, different learning techniques and adaptive methods for model maintenance are developed. Manual re-training, incremental learning, and continuous streaming procedures are explored along with blind and informed adaptive algorithms[11, 15, 16]. Blind adaptations update an HM continuously without realizing that a drift exists, whereas informed algorithms only become effective when predefined triggers are activated[11]. The results will provide insights on how model maintenance for HMs can be performed. Together with system analysis capabilities, this will allow for broader acceptance of HMs for practical use in CPM with the development of Industry 4.0 framework.

1. Ding, B., Pharma Industry 4.0: Literature review and research opportunities in sustainable pharmaceutical supply chains. *Process Safety and Environmental Protection* 2018, *119*, 115-130.
2. Cao, H.; Mushnoori, S.; Higgins, B.; Kollipara, C.; Fermier, A.; Hausner, D.; Jha, S.; Singh, R.; Ierapetritou, M.; Ramachandran, R., A Systematic Framework for Data Management and Integration in a Continuous Pharmaceutical Manufacturing Processing Line. *Processes* 2018, *6* (5).
3. Reis, M. S.; Kenett, R., Assessing the value of information of data-centric activities in the chemical processing industry 4.0. *AIChE Journal* 2018, *64* (11), 3868-3881.
4. Barenji, R. V.; Akdag, Y.; Yet, B.; Oner, L., Cyber-physical-based PAT (CPbPAT) framework for Pharma 4.0. *Int J Pharm* 2019, *567*, 118445.
5. Steinwandter, V.; Borchert, D.; Herwig, C., Data science tools and applications on the way to Pharma 4.0. *Drug Discov Today* 2019, *24* (9), 1795-1805.
6. Thomas, M. C.; Zhu, W.; Romagnoli, J. A., Data mining and clustering in chemical process databases for monitoring and knowledge discovery. *Journal of Process Control* 2018, *67*, 160-175.
7. Liu, Y.; Xu, X., Industry 4.0 and Cloud Manufacturing: A Comparative Analysis. *Journal of Manufacturing Science and Engineering* 2016, *139* (3).
8. Zendehboudi, S.; Rezaei, N.; Lohi, A., Applications of hybrid models in chemical, petroleum, and energy systems: A systematic review. *Applied Energy* 2018, *228*, 2539-2566.
9. von Stosch, M.; Oliveira, R.; Peres, J.; Feyo de Azevedo, S., Hybrid semi-parametric modeling in process systems engineering: Past, present and future. *Comput Chem Eng* 2014, *60*, 86-101.
10. Wang, Z.; Escotet-Espinoza, M. S.; Ierapetritou, M., Process analysis and optimization of continuous pharmaceutical manufacturing using flowsheet models. *Comput Chem Eng* 2017, *107*, 77-91.
11. Gama, J.; Žliobaitė, I.; Bifet, A.; Pechenizkiy, M.; Bouchachia, A., A survey on concept drift adaptation. *ACM Computing Surveys* 2014, *46* (4), 1-37.
12. Webb, G. I.; Hyde, R.; Cao, H.; Nguyen, H. L.; Petitjean, F., Characterizing concept drift. *Data Mining and Knowledge Discovery* 2016, *30* (4), 964-994.
13. Metta, N.; Ghijs, M.; Schäfer, E.; Kumar, A.; Cappuyns, P.; Assche, I. V.; Singh, R.; Ramachandran, R.; Beer, T. D.; Ierapetritou, M.; Nopens, I., Dynamic Flowsheet Model Development and Sensitivity Analysis of a Continuous Pharmaceutical Tablet Manufacturing Process Using the Wet Granulation Route. *Processes* 2019, *7* (4).

14. Boukouvala, F.; Niotis, V.; Ramachandran, R.; Muzzio, F. J.; Ierapetritou, M. G., An integrated approach for dynamic flowsheet modeling and sensitivity analysis of a continuous tablet manufacturing process. *Comput Chem Eng* 2012, *42*, 30-47.
15. Kadlec, P.; Grbić, R.; Gabrys, B., Review of adaptation mechanisms for data-driven soft sensors. *Comput Chem Eng* 2011, *35* (1), 1-24.
16. Pechenizkiy, M.; Bakker, J.; Žliobaitė, I.; Ivannikov, A.; Kärkkäinen, T., Online mass flow prediction in CFB boilers with explicit detection of sudden concept drift. *ACM SIGKDD Explorations Newsletter* 2010, *11* (2), 109-116.

593037 Utilization of Advanced Analytics to Monitor Catalyst Health in an Ethylene Oxide Reactor

Abstract:

Ethylene oxide (EO), a precursor to ethylene glycol (EG), is produced through elective oxidation of ethylene and oxygen in the presence of an Ag supported α-Al2O3 catalyst in the EO/EG plant located in the Petkim Petrochemical Holding complex in Aliaga, Turkey. The EO process takes place in two parallel, fixed-bed multi-tubular reactors at a temperature and pressure range of 240–260 °C and 17–18 bar, respectively. Both ethylene epoxidation and EO combustion take place in the reactors and only the former is desirable. Side reactions result in a reduction of EO yield and an increase in CO2 emission from the plant. Thus, the optimization of the catalyst exploitation continuance becomes an important concern to maximize the primary reaction for product quality and plant profitability.

In this presentation, machine learning methods are utilized in the characterization and monitoring of the industrial catalyst used in the EO reactors. Estimation of the catalyst activity and other indicators has major importance during production due to the five-year lifetime with high catalyst cost and the EO product quality. In order to make accurate estimates, a vast quantity of data is collected from the plant and is processed using several statistical methods in order to capture the major driving signals that provide significant implications. Principal component analysis (PCA) is used for dimensional reduction and to categorize the measurements. The performances of various machine learning algorithms are compared for the prediction of that catalyst status. The machine learning model provides a good insight into the causes of rapid catalyst deactivation. Once these insights are obtained, the operating policies and process conditions that will maximize the catalyst lifetime and other desirable properties of the catalyst in the plant will be discussed in detail.

600749 Collaboration Is Critical for Effective Deployment of Big Data Analytics

Abstract:

Manufacturers across the world have begun their digital transformations with an eye towards utilizing big data analytics. With this effort, process sensors are providing a wealth of data, numerous machine learning algorithms are being developed, and data scientists are being hired at an astounding rate. While these advancements are important, the reality is that a wealth of key information, or context, is typically not captured in any database. That wealth of information is the internal expertise that subject matter experts, like engineers and scientists, rely on to drive effective decision making. Considering this, an effective analytics strategy must utilize technology and workflows that combine both human experience and computer technology.

In collaboration, subject matter experts are relied upon throughout the project to provide relevant context, document assumptions, and explain applicable constraints. Data scientists complement this information with their knowledge of machine learning algorithms and other relevant analytics options. Together, the team finds the hidden patterns in the data that translate to business value for the company.

This session will provide a framework for leveraging analytics applications to promote effective collaboration between subject matter experts and data scientists to accelerate the time to insight and operationalize models and algorithms. The presentation will cover the importance of subject matter expert involvement in data preparation for successful deployment of models and algorithms through a series of examples of effective analytics deployment. Examples include operations selected from a variety of industries, including pharmaceuticals and chemicals.

608672 Dynamic Data Feature Engineering for Process Operation Troubleshooting

Abstract:

Data analytics and machine learning have shown tremendous power in mining and analyzing a large amount of data to reveal useful information and knowledge behind the data. Many industries, including chemicals, petrochemicals, energy, power grids, and pharmaceutical, have turned their attention to the massive data they own for improvement in efficiency and competitiveness. Owing to rich instrumentation and control in engineering and manufacturing systems and the deployment of industrial internet of things (IIoT), data collected from industrial operations are of high dimensions and fast sampling. Therefore, there is a great need to develop analytics tools geared towards analyzing massive process data.

Most process data are collected in the form of time series, which are highly cross-correlated and auto-correlated. In other words, the high-dimensional data are not dynamically excited in all dimensions of the measurement space. This situation is more pronounced with IIoT where sensors are installed with a degree of redundancy which leads to collinearity. Due to collinearity, the dynamic variations in plant data are often concentrated in a low dimensional subspace. On the other hand, the complementary subspace has only random variations which are independent over time. Because of these characteristics, traditional modeling tools such as vector autoregressive moving average (VARMA) analysis are not suitable since they assume full dimensional dynamics (Tsay, 2013).

In this paper we apply the dynamic inner canonical correlation analysis (DiCCA) developed in Dong and Qin (2018) and Dong et al. (2020) to extract low dimensional latent variables. Each of the latent variable models is a self-dependent univariate autoregressive (AR) model. The latent variables are orthogonal or contemporaneously independent of each other, which is convenient for visualizing latent features and troubleshooting abnormal variations in high dimensional data. In addition, each of the latent variables is rank-ordered by the predictability from its own history. This objective promotes self-dependent AR relations to be extracted. For example, integrating components and oscillatory components are favored by this objective.

In this paper, we propose a dynamically engineered latent feature analysis (DELFA) procedure for plant-wide troubleshooting by applying the DiCCA algorithm to decompose high dimensional process data into dynamic latent features. DELFA does not make use of the prediction model of DiCCA. It finds dynamic features of a segment of time series data that contain interesting features, which could be associated with anomalies. We also extend the DiCCA algorithm to deal with exogenous variables, which is referred to the DiCCAX algorithm.

DELFA further identifies measured variables that are best interpreted by the latent features.

The degree of interpretation is represented by the latent variable loadings. Composite loadings and weights are derived to analyze features that appear in multiple latent variables.

The features of interest can be intermittent in time; when they happen their loadings on measured variables are the focus of analysis. We demonstrate the effectiveness of the DELFA troubleshooting procedure on two high dimensional datasets from an industrial plant. One dataset is analyzed with the troubleshooting procedure to find several anomalous features. The other data set, collected after fixing a major anomaly, is analyzed to confirm the fix of the major anomaly and also find other anomalies.

References

Ruey S. Tsay. Multivariate Time Series Analysis: with R and financial applications. John Wiley & Sons, 2013.

Yining Dong and S Joe Qin. Dynamic latent variable analytics for process operations and control. Computers & Chemical Engineering, 114:69-80, 2018.

Yining Dong, Yingxiang Liu, and S Joe Qin. Efficient dynamic latent variable analysis for high dimensional time series data. IEEE Trans. on Industrial Informatics, vol. 16, no. 6, pp. 4068-4076, June 2020.

614025 (Invited Talk) Counting Molecules, Dodging Blood Cells: Continuous, Real-Time Molecular Measurements Directly in the Living Body

Abstract:

The availability of technologies capable of tracking the levels of drugs, metabolites, and biomarkers in real time in the living body would revolutionize our understanding of health and our ability to detect and treat disease. Imagine, for example, a dosing regime that, rather than relying on your watch ("take two pills twice a day"), is instead guided by second-to-second measurements of plasma drug levels wirelessly communicated to your smartphone. Such a technology would likewise provide researchers and clinicians an unprecedented window into neurology and physiology, and could even support ultra-high-precision personalized medicine in which drug dosing is optimized minute-by-minute using closed-loop feedback control. Towards this goal, we have developed a biomimetic,electrochemical sensing platform that supports the high frequency, real-time measurement of specific molecules (irrespective of their chemical reactivity) in situ in the blood and tissues of awake, freely moving subjects. Key to achieving this was the development of sensors small enough to emplace in the body but retaining sufficient s/n in such a complex environment. This was done via the use of nanostructured and nanoporous electrode materials of greatly enhanced microscopic area.

610717 Ultrasensitive, Selective, and Reversible Room-Temperature NO_2 Sensor Based on a Monolayer Transition Metal Dichalcogenide

Abstract:

Chemical sensors are essential for environmental and health monitoring. In particular, gas sensors that can efficiently detect toxic gases emitted from power plants, vehicles, and industrial sources at room temperature, are of substantial importance. Here, we demonstrate a room-temperature NO_2 sensor based on a monolayer transition metal dichalcogenide that is highly sensitive and selective. Depending on the layer number of the sensing devices, their electrical resistances substantially increase or decrease upon exposure to NO_2. We find the atomically-thin sensors to be strongly selective to NO_2 with only minimal responses to NH_3, CH_2O, and CO_2. Importantly, the monolayer sensor shows complete reversibility with fast recovery in ambient condition (*i.e.* at room temperature and under substantial relative humidity). Combined with theoretical calculations, we explore interactions between different molecular species and the monolayer crystal and discuss the underlying sensing mechanism.

605192 Atomically-Thin Sensing Surfaces from 2D Materials for Detecting Cellular Gaps

Abstract:

Cellular layers play critical role as barriers, filters, and membranes in our bodies. The performance of cellular layers is often found to be determined by the properties of individual cellular gaps. Yet these gaps often fall below the diffraction limit, hampering their detection. Here, we develop a novel sensing modality called chemical tomography, where the diffusion of a harmless chemical tracer is used to reveal cellular gaps in human umbilical vein endothelial cellular (HUVEC) layers. For the readout, we employ atomically-thin and optically-active MoS2, whose two-dimensional form-factor enables it to serve as an array of 20,164 wireless sensors over 420x420 μm^2 area. A rigorous screening of 30 analytes identifies several tracer candidates with MoS2 dynamic range over five orders of magnitude in tracer concentration. In a single layer of HUVEC cells, we find local spots with 10^{-8}-10^{-9} cm^2/s diffusivities that are spatially correlated with immunofluorescence of vascular endothelial (VE)-cadherin proteins found on cell membrane. These spots correspond to cellular pores with effective sizes from 50 to 100 nm. This platform enables future studies on properties and dynamics of cellular gaps of various tissues.

607027 Nanocomposites of Multiwalled Carbon Nanotubes and Palladium-Based Nanosheets for Hydrogen Sensing

Abstract:

Thin palladium nanosheets have been gaining attention for various applications due to their high surface area, high surface atom fraction, and excellent electrical conductivity. Palladium has been explored as a resistive hydrogen sensor because of its unique property of forming hydrides in the presence of hydrogen at room temperature. However, most Pd-based sensors suffer from low response with a limit of detection around 300 ppm. This work describes the design and fabrication of multi-walled carbon nanotubes@Pd nanosheets (MWCNT@Pd NS) composites for hydrogen sensing applications. The Pd nanosheets were prepared by a CO confinement strategy using a one-pot solvothermal method. The prepared Pd nanosheets had a high surface area of ~125 m^2/g with a thickness of only ~2 nm. The prepared Pd nanosheets were then dispersed in hexane along with CNT and sonicated to produce a homogeneous suspension. The nanocomposite solution was drop cast onto an interdigitated electrode, and its performance for hydrogen detection in synthetic air was evaluated. Compared with pristine Pd nanosheets, the MWCNT@Pd nanosheets showed enhanced hydrogen detection performance. The resistance change of MWCNT@Pd nanosheets to 2% H_2 was 2%, compared to 0.5% change for pristine Pd nanosheets. The LOD of MWCNT@Pd nanosheets was also found to improve to <10 ppm, far below the 300 ppm LOD of Pd nanosheets. This encouraging result could prove useful in the fabrication of low-cost hydrogen sensors based on 2D Pd nanomaterials.

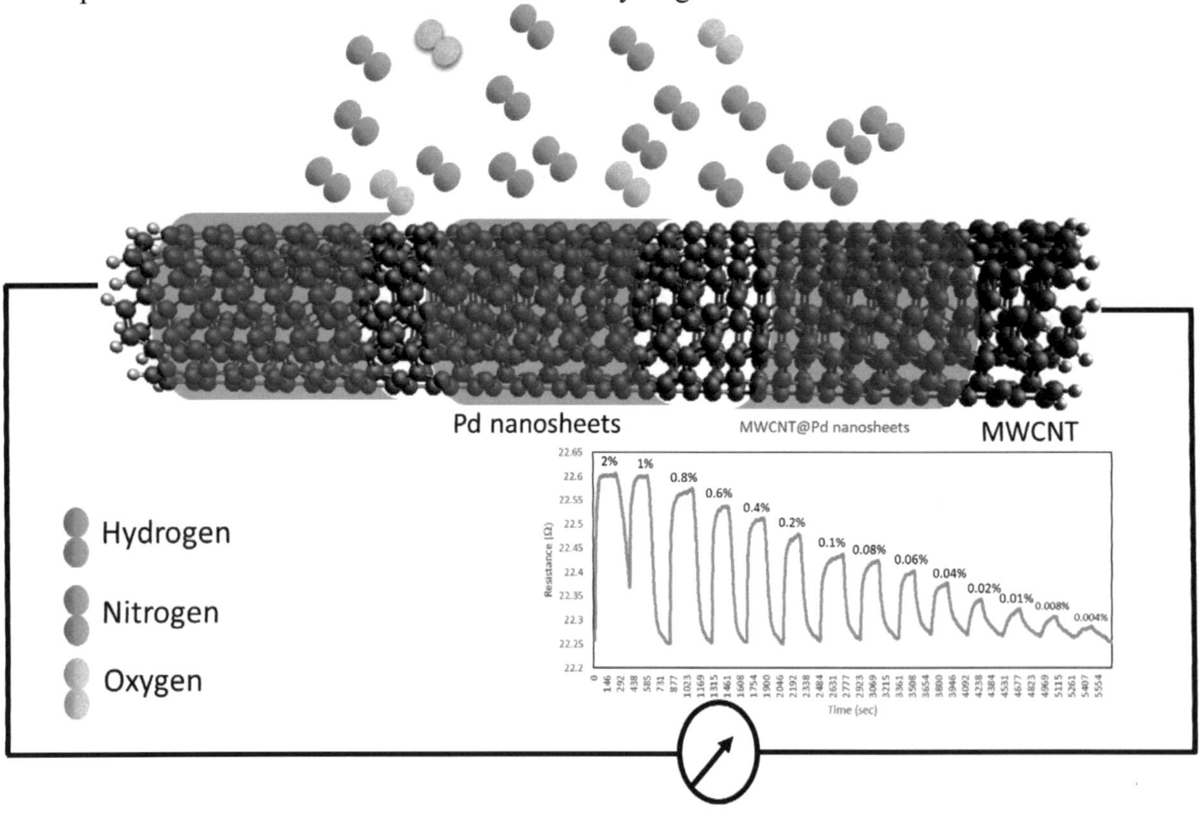

608092 Flame Aerosol Synthesis of Palladium-Decorated Crumpled Reduced Graphene Oxide Nanocomposites for Hydrogen Detection at Room Temperature

Abstract:

Hydrogen (H_2), a green energy source, is highly flammable with a lower explosive limit of just 4% in air. Its leakage during production, transportation, and storage can result in disastrous consequences. Therefore, development of accurate, rapid, and stable H_2 sensors that work in air at room temperature is crucial. In this presentation, we will describe a novel three-dimensional palladium (Pd)-decorated crumpled reduced graphene oxide ball (Pd-CGB) nanocomposite for H_2 detection in air at room temperature. Pd-CGB nanocomposites were synthesized using a flame-driven high temperature reducing jet (HTRJ) process. In this process, a hydrogen-rich flame passes through a converging-diverging nozzle. An aqueous precursor solution containing graphene oxide and palladium nitrate is injected at the throat section of the nozzle and is atomized by the high velocity gas stream. The resulting droplets evaporate, graphene oxide sheets crumple, and the palladium precursor decomposes, initiating nucleation of palladium nanoparticles on crumpled graphene oxide in a reducing environment containing excess H_2. After the reaction zone, nanocomposites are cooled immediately to prevent further growth and coalescence of the Pd nanoparticles. Hydrogen sensors produced by simple drop-casting of these nanocomposites onto electrodes performed well over a wide range of H_2 concentrations (0.01-2%) with response (resistance change) value, response time, and recovery time of 14.8%, 73 s, and 126 s, respectively, at 2% H_2. Compared with other reports on Pd-graphene sensors, the present study shows superior response/recovery time. The HTRJ process enables the single-step, continuous production of Pd- and other metal-decorated graphene nanocomposites with great potential for creating various gas sensors by simple drop-casting onto low-cost electrodes.

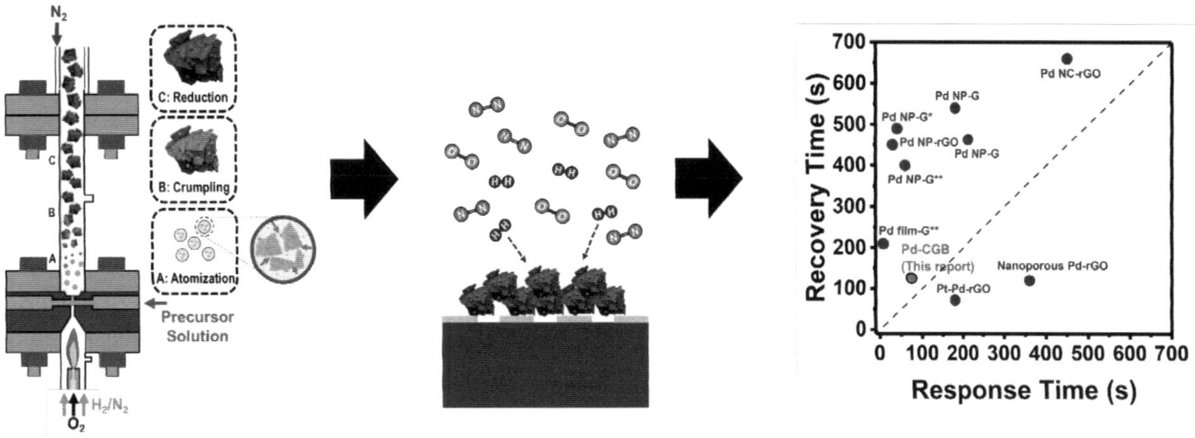

601668 Carbon Black-Gold Nanoparticles for Detection of Analytes Using Surface Enhanced Raman Scattering

Abstract:

Surface-enhanced Raman scattering (SERS) is a powerful technique for the detection of a variety of molecules at very low concentrations. We have developed irregularly shaped gold nanoparticles using a carboxyl-terminated carbon black (CB) nanoparticles as a template. The CB particles were coated with the cationic polyelectrolyte poly-L-lysine (PLL) and exposed to a tetrachloroauric acid solution. Gold - carbon black (AuCB) particles were formed by the reduction of gold chloride ions that concentrated on the surfaces of the PLL. The morphology of the gold on the AuCB can be modified using different PLL concentrations,. The plasmon resonance can then be matched to the incident laser wavelength, thus greatly enhancing Raman signals coming from analytes in the vicinity of the gold surface. We used these AuCB particles to detect a wide range of analytes, including 4-nitrobenzene thiol, and nitrate and sulfate ions in a mixture. The limits of detection were well below that obtained by other SERS particles.

609081 3D Printing for High Performance Polymers and Nanocomposites

Abstract:

There is high interest in the use of 3D printing to create objects from polymeric nanocomposite materials. It has appended the design functionality of polymeric materials enabling wide uses and rapid scale development in new and demanding applications. Other materials based on metallic, ceramic, hydrogels and other non-metallic do not have as wide a range of properties leading to new structure-composition-property relationships. Polymers can be classified into thermoplastics, thermosets, and elastomers based on their thermo-mechanical properties. However, it is not as simple as the classification into two groups when it comes to the use of pre, real-time, and post processing and curing when the transition to a final phase or cross-linked structure results in new properties. However, each class of polymers when paired with a corresponding nanomaterial can become a class of its own. This is also more evident with the choices of 3D printing methodologies (FDM, SLA, SLS, VSP) which can make use of blended or formulated compositions. By incorporating nanomaterials with 3D printing it allows the design of new and toughened materials and applications based on nanostructuring and integrating the chemistry of conversion with the printing mode. In this talk, we demonstrate the design principles and fabrication of higher performance and toughened 3D printed materials based on the use of nanomaterials and their nanostructuring. Other work based on the use of SLA, SLS, FDM, towards high strength nanocomposite materials and their applications will be discussed. 4D Printing will be demonstrated based on stimuli-responsive properties to be observed including supercompressibility and shape-memory properties.

604005 Room-Temperature 3D Printing of a Super-Soft and Solvent-Free Elastomer

Abstract:

Super-soft elastomers derived from bottlebrush polymer networks show promise as advanced materials for biomimetic tissue and device applications, but current processing strategies are restricted to simple molding. Here, we introduce an alternative approach rooted in molecular design that enables direct ink write (DIW) 3D printing of super-soft bottlebrush elastomers at room temperature without solvent. The key advance is a new class of shear-thinning inks comprising statistical bottlebrush polymers that self-assemble into well-ordered body-centered cubic spheres at remarkably asymmetric compositions. These single-component soft solids undergo sharp and reversible yielding at 20 °C in response to shear with a yield stress and structural modulus that can be tuned by manipulating the length scale of microphase separation. Addition of a soluble photo-crosslinker containing benzophenone end-groups allows complete UV curing after extrusion to form super-soft elastomers (shear moduli ≈ 1–50 kPa) with unusual mechanical properties, namely near-perfect recoverable elasticity well beyond the yield strain. In summary, the unique bottlebrush polymer structure–property design rules elucidated herein create new opportunities to tailor the performance of super-soft elastomers in ways that are not possible with current materials and processes.

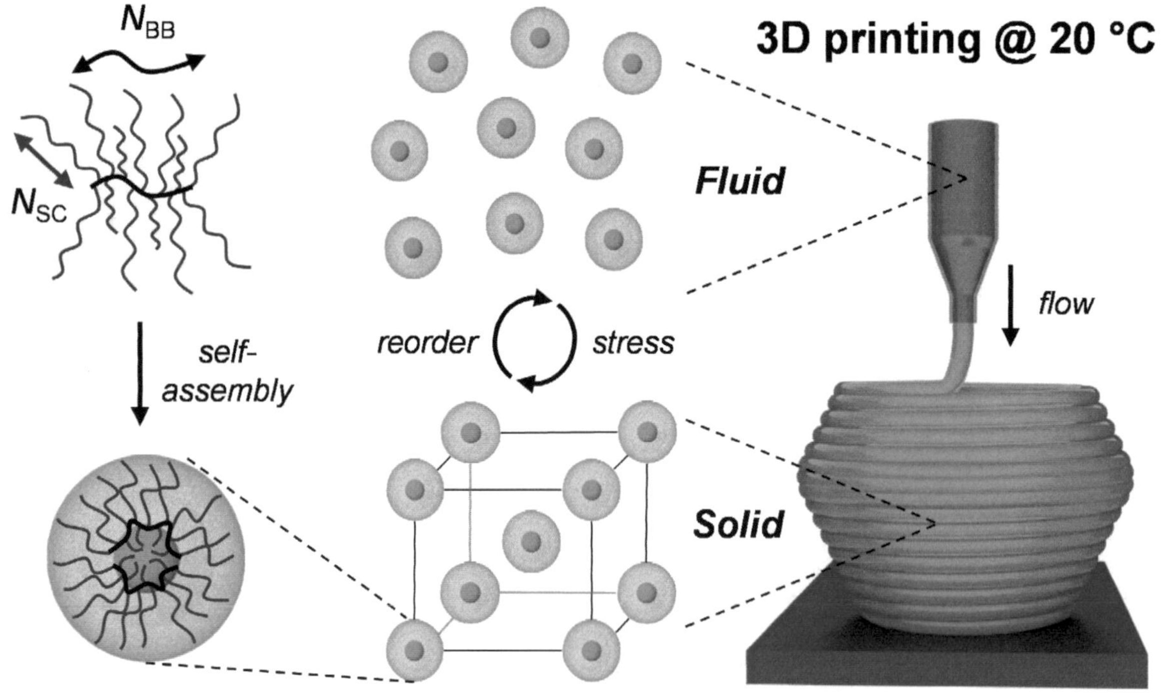

599674 High T_g, Low Viscosity Resin Systems Based on Isosorbide Methacrylate (IM) for Additive Manufacturing

Abstract:

Additive manufacturing is currently at the forefront of the seemingly impending industrial revolution as researchers are investigating new ways to improve and optimize the manufacturing of machine/tooling components. Widely understood as Industrialized 3-D printing, additive manufacturing provides the promise of manufacturing entire machines and machine parts such as jet engines, hearing aids and gas turbines. According to studies conducted by the Fraunhofer Additive Manufacturing Alliance, approximately 150 companies currently operate in this services market; an interesting prospect as market researchers forecast a 300% growth in additive manufacturing niches over the next decade. Although currently used for rapid prototyping, additive manufacturing is expected to grow past conventional production methods as it is projected to be 50% cheaper and 400% faster in the next five years. This work aims to show the synthesis and characterization of isosorbide based low viscosity crosslinkers that can potentially be integrated into additive manufacturing resins. Isosorbide based crosslinkers have the potential to impart high thermal properties (T_g~ 220 °C) and high modulus (> 4 GPa) on bio-based polymers for additive manufacturing.

604748 Investigation of the Physiochemical Effects of Heat-Induced Aging on 3D Printed Photopolymers

Abstract:

In three-dimensional (3D) digital light processing (DLP), ultraviolet light (typically at 405 nm) is emitted from a digital projector and directed at a mobile build area in a layer-by-layer fashion to create solidified cross-linked networks that are composed of rectangular voxels via free radical polymerization of photopolymer resins. The photopolymers that are typically used in DLP (e.g., acrylate- or epoxy-based thermoset resins) can contain different properties after printing that are dependent on resin formulation, printing parameters, and post-processing conditions; hence, the need exists to appropriately engineer and characterize a resin formulation that is based on the specific requirements of the end-use application in both a time-efficient and cost-effective manner. By comparing results of post processing (e.g., UV cure, convection oven) to a set of green DLP prints, protocols for determining and predicting the dependence of mechanical performance on chemical and physical aging has been established. Surface profilometry shows the topological effects of different environments in 3D-DLP parts over time, which has proven to be the most time-efficient and reliable method in anticipating changes in mechanical performance of DLP specimens. Dynamic mechanical analysis (DMA) was found to be very useful in evaluating the mechanical performance, utilizing time-temperature superposition (TTS) to measure the modulus over more than eight (8) orders of magnitude in frequency. The activation energy that is associated with the glass transition and the magnitude of the constants that are associated with the shift factors for each test were further analyzed using the Arrhenius and William-Landel-Ferry (WLF) empirical equations as tools for predicting the effects of temperature-induced physical changes of the DLP specimens. The successful completion of the tasks that are described in this work provided the necessary experimental framework for predicting and evaluating prolonged mechanical functionality of DLP parts that otherwise may be unattainable.

599251 Continuum Simulations of Polymer Deposition in Fused-Filament Fabrication (FFF)

Abstract:

We utilize the OpenFOAM tool box to simulate the viscoelastic flow of polymer deposition in Fused-Filament Fabrication (FFF). In FFF, molten polymer extruded from a nozzle is built up layer-by-layer to form three-dimensional structures; the ultimate mechanical properties of the resultant structures depend on the material properties of the polymer as well as the microstructural changes of the polymer molecules induced by the processing history of the melt. Prior theory has examined polymer conformational changes during deposition, however, these theories relied on an "advection" calculation with no dynamics during deposition, suggesting the need for simulations to access the full dynamics. We simulate the non-uniform flow within nozzle and the deposited material; the polymer microstructure is described by coarse-grained, continuum-level structural variables (orientation, stretch, and entanglement fraction). The evolution equations for these structural variables, as well as conservation laws of mass, momentum, and energy, are solved to obtain the polymer conformation, velocity, and temperature within the nozzle as well as within the extruded polymer. We compare different entanglement dynamics schemes, and find that models with shear-induced disentanglement schemes overpredict disentanglement at high shear rates. We investigate the conformational changes induced by regions of local shear and extensional flows within the nozzle. The influence of processing conditions (e.g. volumetric flowrate, print speed, printer height) are investigated; it is found that these processing conditions may be tuned to produce desired conformational changes in the polymer molecules.

601834 Systematic Application of 3D Bioprinting in the Fabrication of a Composite Full Thickness Human Skin Model

Abstract:

Considering the increasing aversion to the use of animal models in assessing the safety and efficacy of cosmetic and drug products, the use of tissue models is a promising alternative. While *in vitro* skin models are conventionally fabricated via manual deposition of cells onto a suitable extracellular matrix, a drawback of such a method is its inability to produce irregular tissue shapes and difficulties in scale up. The use of 3D bioprinting techniques has the potential to overcome these limitations through the efficient and uniform fabrication of skin models at a large scale. This study thus focuses on the development and validation of a novel full-thickness biomimetic skin model via extrusion-based bioprinting. A composite PCL collagen acellular layer was first constructed before a layer of fibroblast encapsulated collagen and a layer of cell culture medium containing keratinocytes were sequentially printed above it. The construct was then airlifted after a few days to allow for stratification and differentiation of keratinocytes. Bioprinted skin constructs were then compared with full thickness skin models produced via manual seeding of cells both qualitatively and quantitatively. Compared to full-thickness skin equivalents produced via manual seeding of cells, skin equivalents fabricated via 3D extrusion bioprinting had comparable qualitative and quantitative features when evaluated via cell viability assays, H&E staining and biomarker analysis. Permeability tests also provide evidence of good barrier function of skin models produced via 3D extrusion bioprinting. Our studies thus demonstrate the capability of dynamic extrusion PCL scaffold based printing in producing reconstructed human bi-layered skin equivalents with better epidermal morphogenesis, differentiation and physiological barrier functions. While promising, more studies are needed to optimize this 3D bioprinting methodology in order to improve recovery time and biomimeticity of human skin fabricated.

607912 Constrained Control of Dissipative Distributed Parameter Systems Via on-Demand Data-Driven Model Reduction

Abstract:

Distributed parameter systems (DPSs) that frequently appear in fluid flows and diffusion-convection-reaction processes can be mathematically modeled by dissipative partial differential equations (DPDEs). The control problem of such systems is challenging due to the spatial variation and complex temporal dynamics of the state variables [1]. Model reduction is known as an auspicious method to address the control and optimization problems for dissipative DPSs, where a reduced-order model representation of the governing DPDEs is used as the basis for the optimization formulation and control design [2]. Given the required spatial basis functions, the reduced-order models can be derived by employing weighted residual methods in the form of ordinary differential equations describe the temporal dynamics [1].

Analytical methods for basis functions computation are applicable only to linear systems with simple boundary conditions. Data-driven methods such as proper orthogonal decomposition (POD) have been applied as an alternative solution to compute dominant empirical basis functions [2-5]. Specifically, POD analyzes spatially distributed profiles of the system states (aka process snapshots) obtained by high-fidelity simulations or measured by spatially distributed sensors to compute the basis functions. POD may require a large number of process snapshots to guarantee the optimality of representing the process dynamics, specifically in the presence of model uncertainty and time-varying parameters [6]. Then, the resulting empirical basis functions may need to be revised as needed due to the appearance of new trends.

In previous work, adaptive proper orthogonal decomposition (APOD) was developed to recursively update the empirical basis functions using the new process snapshots [6]. Once the empirical basis functions are updated, the current reduced-order model switches to a new one and the controller design is revised. These operations increase measurement, communication, and computation costs [7].

In this work, we develop an on-demand recursive POD to minimize the frequency of reduced-order model revisions. The proposed method hinges on an event-triggered mechanism which evaluates the constrained control Lyapunov function (CLF) of system modes estimated by a static observer. The input-constrained controller based on the reduced-order model is synthesized to force the CLF to decrease between the reduced-order model revisions to guarantee the asymptotic stability of the closed-loop switching system [8]. Then, an increase of the CLF at any time implies that the reduced-order model is inaccurate. We identify the periods when the reduced-order model must be revised by tracking the time derivative of the CLF. The recursive revision continues until the CLF value satisfies the switched system stability criteria. The proposed approach is applied to stabilize the Kuramoto-Sivashinsky equation, which exemplifies wave motion of falling liquid thin films, instability at the interface of two viscous fluids, and phase turbulence in transport-reaction processes.

1. Christofides, P.D. Nonlinear and robust control of PDE systems; Birkhauser: New York, NY, 2000.
2. Graham, M.D.; Kevrekidis, I.G. Alternative approaches to theKarhunen-Loeve decomposition for model reduction and data analysis. Comp. & Chem. Eng. 1996, 20, 495-506.
3. Armaou, A.; Christofides, P.D. Finite-dimensional control of nonlinear parabolic PDE systems

with time-dependent spatial domains using empirical eigenfunctions. Int. J. Appl. Math. & Comp. Sci. 2001, ,11, 287–317.
4. Tarman, I.H.; Sirovich, L. Extensions to Karhunen-Loevebased`approximation of complicated phenomena. Comput. Methods. Appl. Mech. Eng.1998, 155, 359-368.
5. Xu, C.; Ou, Y.; Schuster, E. Sequential linear quadratic control of bilinear parabolic PDEs based on POD model reduction. Automatica 2011, 47(2), 418-426.
6. Pourkargar, D.B.; Armaou, A. Design of apod-based switching dynamic observers and output feedback control for a class of nonlinear distributed parameter systems. Chem. Eng. Sci. 2015, 136, 62-75.
7. Pourkargar, D.B.; Armaou, A. APOD-based control of linear distributed parameter systems under sensor/controller communication bandwidth limitations. AIChE J. 2015, 61, 434-447.
8. Christofides, P.D.; El-Farra, N. Control of nonlinear and hybrid process systems: Designs for uncertainty, constraints and time-delays; Springer-Verlag, Berlin, Germany, 2005.

602486 Design of Optimal Multi-Size Proppant Pumping Schedule to Enhance Shale Gas Production

Abstract:

Hydraulic fracturing is a technique extensively used in the oil and gas industry, where water, proppant (sand) and additives are injected into unconventional reservoirs to enhance the recovery of shale hydrocarbon. Although some previous studies have developed pumping schedules that maximize gas production for a single-size proppant [1, 2], they did not consider the effect of varying proppant diameters across pumping stages on shale gas production. For determining the oil and gas production from unconventional reservoirs, average propped surface area (PSA) and average fracture conductivity (FC) of the propped hydraulic fractures play an important role, all of which depend largely on the injected proppant diameter. Unfortunately, existing studies on average PSA or average FC considered these two effects separately, and did not consider their combined effect on shale gas production [3, 4].

Motivated by this lack of fundamental studies on this important topic, we first conducted a sensitivity analysis to study the effect of multi-size proppant during hydraulic fracturing on shale gas production from unconventional reservoirs. From the sensitivity analysis, we found out that the pumping schedule used for hydraulic fracturing determines the average PSA and average FC of the propped fractures, which in-turn are two useful parameters that determine the cumulative shale gas production volume from an unconventional reservoir. Therefore, we developed a framework called Sequentially Interlinked Modeling Structure (SIMS) to predict the average PSA, average FC and cumulative shale gas production volume for a given pumping schedule. The first model in SIMS is a multivariable output error state space (MOESP)-based reduced-order model (ROM) of the fracturing process, which provides the pre-shut-in average PSA and pre-shut-in average FC as the outputs. The second model in SIMS is an artificial neural network (ANN) that has been used to accurately simulate the gravity-induced proppant settling process, which provides the post-shut-in average PSA and post-shut-in average FC as the outputs. The third model in SIMS is a map that links the average PSA and the average FC of the created hydraulic fractures to the cumulative shale gas production volume from an unconventional reservoir. These models are interlinked in the sense that the output from the first model is the input to the second, and the output from the second model is the input to the third. Since this approach also provides us with average PSA and average FC values in the intermediate step after the hydraulic fracturing operation that typically takes a few hours, these parameters are usually more important to the oil and gas industries instead of just the cumulative shale gas production volume at the end of 10 years. Then, we used the SIMS framework to obtain a multi-size proppant pumping schedule that maximizes shale gas production. Finally, we demonstrated that obtained pumping schedule gives a gas production volume greater than the values obtained from the existing pumping schedules which consider only single-size proppant.

References:

[1] Siddhamshetty, P., Kwon, J.S., Liu, S., & Valkó, P.P. (2017). Feedback control of proppant bank heights during hydraulic fracturing for enhanced productivity in shale formations. AIChE Journal, 64, 1638-1650.

[2] Siddhamshetty, P., Bhandakkar, P., & Kwon, J.S. (2020). Enhancing total fracture surface area in naturally fractured unconventional reservoirs via model predictive control. Journal of Petroleum Science and Engineering, 184, 106525.

[3] Wahl, H.A. (1965). Horizontal fracture design based on propped fracture area. Journal of Petroleum Technology, 17, 723 – 730.

[4] Ramurthy, M., Barree, R.D., Kundert, D.P., Petre, J.E., Mullen, M.J. (2011). Surface-area vs. conductivity-type fracture treatments in shale reservoirs. In SPE Hydraulic Fracturing Technology Conference, The Woodlands, TX.

610803 Molecules to Systems Design of Advanced Membrane Materials and Technologies

Abstract:

The synthesis of many advanced materials in the past three decades have failed to produce a corresponding revolution in sustainable separation applications.[1] For example, new technologies for carbon capture can take up to 30 years to progress from a laboratory to commercial deployment.[2] To accelerate technology development, there is a critical need to move away from classical Edisonian iterative methods of materials discovery. Principled and data-driven frameworks will overcome the gaps in knowledge which inhibit the translation of new materials and devices from laboratory scale to sustainable separation technologies. In this talk, we describe our work using superstructure optimization at the systems scale and nonlinear parameter estimation at the materials scale in designing next generation membrane technologies and materials. We highlight how the current work lays the foundation for a holistic molecules-to-infrastructure computational design framework for the development of sustainable separation technologies.[3]

Relative to conventional technologies, membrane separations have demonstrated significant advantages in sustainability and energy efficiency.[4] For example, chemically selective separations enabled by polymer membranes have the potential to replace hazardous and toxic solvents used in conventional lithium-ion battery (LIB) recycling technology. However, there is no clear understanding of the interfacial and thermodynamic phenomena underlying the chemically-selective transport mechanisms needed to separate molecules with comparable sizes such as lithium (Li) and cobalt (Co) which hinders the development of membrane materials for this task.[5] Another challenge is the recovery of high purity products from a membrane separations unit. The formation of salt precipitate layers at the membrane interface when the feed side solution approaches the solubility limit prevents the extraction of high purity products, which also prevents the direct staging of these units.

In the first part of this talk, we demonstrate how a continuous, membrane based diafiltration cascade[6] enables the recovery of high purity Li and Co rich solutions for a LIB recycling case study at purities comparable with existing technologies. In diafiltration, the effects of salt precipitation encountered while staging membrane units are offset by adding a pure or dilute solvent, known as the dialysate to the feed side of the membrane filtration process. We show how the unique configuration of our system enables the utilization of the retentate side product, which is dilute in Li, to be recycled as dialysate to preceding stages of the process. We highlight how superstructure optimization searches multiple, complex process configurations postulated by the modeler to identify promising applications for emerging materials. We treat flow and concentration of all streams as decision variables in our nonlinear optimization model. We then characterize the Pareto set using an epsilon-constrained method to understand the tradeoffs between purity and recovery objectives for the diafiltration cascade. We observe interesting regions of the Pareto set with unique flow and concentration profiles but having similar recycle strategies. We use this information to discern physical insights into the phenomena governing the underlying process and demonstrate the top-down capabilities of this framework to identify property targets for emerging membrane materials for LIB recycling.

The superstructure optimization framework developed so far uses lumped parameter models in optimizing the diafiltration cascade. In the second part of this talk, we apply parameter estimation and design of experiments formalisms to analyze novel dynamic diafiltration. We use timeseries data to characterize membranes by estimating model parameters governing interfacial and transport phenomena across membranes. We argue that analyzing timeseries data from dynamic experiments will allow for faster characterization of membranes and capture non-equilibrium effects characteristic of the process. Moreover, dynamic modeling and parameter estimation will provide first pass insights into realizing model predictive control for diafiltration systems. We formulate a multiobjective parameter estimation problem, to characterize the trade-offs between minimizing errors for three different experimentally measured quantities (permeate concentration, retentate concentration and timeseries filtration cell mass). We perform localized and Monte Carlo analysis to identify the uncertainty in fitted parameters. Finally, we discuss how to optimize sequential experimental campaigns to elucidate transport and thermodynamic mechanisms that govern multicomponent separations via diafiltration.

We conclude this two-part talk by demonstrating how superstructure optimization and advanced data analytics form the pillars for comprehensive molecules-to-infrastructures computational design framework. We hypothesize that embedding detailed membrane models into the superstructure optimization framework will reveal optimum process network configurations that enable the identification of operating parameters and control strategies that influence the behavior of the diafiltration cascade (bottom-up design). Detailed membrane models will also inform how to engineer materials with desirable properties for specific application requirements (top-down design). Uncertainty information quantified in the statistical learning framework will allow the optimal design of future experiments using techniques such as model-based design of experiments. Finally, we provide brief comments on materials informatics and the ubiquitous role of uncertainty quantification which make up the rest of our holistic framework.

References

[1] Luo, J. & Crittenden, J. C. Nanomaterial Adsorbent Design: From Bench Scale Tests to Engineering Design. *Environmental Science & Technology* **53**, 10537 - 10538 (2019). 10.1021/acs.est.9b04371.

[2] Miller, D. C., Syamlal, M., Mebane, D. S., Storlie, C., Bhattacharyya, D., Sahinidis, N. V., Agarwal, D., Tong, C., Zitney, S. E., Sarkar, A., Sun, X., Sundaresan, S., Ryan, E., Engel, D. & Dale, C. Carbon Capture Simulation Initiative: A Case Study in Multiscale Modeling and New Challenges. *Annual Review of Chemical and Biomolecular Engineering* **5**, 301-323 (2014). 10.1146/annurev-chembioeng-060713-040321.

[3] Eugene, E. A., Phillip, W. A. & Dowling, A. W. Data science-enabled molecular-to-systems engineering for sustainable water treatment. Current Opinion in Chemical Engineering 26, 122-130 (2019). 10.1016/j.coche.2019.10.002.

[4] Phillip, W. A., & Elimelech, M. The Future of Seawater Desalination: Energy, Technology, and the Environment. *Science*, **333**, 6043, 712–717 (2011). 10.1126/science.1200488.

[5] Park HB, Kamcev J, Robeson LM, Elimelech M, Freeman BD. Maximizing the right stuff: the trade-off between membrane permeability and selectivity. *Science* **356**, 1-10 (2017). 10.1126/science.aab0530.

[6] Eugene, E., Phillip, W. & Dowling, A. Material Property Goals to Enable Continuous Diafiltration Membrane Cascades for Lithium-ion Battery Recycling. In *Computer Aided Chemical Engineering* **47**, 469-474 (2019). 10.1016/B978-0-12-818597-1.50075-8.

600938 Development of Condition Monitoring and Prognostic Capabilities for a Vanadium Redox Flow Battery

Abstract:

The vanadium redox flow battery (VRFB) is a rechargeable flow battery that is one of the most promising large-scale energy storage systems making them suitable for grid energy storage. However, the self-discharge reactions along with the undesired side reactions and water transfer through the membrane causes imbalance in electrolyte and state-of-charge (SOC), which can significantly reduce the capacity of VRFBs. Self-discharge reactions, caused by the diffusion, convection, and migration of the vanadium ions from one half-cell to the other, lead to an imbalance between the state-of-charge of the two half-cell electrolytes and consequently cause a capacity fade. As a result of evolution of hydrogen or air oxidation of V^{2+}, the side reactions also affect the capacity of the VRFB. Likewise, inadequate flow rate reduces the cycling time, thus affecting the capacity. To overcome the effect of side reactions, it has been suggested that the operating range of the cell be maintained between 10% and 90% SOC[1]. However, even when the overall SOC is constrained in this desired region, there can be considerable local inhomogeneity causing capacity loss. For maximizing the reliability of VRFBs under load-following applications, capacity fade needs to be estimated along with a prognostic capability that can estimate the remaining useful life (RUL) of the battery.

A first-principle non-isothermal model[2] is developed based on the physical processes and electro-chemical reactions taking place inside a cell to capture the dynamics of the model. This model is based on momentum and mass transfer, ion conservation, and energy transport phenomena. This is combined with a kinetic model for electro-chemical reactions involving vanadium species. Bubble formation due to the gas evolving reactions at the positive and negative electrodes is included to account for the effect of momentum transfer between gas and liquid phases and also the reduction in liquid volume. Heat generation due to activation losses, electrochemical reaction, and Ohmic resistance is modeled. Water transport through the membrane is also modeled. Due to the different diffusion coefficients of the vanadium species, their cross-over causes capacity decay. In addition, the gas evolving reactions are modeled. These reactions lead to an imbalance in the SOC leading to capacity loss.

A model-based approach is developed for condition monitoring and prognosis. The detailed first-principles model is computationally intractable for prognosis. Thus, an adaptive neural network (NN) model-based approach is developed to estimate SOC and capacity fade. Given the current state and quantified uncertainty in the NN model and model parameters, the probability density function (PDF) for the RUL is calculated. The PDF for the RUL is conditional and is updated at discrete time intervals taking into consideration the accumulated damage and expected future operating profile.

Reference

1. Tang, Ao, Bao, Jie, and Maria Skyllas-Kazacos, "Dynamic modelling of the effects of ion diffusion and side reactions on the capacity loss for vanadium redox flow battery", Journal of Power Sources 196 (2011) 10737– 10747.
2. Hassan Al-Fetlawi. "Modelling and Simulation of All-Vanadium Redox Flow Batteries". Thesis for the degree of Doctor of Philosophy. University of Southampton. November 2010.

606584 Data-Driven Optimization of Manufacturing Systems Under Supply and Demand Volatilities in an Industrial Symbiosis Network

Abstract:

The concept of industrial symbiosis (IS) draws inspiration from circular economy, and is a promising approach for resource exchange, energy conservation and emission reduction in the global manufacturing industry (1). However, uncertainties in supply, demand and production environments render significant complexities for decision support systems (DSS) within individual entities of the IS network (2), thereby introducing the risk of suboptimal decisions when operations are performed deterministically . Traditionally, such problems have been solved using the classical stochastic or chance-constrained optimization under uncertainty.

In contrast to previous studies, this research investigates the application of data-driven optimization by coupling the concepts of machine learning and conventional mathematical optimization (3), to devise a DSS. Specifically, we consider uncertainties associated with the supply of resources and demand – applied to a real-world wood manufacturing company in an IS network. For its production requirements, the company primarily depends on the supply of waste-wood (material resource) from other entities of the IS network, which are highly volatile in terms of quantity and quality. The company also relies on a hybrid power system (energy resource) comprising of solar panels in addition to the main power grid to meet the energy demands, wherein the volatility in weather conditions and time-of-use power pricing significantly impacts the production.

To this aim, based on historical data, we first develop hybrid machine learning (4) models to predict supply of material and energy resources and incorporate the concept of prediction interval (PI) to the outcome of the model to quantify the degree of uncertainty associated with the respective volatile resources. This prediction interval serves as an ambiguity set for an optimization formulation that accounts for the optimal utilization of wood-waste as well as optimal combination of power sources to meet the production demands; with multi-objectives to minimize the overall production cost, emissions and optimize inventory utilization. An optimal DSS based on data-driven optimization framework as described in this study, is computationally less intensive (5) and offers greater flexibility in decision making in the face of uncertainties (5) to aid the overall productivity, profitability and sustainability of individual networks in an IS network.

References

1. Valenzuela-Venegas G, Vera-Hofmann G, Díaz-Alvarado FA. Design of sustainable and resilient eco-industrial parks: Planning the flows integration network through multi-objective optimization. Journal of Cleaner Production. 2020 Jan 10;243:118610.
2. Cao X, Wen Z, Xu J, De Clercq D, Wang Y, Tao Y. Many-objective optimization of technology implementation in the industrial symbiosis system based on a modified NSGA-III. Journal of Cleaner Production. 2020 Feb 1;245:118810.
3. Ning C, You F. Data-driven Wasserstein distributionally robust optimization for biomass with agricultural waste-to-energy network design under uncertainty. Applied Energy. 2019 Dec 1;255:113857.
4. Chandra R, Ong Y-S, Goh C-K. Co-evolutionary multi-task learning for dynamic time series prediction. Applied Soft Computing. 2018 Sep 1;70:576–89.
5. Ning C, You F. Optimization under uncertainty in the era of big data and deep learning: When machine learning meets mathematical programming. Computers & Chemical Engineering. 2019 Jun 9;125:434–48.

605830 Comprehensive Dynamic Simulation Model of a Cryogenic Storage Tank

Abstract:

Many industrially useful gases such as hydrogen, argon, helium, oxygen, nitrogen, methane, ethane, ethylene, propane, propylene, butane, and ammonia are gases at ambient conditions. Hydrogen is becoming more important, as the world moves towards clean and renewable energy fuels. The demand for natural gas (NG) as an energy source with the lowest carbon, sulfur, and particulate emissions is growing at twice the rate of global energy demand. Ammonia is gaining attention as a potential hydrogen carrier. These gases are normally stored and transported in their cryogenic liquid states in huge cylindrical or spherical, heavily insulated, and auto-refrigerated tanks. In such tanks, evaporation of saturated liquid, or condensation of boiled off vapors (called Boil-Off Gas or BOG) are inevitable. While BOG generation auto-refrigerates the liquid, BOG must be removed periodically to keep the tank pressure below a safe limit. BOG also represents a loss of valued product, unless reliquefied or reused. However, BOG reliquefaction is energy intensive. Therefore, it is important to understand, quantify, and minimize BOG generation in an industrial cryogenic storage tank, which is a 2-phase system with temporal liquid level, complex spatiotemporal physical phenomena (heat transfer, evaporation, and condensation), and spatiotemporal state variables (temperature, pressure, and composition).

Several studies in the past have attempted to simulate cryogenic tanks using either simplified lumped parameter models or complex full-fledged CFD (Computational Fluid Dynamics) models. While the latter are computationally too expensive for practical applications (Saleem et al., 2018), most of the former lack the rigor. Most models (Migliore et al, 2015) assume vapor and liquid phases to be in Vapor-Liquid Equilibrium (VLE), which is not the reality (Effendy et al., 2017; Migliore et al., 2017) in an industrial tank. Process simulators such as Aspen Hysys® and Unisim® have inbuilt tank modules for dynamic simulation, but with serious limitations. These lumped parameter models cannot predict spatial temperature and pressure gradients, do not allow independent non-equilibrium initial hold-up conditions, have no provisions for some heat leaks, and have non-intuitive ways to deal with mixing in the tank. Therefore, it is highly desirable to develop a simpler, faster, and more rigorous dynamic simulation model for this heart of most energy systems.

Modeling the spatiotemporal behavior of a cryogenic storage tank poses two key challenges. One is the handling of the moving vapor-liquid interface, and the other is the accurate prediction of the BOG generation rate. While many existing models consider only the evaporation, condensation cannot be ignored. In a multicomponent system, some components may evaporate, and others may condense at the same time. In this work, we will present a practically useful, comprehensive, and spatiotemporal mathematical model for simulating the dynamic behavior of a generic cylindrical tank holding a multi-component cryogenic liquid. The tank can be completely or partially insulated and allows multiple feed and product streams. The model uses a novel, flexible, and moving spatial grid instead of the fixed grid used in most models and CFD. It does not assume well-mixed liquid and vapor phases and allows non-equilibrium behavior. It also offers a practically meaningful and intuitive treatment for mixing feeds with the tank contents and withdrawing products. It models evaporation/condensation at the vapor-liquid interface using appropriate driving forces based on sound interphase mass transfer principles. The mass, energy, and component balances describing the spatiotemporal variations of pressure, temperature, and composition in the vapor and liquid holdups result in a complex differential algebraic (DAE) equation system. The system is further reduced using clever mathematical manipulations to improve solution speed. The solution requires repeated iterative and nonlinear vapor-liquid equilibrium or flash calculations that need accurate physical properties. For these, we have designed a series of ANN (Artificial Neural Network) models with help from Aspen Hysys to reduce computational burden without sacrificing accuracy in physical property estimation. Our simulation model can predict the dynamic behavior of any cryogenic storage tank in any mode of operation (loading, unloading, supplying, etc.) It estimates BOG generation accurately by capturing the effect of competitive multi-component evaporation and condensation occurring at the interface. The developed model is thus computationally cheaper and comprehensive in capturing all major tank dynamics. It provides a sound basis for the design, operation, and control of these tanks.

References

Saleem, et al., A CFD simulation study of boiling mechanism and BOG generation in a full-scale LNG storage

tank, Comput. Chem. Eng. 115 (2018) 112–120.

Migliore, C. Tubilleja, V. Vesovic, Weathering prediction model for stored liquefied natural gas (LNG), J. Nat.

Gas Sci. Eng. 26 (2015) 570–580.

Migliore, A. Salehi, V. Vesovic, A non-equilibrium approach to modelling the weathering of stored Liquefied

Natural Gas (LNG), Energy 124 (2017) 684–692.

Effendy, et al., Dynamic modelling and optimization of an LNG storage tank in a regasification terminal with

semi-analytical solutions for N2-free LNG, Comput. Chem. Eng. 99 (2017) 40–50.

602216 Ontology Engineering: Support to Decision Making in Biorefining

Abstract:

Biorefining is a dynamic field with ever grooving number of processes operating in different environment, using different feedstocks and providing expanding range of products. Option scoping and process synthesis and process design are becoming increasingly complex [1]. While development and integration of process and feedstock models and datasets from laboratory experiments are crucial at design stage [2], process and feedstock discovery and valorisation are important at operational stage [3]. This in turn necessitates efficient decision support tool addressing stated complexity and assuring timely and accurate decisions made at increasingly shorter time span.

This paper proposes a knowledge based decision support tool for biorefining which:

- Coordinates operation of biorefining repositories of process and feedstock models and datasets, as well as repositories of existing biorefining processes and available feedstocks, all with fast and precise discovery and audited updates;
- Supports model and data integration used for process synthesis and design [4];
- Enables screening opportunities and integration of concomitant networks based on technological capability and resource availability for targeted products [3, 5];
- Assessment of economic and environmental potential and effects [6].

The proposed decision making tool employs ontologies to embed tacit knowledge in the domain of biorefining, knowledge gained from past experience but also from the latest research and otherwise advances in biorefining [1, 2]. More specifically, a set of integrated ontologies address i) functionality and otherwise properties of biorefining models and datasets from laboratory experimentation and biorefining processes, or simply from analytical assessments, ii) nature of feedstock, i.e. variability in composition, dynamics in availability and pricing, as well as economic and environmental properties including hazardousness, iii) processing technologies in terms of processing capabilities, but also range of type of inputs, conversion rates, water and energy requirements, range of capacities, emissions as well as fixed and operational costs and environmental effects, and iv) general knowledge and past experience in process design and synthesis. While tacit knowledge associated with all four elements is embedded in the ontology structure, the explicit knowledge associated with particular process models and datasets, as well as particular process and feedstock is enquired during the ontology instantiation with actual data collected from the model, data, process and feedstock owners during the registration. The ontologies are designed using ontology web language and hence prepared to grow and to share. In the current implementation more than 400 different model and datasets have been included on the top of more than 150 different type of feedstock and over 40 different biorefining technologies, all organised in respective repositories.

Purpose designed matchmaker [5] is used to i) provide a fast and accurate discovery of elements in all repositories and enable repository audited updates, ii) enable model and data integration, and iii) formation of processing networks with assessment of economic and environmental benefits. Semantic relevance of integrated models and datasets, as well as integrated processes into a network solution defines their mutual suitability and it is calculated from distance between the two instances in the respective ontology. Explicit relevance is calculated using vector similarity algorithm for respective properties, such as quantity, availability, geographical location and environmental and economical potential etc. More intuitive and complex model and data networks, as well as processing networks are proposed by reclusively repeating matches between two participants which in turn gives an opportunity for even better economic and environmental savings and/or targeted production. Both semantic and explicit matching relevance are aggregated in into a numerical values use for match ranking and providing beter but flexible decision support to the user.

The proposed decision making platform has been implemented as a web service with performance validated verified using several real-life cases for model and data integration, as well as real-life biorefining scenarios.

1. Bussemaker, M., et al., *Ontology Modelling for Lignocellulosic Biomass: Composition and Conversion.* Computer Aided Chemical Engineering: 28th European Symposium on Computer A, 2018.
2. Koo, L., N. Trokanas, and F. Cecelja, ***A semantic framework for enabling model integration for biorefining***. Computers & Chemical Engineering, 2017. **100**: p. 219-231.
3. Raafat, T., et al., *An Ontological Approach Towards Enabling Processing Technologies Participation in Industrial Symbiosis.* Computers & Chemical Engineering, 2013. **59**(1): p. 33-46.
4. Cecelja, F., et al., *Integration of CAPE Models and Data for the Domain of Biorefining: InterCAPEmodel Ontology Design.* Computers and Chemical Engineering, 2019. **- accepted for publication**.
5. Trokanas, N., F. Cecelja, and T. Raafat, *Semantic Input/Output Matching for Waste Processing in Industrial Symbiosis.* Computers & Chemical Engineering, 2014. **66**(1): p. 259 - 268.
6. Trokanas, N., F. Cecelja, and T. Raafat, *Semantic Approach for Pre-assessment of Environmental Indicators in Industrial Symbiosis.* Journal of Cleaner Production, 2015. **96**(1): p. 349-361.

612116 Hydrothermal Liquefaction of Unhydrolyzed Solids for Fuels and Value-Added Products

Abstract:

Integrated biorefineries utilizing lignocellulosic biomass feedstock for fuel production will require additional revenue generation from the derived waste streams to demonstrate technoeconomic feasibility. One of the ways of revenue generation will be the efficient valorization of waste streams for the production of value added products. To understand technoeconomic feasibility of integrated bioprocessing, we investigated hydrothermal liquefaction (HTL) of unhydrolyzed solids (UHS) derived from some preprocessing of a lignocellulosic biomass and subsequent valorization of biocrude waste streams at a laboratory scale. Both solid and liquid side products were selectively enriched and their yields were taken into account to estimate revenue generation. A scale-up of the envisioned pilot scale integrated bioprocessing plant, and preliminary process and cost analysis were performed. Biomass processing steps, energy and mass balances, energy recovery, waste stream valorization, and initial assessment of technoeconomic feasibility will be presented.

608027 Understanding of Metabolic Pathways in *Pseudomonas Putida* for Biosynthesis of PHA Using Sugar, Glycerol, Benzoate, and Their Combinations

Abstract:

Efficient utilization of all available carbons from lignocellulosic biomass is critical for economic efficiency of a bioconversion process to produce renewable bioproducts. However, the metabolic phenotypes enabling bacteria such as *Pseudomonas* to utilize mixed carbon sources to generate energy and stored bioproducts remain poorly understood. In this study, *Pseudomonas putida* KT2440 was employed to investigate the polyhydroxyalkanoates (PHA) production from mixed carbon source of glucose/benzoate or glycerol/benzoate. A multi-omics approach involving label-free proteomics and NMR metabolomics was taken to compare the metabolic profiles of different mixed carbon sources. Results showed different patterns of reductant generation and fatty-acid metabolism pathways with different mixed carbon sources, which resulted in enhancing or inhibiting biomass and PHA productions. Overall, our study provides new insights into the metabolic architecture of *P. putida* with different mixture carbon substrates and reveals the potential regulatory metabolic nodes that may guide further system biology design to improve the PHA production.

594776 Lignin-Based Drop-in Jet Fuel Production

Abstract:

Sustainable aviation fuels remain the only near-, mid-, and likely long-term solution for lowering the carbon emissions of commercial and military aviation. Utilizing lignin feedstock for production of high energy density jet fuel along with cellulosic ethanol offers an unparalleled opportunity to enhance the overall operation efficiency, carbon conversion efficiency, economic viability, and sustainability of biofuels and chemicals production. In addition, determination of sustainable aviation fuel's chemical compositions and prediction of their properties is the critical first step for further research and development that eventually lead to the final certification process. Several catalytic processes with bifunctional catalyst Ru/H^+-Y, Ru/Super Lewis acid, and Ru-M/H^+-Y (M= Fe, Ni, Cu, Zn) were recently developed by our lab. Biomass-derived lignin was successfully converted into the lignin-based jet fuel, which consists of mainly paraffinic hydrocarbon species. They can be further classified into several classes, including n-paraffins, iso-paraffins, mono-, di-, and tri-cycloparaffins of which the majority contains carbon numbers in the range of 7 to 20. However, the "impurities" in the lignin wastes such as sugars, furans, and their derivatives were found to suppress metal catalysts and inhibit lignin depolymerization and hydrodeoxygenation reactions. Furthermore, the possibility of meeting drop-in specifications such as the low aromatic concentrations, which illuminates favorable energy content, energy density, possible low emissions, and high-performance characteristics, will be discussed in the presentation.

613064 Toward Feedstock Design for Biomaterial: Novel Biomass Structure Characteristics Determining Properties of Lignin-Based Carbon Fiber

Abstract:

Lignin has been explored extensively as a renewable feedstock for biomaterials, considering its abundance as a major component of plant cell wall and its sustainability as a waste of lignocellulosic biorefinery and paper-making industries. Despite the extensive efforts on defining process-property relationship, it remains largely unknown how lignin biosynthesis and precursor chemistry in planta would impact on the properties of biomaterials. Current knowledge about reducing lignin content to improve saccharification efficiency does not necessarily inform the feedstock for enhanced properties of biomaterials. Such inadequate understanding fundamentally limited the feedstock design and selection to improve carbon material properties toward broader commercial application. Using lignin from a broad range of biomass feedstock for carbon fiber manufacturing, we have fundamentally explored the structure-function relationship between lignin chemistry and carbon fiber properties, for both mechanical and electro-conductive performance. Specifically, lignin extracted from hardwood (sugar maple), softwood (loblolly pine and red cedar), and herbaceous plant (corn stover and switchgrass) were used for carbon fiber manufacturing, considering the very different lignin and biomass structures from these feedstock. Linear regression models were established to define the relationship between carbon fiber mechanical properties (Elastic Modulus, Tensile Strength, and Elongation) with lignin structure characteristics (content, composition, linkage profile and others), along with carbon fiber microstructures (crystallite content and size). The results highlighted that traditional biomass characteristics such as lignin content and composition don't impact the properties of lignin-based carbon fiber. Instead, the content of β-O-4 linkage correlates significantly with tensile strength and elastic modulus of lignin carbon fiber, indicating that the more linear β-O-4 linkage would promote carbon fiber performance. Besides the mechanical property, electro-conductive property is essential for broader energy application of lignin-based carbon fiber, yet the mechanisms controlling the electrocoductive properties of lignin-based carbon fiber is largely unknown. We hereby demonstrated that a higher β-O-4 content also promotes electroconductive performance of carbon fiber. Moreover, we also found that higher β-O-4 linkage profile defines the crystallite content and size of the resultant carbon fiber, which could be due to the better miscibility with the guest polymer. Overall, the study established lignin linkage profile, in particular, β-O-4 content as the new structure determinants for biomaterial property, which could guide the future biomass feedstock development.

611655 3D-Printed Zeolite Monoliths with Metal Dopants for Methanol Conversion

Abstract:

Additive manufacturing by 3D printing has gained worldwide attention in formulation of porous materials into practical contactors as a result of flexibility in design alteration and preparation of robust and complex geometries which requires fewer steps and minimal resources in comparison to the previously established extrusion process.[1-4] In this study, 3D-printed ZSM-5 monoliths doped with Ga_2O_3, Cr_2O_3, CuO, ZnO, MoO_3, and Y_2O_3 were synthesized using the state-of-art 3D printing technique. The physicochemical properties of the catalysts were characterized by X-ray diffraction, N_2 physisorption, NH_3 and CO_2 temperature-programed desorption and H_2 temperature-programmed reduction. The promotional effect of doped metals on catalytic performance of 3D-printed ZSM-5 monoliths in methanol to hydrocarbon (MTH) reaction in the presence and absence of CO_2 was investigated. Results indicated that both metal dopants type and reaction atmosphere greatly influence catalyst stability and product distribution. The yield of light olefins was enhanced over all metal-doped 3D-printed ZSM-5 monoliths in N_2 atmosphere (absence of CO_2), however, CO_2 atmosphere did not favor the production of light olefins. Although selectivity toward ethylene slightly decreased, the propylene yield was almost constant after switching N_2 to CO_2 in MTH reaction at 400 °C. Furthermore, it was found that Y- and Zn-doped ZSM-5 monoliths exhibited higher yield of light olefins and BTX compounds in the in the absence and presence of CO_2, respectively.

References:

[1] F. Magzoub, X. Li, J. Al-Darwish, F. Rezaei, A.A. Rownaghi, 3D-printed ZSM-5 monoliths with metal dopants for methanol conversion in the presence and absence of carbon dioxide, Appl. Catal. B Environ. 245 (2019) 486–495.

[2] X. Li, A. Alwakwak, F. Rezaei, A.A. Rownaghi, Synthesis of Cr , Cu , Ni , and Y - Doped 3D-Printed ZSM-5 Monoliths and Their Catalytic Performance for n-Hexane Cracking, ACS Appl. Energy Mater. 1 (2018) 2740–2748.

[3] X. Li, F. Rezaei, A.A. Rownaghi, 3D-printed zeolite monoliths with hierarchical porosity for selective methanol to light olefin reaction, React. Chem. Eng. 3 (2018) 733–746.

[4] X. Li, F. Rezaei, A. Rownaghi, Methanol-to-olefin conversion on 3D-printed ZSM-5 monolith catalysts: Effects of metal doping, mesoporosity and acid strength, Microporous Mesoporous Mater. 276 (2019) 1–12.

596901 Fluid Flow Control Devices with 3D-Graded Permeability

Abstract:

Chemical engineering systems often rely on components that involve intimate contact between a fluid and a solid, such as in catalyzed reactive flows, fluid purifiers, and chromatographic separations. To accomplish this, it is often necessary to distribute the fluid from a relatively narrow tube to a broader cross section, to an array of tubes, or into a porous medium that can be modeled by Darcy's Law. (Darcy's Law states that the volumetric flow rate is proportional to the permeability, an intrinsic material property related to its porosity.) Ideally, the flow rates in the tube array or throughout the porous medium are uniform, and the size, weight, and cost of this portion of the system are a small fraction of those of the chemically reactive portion.

Additive manufacturing techniques raise the possibilities that porous media can be fabricated in which the permeability can be arbitrarily specified in three dimensions, and that a broader range of permeabilities can be achieved than by traditional methods used to manufacture porous media.[1] We are using optimization algorithms[2] to design devices that distribute a fluid from a narrow inlet to a broad outlet, where the outlet flow rate is spatially uniform, and device geometry and/or pressure drop are constrained, by spatially varying the permeability in the device. We have considered a Darcy's law model, as well as a modification of the Navier-Stokes equations with a term representing permeability.[3] Numerical models show that designs varying permeability in three or two dimensions can achieve greater uniformity than designs that vary permeability in only one dimension (such as with stacked flow elements).

Meanwhile, we are establishing and evaluating methods to fabricate stainless steel structures with spatially varying porosity on conventional additive manufacturing tools, as a path to build the designed devices for future experimental testing.

Our work is supported by the Laboratory-Directed Research and Development program at Sandia National Laboratories, a multimission laboratory managed and operated by National Technology & Engineering Solutions of Sandia, LLC, a wholly owned subsidiary of Honeywell International Inc., for the U.S. Department of Energy's National Nuclear Security Administration under contract DE-NA0003525.

SAND2020-3764 A

[1] V. P. Palumbo et al. "Porous Devices Made by Laser Additive Manufacturing." US Patent Application 2017/0239726 A1, Mott Corporation, 2017.

[2] M. A. Heroux, R. A. Bartlett et al. "An overview of the Trilinos project" ACM Trans. Math. Softw. 31(3), 397-423, 2005.

[3] T. Borrvall, J. Petersson. "Topology Optimization of Fluids in Stokes Flow." Int. J. Numer. Meth. Fluids 41, 77–107, 2003.

605743 3D Printed Gyroid-like Structured Packings

Abstract:

Structured packings are an integral part in distillation columns, absorption columns, and chemical reactors. An effective packing requires large surface area, low pressure drops, and even liquid distribution. Standard structured packings have been established for decades and typically consist of corrugated sheets stacked in alternating arrangements. Recently, triply periodic minimal surfaces (TPMS) have received attention for potential use in heat exchangers, mass exchangers, and chemical reactors. This class of structures, including the familiar 'Gryoid', uniquely requires additive manufacturing for production. Here, we demonstrate the use of several TPMS geometries as structured packings fabricated using additive manufacturing. These packings exhibit improved liquid distribution, reduced wall channeling, and increased fluid mixing in comparison to conventional structured packings.

This work was performed under the auspices of the U.S. Department of Energy by Lawrence Livermore National Laboratory under Contract DE-AC52-07NA27344. LLNL-ABS-809142.

610417 Coupled Numerical and Experimental Analysis of Transport Properties in 3D-Printed Periodic Cellular Media to Enable Catalytic Reactors Design

Abstract:

Process Intensification is a key topic in academic and industrial research aiming at the development of more compact and energy-efficient processes [1]. In this context, structured catalysts have been considered as a promising solution to overcome the limitations of conventional catalytic supports. Currently, cellular materials (foams, Periodic Open Cellular Structures – POCS) are envisioned as potential candidates for enhanced catalyst substrates. Thanks to the possibility of advanced manufacturing and design offered by 3D printing, POCS allow for extreme flexibility in the geometrical optimization (cell shape, strut size, solid fraction) to tailor the design to the process needs [2]. In this view, the fundamental understanding of the coupling between the geometrical properties and the transport phenomena is essential for the optimization of the supports and their industrial application.

In this work, we propose a combined experimental and numerical investigation of the transport properties of POCS aimed at the derivation of engineering correlations to enable the development and optimization of enhanced catalytic reactors. Computational Fluid Dynamics (CFD) allows for the analysis of the transport properties as a function of the geometrical parameters, enabling the understanding of the effect of single morphological parameters (e.g. cell shape, porosity, cell diameter) on the overall performances and providing deep insight in the transport mechanisms.

The significant computational burden of CFD simulations hinders their direct application to the industrial reactor scale. Hence, a hierarchical approach following the procedure proposed by Rebughini et al. [3,4] is employed to investigate representative structure volumes by means of CFD.

While numerical simulations allow to evaluate the transport properties of structures regardless of the limits imposed by their manufacturability, they are unable to provide a broad range of data capable of completely define the performances of a specific structure. Contextually, experimental analysis performed on 3D-printed replicas, while requiring specialized manufacturing infrastructures, can provide enough data for the complete characterization of the most promising geometries and cross-validation with CFD simulations.

In this view, high precision 3D printing technology enables the fabrication of highly accurate structures. The precision and accuracy of currently available SLA 3D printers, combined with the wide range of available materials, allows for their effective deployment in the experimental analysis of fluid-solid interactions. The possibility to create faithful physical replicas of digital structures allows for the direct comparison between simulations and experiments hence providing a reliable tool for the cross-validation of experimental and computational results.

As a result, the numerical and experimental analyses can be combined and exploited to formulate engineering correlations able to predict the structures behaviour as a function of the geometrical and physical parameters [5].

As an example of this approach, an analysis of pressure drops and mass transfer over TKKD cellular structures is presented. Experimental investigation is performed both in cold flow (pressure drops) and reactive conditions (mass transfer) over 3D printed resin samples with different geometrical characteristics. Pressure drops are evaluated by flowing air and measuring the differential pressure across 3D printed resin samples. Transport properties are assessed by reactive runs performed on samples realized with a High Temperature 3D printed resin ($HDT_{@0.45MPa}$ = 289°C). The oxidation of H_2 over Pd/CeO_2 in rich conditions is used as the test reaction to allow for the onset of external mass transfer limited regime at temperatures compatible with the resin. The conversion of O_2 (the limiting reactant) in external mass transfer limited regime is employed to evaluate the mass transfer coefficients due to the similarity between its diffusivity and CO while the large excess of H_2 prevents back-diffusion problems.

In parallel, CFD simulations are performed on the same geometry. The working fluid is CO diluted in air at 298K, 1 atm, the test reaction is the total oxidation of CO. Hence, the mass transfer properties are assessed in steady state conditions by evaluating CO conversion (limiting reactant). Pressure drops and conversion are together estimated for a representative elementary volume (REV, [6]) made of a set of unitary cells.

Figure 1 (left) shows the result of the numerical analysis (lines) along with experimental results (dots) for the TKKD geometry, showing a very good agreement between CFD and experiments. The analysis shows the effects of the different geometrical parameters (cell size, porosity) on the pressure drops. In this view, the envisioned approach offers a reliable tool for assessing the pressure drop functional dependency on the structures geometrical properties. Figure 1 (right) shows the outcome of the analysis of the gas-to-solid transport properties of the same geometry. Experimental and numerical data (dots) are exploited to develop a gas/solid mass transfer correlation for the TKKD unit cell able to predict its behaviour as a function of the geometrical and the physical parameters. The proposed dimensionless correlation (line) shows a good agreement with respect to the available data.

In conclusion, we propose a methodology for the fundamental investigation of transport properties in periodic cellular substrates based on coupled numerical and experimental analysis. The combined use of CFD and experiments allows for the fast screening of a wide variety of different structures while enabling the in-depth characterization of the most promising geometries without the limitations of either numerical or empirical independent approaches The illustrated case study highlights the potential of such approach in the development of engineering correlations able to describe substrates transport properties regardless of the specific geometries. Thanks to the gathered knowledge and to commercial 3D printing solutions, it is then possible the manufacture catalytic substrates with tailored properties for each specific applications with materials such as metals (Steel FeCrAlloy, Copper, Titanium, and others) or ceramics, with details as small as 200μm. In this view, the proposed methodology allows for the proficient exploitation of the extreme design flexibility of advanced 3d printing, thus enabling effective process intensification.

References

[1] A.I. Stankiewicz at al. Chem. Eng. Prog. 96 (2000) 22–33.

[2] C. Busse at al. Chem. Eng. Process. Process Intensif. 124 (2018) 199–214.

[3] S. Rebughini, et al., React. Chem. Eng. 3 (2018) 25–33.

[4] S. Rebughini,et al., Chem. Eng. J. 289 (2016) 471–478.

[5] M. Bracconi, et al., Chem. Eng. J. 377 (2019.)

[6] M. Bracconi et al., Chem. Eng. J. 352, pp. 558-571 (2018).

Acknowledgements

This project has received funding from the European Research Council under Grant Agreement no. 694910 (INTENT).

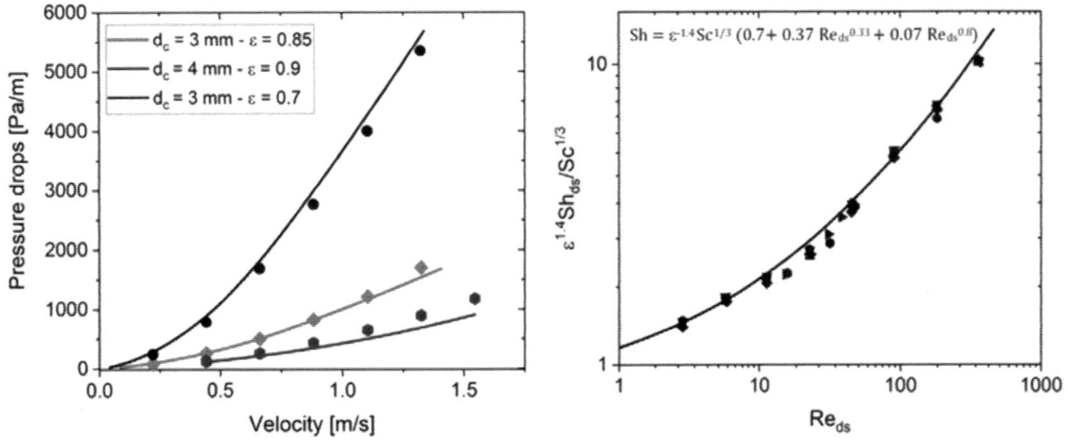

Figure 1: Left – Pressure drops of TKKD cells evaluated with experiments (points) and CFD (lines); right: correlation for Sh number of TKKD unit cells (line) derived from CFD simulations (points)

605946 Development of Polymer Monolith Pd Catalysts for Batch, Modular Continuous Flow and Novel 3-D Printed PEEK Reactor Platforms for Hydrogenation Reactions

Abstract:

Direct Hydrogenation is one of the most heavily utilised and essential reactions within the synthetic routes of many active pharmaceutical ingredients (APIs). Continuous hydrogenation as a result is becoming a topic of significant interest throughout the chemical industry with the development of novel reactor technologies[1] and catalysts[2], to optimising mass transfer, residence time distribution, and surface area limitations often associated with heterogeneous catalysis.

In this study, novel polymer monolith Pd catalysts have been developed for both batch and continuous flow hydrogenations. Modular reactors based on standard tubes and fittings utilised the commercially available H-Cube. Comparison with 5% Pd/C for the direct hydrogenation of nitro, alkene, and imine functional groups were investigated. Furthermore, novel 3-D printed polymer reactors fabricated in the chemically and mechanically resistant material PEEK (poly(etheretherketone)) have been designed and printed for heterogenous hydrogenation reactions[3]. The monolithic catalysts developed are ideally suited for incorporation into Fused Filament Fabrication printed reactors as they can be covalently attached to the reactor walls, grown within complex printed geometries and catalyst synthesis conditions are compatible with standard PEEK processing conditions.

Acknowledgements:

This publication is supported by Science Foundation Ireland (SFI) through SSPC, The SFI Research Centre for Pharmaceuticals (Grant number 12/RC/2275_P2) and I-Form, the SFI Research Centre for Advanced Manufacturing (Grant number 16/RC/3872)

References:

[1] M. O'Brien, N. Taylor, A. Polyzos, I. R. Baxendale, S. V. Ley, *Chem. Sci.* **2011**, *2*, 1250–1257.

[2] CH. Pélisson T. Nakanishi, Y. Zhu, K. Morisato, T. Kamei, A. Maeno, H. Kaji, S. Muroyama, M. Tafu, K. Kanamori, T. Shimada, K. Nakanishi, *ACS Appl. Mater. Interfaces.*, **2017**, *1*, 406–412 .

[3] M. J. Harding, S. Brady, H. O'Connor, R. Lopez-Rodriguez, M. D. Edwards, S. Tracy, D. Dowling, G. Gibson, K. P. Girard, S. Ferguson, *React. Chem. Eng.*, **2020**, *5*, 728-735

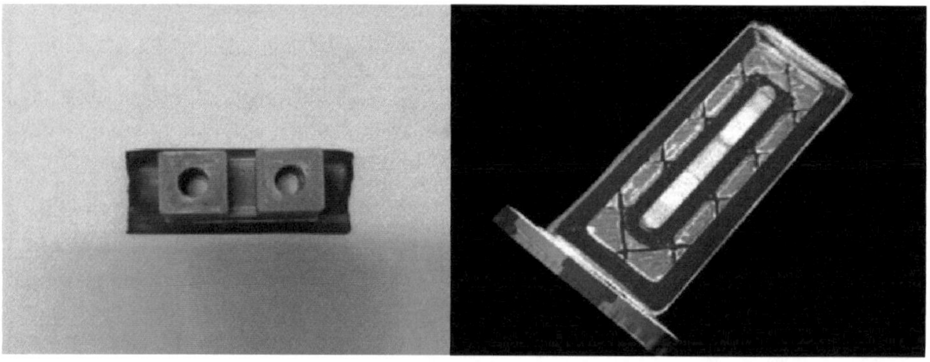

Figure 1: (left) PEEK 3-D printed reactor with a straight channel (2 mm ID) housing polymer monolith. (right) X-Ray scan of polymer monolith grown throughout the inner channel of the reactor.

610202 Molecular Recognition and *In Vivo* Detection of Temozolomide and 5-Aminoimidazole-4-Carboxamide for Glioblastoma Using Fluorescent Nanosensors

Abstract:

The efficacy of chemotherapeutics such as temozolomide (TMZ) and its metabolic product 5-aminoimidazole-4-carboxamide (AIC) is often affected by the timing, quantity and frequency of dosages. There is strong interest in facilitating the ability to monitor efficacy in individual patients for specific subtypes of cancer. Real-time, dynamic measures of potency may also supplement or in some cases replace reliance on bio-imaging. Towards this end, in this work we develop new, synthetic molecular recognition sites for TMZ, and its decomposition product AIC, grafting them onto near infrared fluorescent nanoparticles capable of forming optode or other biosensor interfaces to monitor drug efficacy in real-time. Infrared fluorescent single-walled carbon nanotubes, wrapped using specific DNA oligonucleotides, are encapsulated within biocompatible, poly(ethylene glycol) diacrylate hydrogels and enable the selective recognition of an anti-cancer drug, TMZ, on U-87 MG human glioblastoma cells. In both solution phase and hydrogel form, the sensors exhibit a fluorescence response to TMZ with a detection limit of 30 M. Furthermore, U-87 cells exhibit no changes in viability for 7 days in contact with the hydrogel. The sensors were used to track the progression of glioblastoma death following TMZ administration. In addition, the sensors were able to monitor *in vivo* AIC and TMZ dynamics in SKH-1E mice. These results enable real-time detection of chemotherapeutic concentration and metabolism at the cellular and subcellular levels, with potential to increase the efficacy of cancer treatments.

606459 An Enzymatic Electrochemical Biosensor for Real-Time Detection of Physiologically Relevant Nicotine Concentrations

Abstract:

The glucose biosensor, built upon the redox enzyme glucose oxidase, is the most commercially successful and studied enzymatic biosensor. However, the lack of available and functionally validated enzymes is prohibiting the development of redox-based sensors for other important analytes. Herein, we present the development and assessment of an electrochemical nicotine biosensor, using genomic screening to identify the gene for a known nicotine catabolizing redox enzyme. The resulting nicotine biosensor demonstrated a specific, sensitive, and stable operational profile with a limit of detection of 27 μM over the range of 0-200 μM. This range is well within the physiological concentrations of nicotine present in smoker urine. Specificity and cross-reactivity were measured against structurally similar compounds to nicotine as well as to known physiological by-products. Our results highlight that this novel enzymatic electrochemical nicotine biosensor possesses operational capabilities for monitoring of nicotine in physiologically relevant conditions. The screening methodology can be generalized for the discovery of enzymes for novel sensor development.

608267 Simulations Guide Optimization of Electroenzymatic Biosensors for Neurotransmitters and Enable Proper Interpretation of Sensor Response *In Vivo*

Abstract:

Electroenzymatic biosensors for neurotransmitters are uniquely compact, sensitive, and selective devices that provide the neuroscience community with a method to monitor fluctuating chemical concentrations in the brain. Many studies primarily consider glutamate, the most common excitatory neurotransmitter, and study the relation between sub-second neurotransmitter dynamics and various stimuli, behaviors, or diseases. Challenges, however, have remained in the optimization of sensor design and in clarifying the significance of assumptions made during *in vivo* use. Accordingly, a number of simulations were carried out to guide the development of glutamate sensors and to test the limitations of the most prominent assumption, that sensor response can be linked to the immediate extracellular neurotransmitter concentration using a predetermined calibration factor. Recommended modifications to sensor design were developed and successfully implemented, improving sensitivity six-fold (up to 320 nA/μM/cm^2) with an accompanying order of magnitude decrease in response time (down to ~80 ms). Since even this improved response time is an order of magnitude slower than the timescale of synaptic neurotransmitter release, it is clearly wrong in many cases to assume that the use of a calibration factor to translate sensor response to neurotransmitter concentration without any signal deconvolution is accurate. To investigate the significance of this assumption and the conditions for which this practice remains reliable, the model was expanded with a new set of partial differential equations to describe the extracellular space of the brain. Considering the effects of variations in the local biological environment bordering a sensor, a range of representative parameters for some biological processes were selected. Comparisons were made between the theoretical responses expected from *in vitro* calibrations, tonic steady-state glutamate presence, and bolus-type glutamate releases that mimic synaptic release. Surprisingly, in steady-state cases, the sensor response could indicate a concentration over twice the actual value, while in cases where sensors were 10 μm or more from the source of glutamate being detected, sensor response was < 50% of the expected value. In some cases, transient concentrations due to a single synaptic-type release of neurotransmitters at a moderate distance from the sensor were not detected by the sensor at all, and when they were, the time course of signal decay did not match the rates of reuptake expected in the brain without significant signal deconvolution. Ultimately, results show that additional characterization of sensor performance *in vivo* may be necessary to enable proper interpretation of sensor data.

611068 Theoretical and Experimental Studies on an Electrochemical Enzyme Immunosorbent Biosensor

Abstract:

Electrochemical immunosensors integrate biological recognition molecules (e.g., antibodies) with redox enzymes (e.g., horseradish peroxidase (HRP) to combine the advantages of immunoassays (extremely high sensitivity and selectivity) with those of electrochemical biosensors (reproducible, quantitative electrical output). Mechanistic mathematical models that describe the multiple mass-transfer and chemical reaction steps that give rise to the electrical output are needed to help design, optimize, and validate electrochemical immunosensors for medical applications. In this study, we developed such a model of an electrochemical immunosensor that used HRP as a reporter enzyme on the secondary and validated it experimentally.

The model consists of a system of non-linear mass-balance equations that describe simultaneous mass transfer and chemical reactions of the reactants (catechol and hydrogen peroxide) that HRP oxidizes, as well as the HRP product, O-quinone, which is reduced at the working electrode to give the amperometric output. A bi-substrate, ping-pong reaction mechanism was used to describe the enzymatic reaction, and the *Butler-Volmer* kinetic model was used to describe the reduction of o-quinone back to catechol at the working electrode's surface. The system of equations was solved numerically using the BVP4C function in MATLAB.

Experimental data for model validation was obtained using electrochemical immunosensors for a surrogate protein antigen (mouse IgG). The immunogens interface was fabricated on the gold working electrode of DropSens screen printed electrode arrays. The steady-state amperometric output of the resulting immunosensors was measured over a range of four independent variables: the concentrations of the two reactants, the voltage of the working electrode, and pH. The experimental results were used to determine optimal operating conditions and to validate the model. The validated model was then used to calculate Damkohler numbers and flux-control coefficients to identify mass-transfer and reaction steps that most strongly affected the magnitude of the immunosensors' output and the output's sensitivity to analyte concentration.

610485 HTL Derived Biochar and Graphene Nanoplatelets for Biosensor Applications

Abstract:

HTL Derived Biochar and Graphene Nanoplatelets for Biosensor Application

Bharath K Maddipudi, Vinod Amar, Hope Dosch, Anuradha Shende and Rajesh V Shende*Department of Chemical and Biological Engineering, South Dakota School of Mines and Technology, Rapid City, South Dakota 57701, United States of America

***Corresponding Author:** Rajesh Shende, Department of Chemical and Biological Engineering, South Dakota School of Mines and Technology, Rapid City, South Dakota 57701, United States of America (E-mail: *Rajesh.Shende@sdsmt.edu*)

Abstract

Biosensors with Au electrode layer immobilized with peptides are currently being developed for the detection of microbial contamination. It is anticipated that the use of a low-*k* dielectric layer modified with graphene will improve the sensor sensitivity. The low-*k* dielectric layer can be modified with the biochar derived from hydrothermal liquefaction of different biomass substrates. In this investigation, HTL of lignocellulosic substrates was performed under non-catalytic and catalytic conditions and by-products were analyzed. The biochar was thermally treated in a tubular reactor under inert environment from 400-1100°C without any pretreatment. In other set of experiments, HTL derived biochar was acid treated and subsequently, thermal activation was performed. After the acid pretreatment and thermal treatment steps, the porosity, the specific surface area (SSA) and the morphology of the char was analyzed using BET and SEM/EDX. The char obtained after thermal treatment was combined with low-*k* dielectric layer and biosensor was fabricated. The sensor output was compared with the biosensor fabricated with low-*k* and graphene nanoplatelets. In addition, the sensor output was correlated with the char materials obtained under different thermal conditions. The results obtained on hydrothermal liquefaction of biomass substrates, characteristics of biochar before and after thermal treatment, biosensor fabrication and sensor output will be presented.

604780 Construction of a Red Emission Bodipy-Based Probe for Tracing Lysosomalviscosity Changes in Culture Cells

Abstract:

A red emission BODIPY-based lysosome-targeted viscosity probe (Lys-VBOD) was synthesized for monitoring lysosomal viscosity in live cells. The increase of viscosity can restrict the twisting of double bond between indole
and BODIPY fluorophore, also restrict the rotating of single bond between quinoline and BODIPY, which lead to the enhancement of fluorescence intensity in high viscosity environment. The changes of fluorescence intensity
exhibit a good linear relationship with viscosity values. Besides, Lys-VBOD is quite stable in various pH solvents with given viscosity, suggesting that Lys-VBOD is suitable for applying in a wide pH range. Moreover, Lys-VBOD
can visualize the changes of lysosomal viscosity in live cells under the stimulus of dexamethasone. BSA titration experiment demonstrated that Lys-VBOD is stable in the presence of macromolecules and proteins. Notably, the
fluorescence lifetime of Lys-VBOD in large viscosity solvents is extremely high. These results indicate that LysVBOD is an excellent tool for detecing viscosity changes in vitro and live cells.

610593 Super Sensitive Cerium Oxide-Based Composite Sensor for the Detection of Hydroxyl Radicals

Abstract:

Free radicals are recognized as essential molecules for upholding the function of normal cells. Immune systems, such as bacteria elimination and virus inactivation, are examples of the crucial role that free radicals play in human cells. Even though free radicals have benefits, an optimum production is key to maintain their advantages, as an excess can cause the development of oxidative stress conditions, such as cancer, Alzheimer's, and Parkinson's diseases. Among all free radical species, hydroxyl radicals (•OH) are the most reactive and dangerous, and they can be used as biomarkers for the detection of those diseases at their initial stages. Thus, a very sensitive analytical sensor is needed to detect and monitor low concentrations of •OH. The integration of an electrochemical technique with a sensing element is regarded as a promising method for •OH detection due to a rapid and direct measurement without the pretreatment of samples. Thus, in this study, a glassy carbon electrode was modified with a sensing element comprised of ultrasmall cerium oxide nanoclusters (<2 nm) grafted onto a superconductive carbon (Figure 1a). XPS results showed 46.3% Ce^{3+} sites, which are those able to react with •OH. Cyclic voltammetry (CV) was implemented to characterize the interaction of the composite sensor with •OH, and they showed that the composite sensor was able to differentiate between solutions with and without •OH by an increased redox response in presence of •OH (Figure 1b). Furthermore, we observed a linear relationship between the redox response and the concentration of •OH in the range of 0.1 mM to 10 mM, with a limit of detection as low as 0.1 μM when using only 0.5 wt% CeO_x loading.[1]

(1) A.C. Alba-Rubio, D.S. Kim, S. Duanghathaipornsuk; Super sensitive sensor for the detection of hydroxyl free radicals with scavenging properties. Application number 62/977,945 (2020).

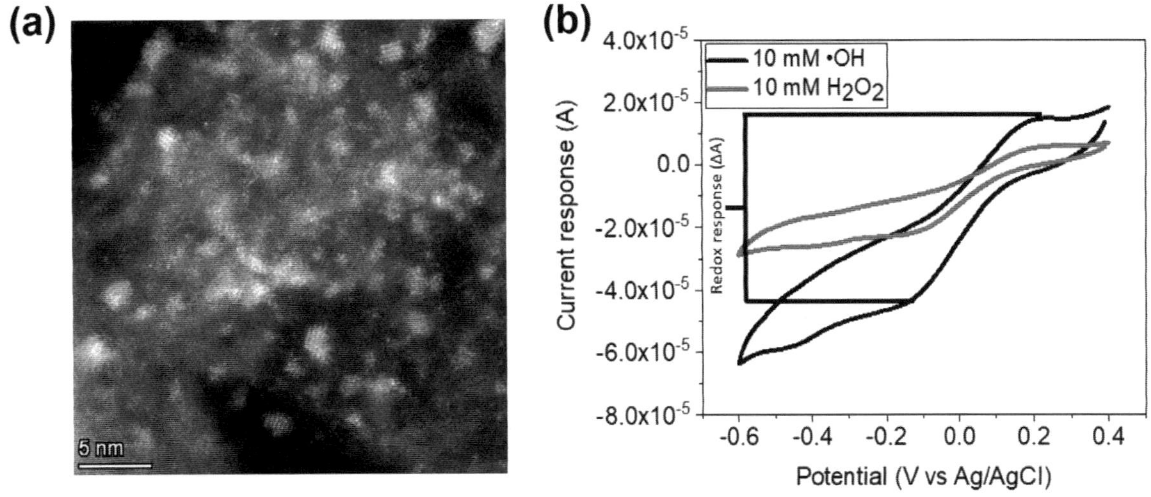

Figure 1. (a) STEM image of 0.5 wt% CeO_x/Carbon and (b) Current response to hydroxyl radicals.

611380 High Throughput Antibiotic Susceptibility Testing with Optical Nanosensors

Abstract:

Infections in a clinical setting can result from biofilm-forming species, which presents two key challenges: increased antimicrobial resistance, and poor diagnostic approaches to categorize antimicrobial susceptibility. Traditional antimicrobial susceptibility assay approaches are built upon planktonic bacteria (bacteria that are growing in solution). Biofilms are bacteria that grow in a community on a surface, and this community structure changes resistance to antimicrobials that is not reflected in these planktonic assays. While current biofilm-based assays exist, they are both time and material intensive, and can still provide antibiotic susceptibility results that may not reflect the actual biofilm response. Here we present our nanosensor based approach to monitor biofilm metabolic response during the administration of antibiotics. We grow biofilms with oxygen-monitoring nanosensors embedded throughout the film, and these nanosensors continuously report out the metabolic state of the biofilm. As we add antibiotics, we can measure what concentration can stop metabolism, and infer the minimum biofilm inhibitory concentration (MBIC) of the specific strain. Our diagnostic approach functions in 96 well plates with minimal materials and time (important for high throughput screening), and works with both lab and clinically relevant strains (demonstrated with Pseudomonas aeruginosa).

602734 Renewable NH_3 As Alternative Energy Fuel for Ocean Exploration

Abstract:

Renewable NH_3 as Alternative Energy Fuel for Ocean Exploration

Jian Liu, Yuyan Shao, Vanda Glezakou, Robert Cavagnaro, and Li-Jung Kuo

Pacific Northwest National Laboratory, Richland, WA 99352, USA.

Wind and wave energy constitute an abundant renewable energy source readily available to the United States. However, their practical use is limited by their fluctuating power output, expensive transmission line infrastructure and complicate integration systems to the existing grid. We propose a modular energy storage solution to convert wave or tidal energy into a carbon-neutral NH_3 fuel using an electrochemical method for ocean exploration and applications at sea. Currently, state-of-art electrocatalytic conversion using precious metals favors the competing hydrogen evolution reaction leaving insufficient vacant sites (Faraday efficiency<1%) for nitrogen reduction to occur. A high surface carbon catalyst with pyridinic N derived from metal-organic frameworks is developed under the guidance from the theoretical simulation. The enhanced N_2 adsorption and dissociation and the potential ability to suppress the competitive hydrogen evolution reaction will help to make the electrochemical synthesis of NH_3 from seawater and air more viable. An economic assessment of making and using NH_3 prepared through this route as the alternative energy fuels will be discussed. The performance of the electrocatalyst will be characterized using the rotating ring disk electrode method and compared with that of the conventional electrocatalysts.

611810 Mixed-Phase Perovskite Materials for H_2 Generation and Hybrid Charge Storage

Abstract:

Perovskite (ABO_3) materials have applications both in fuel production and energy storage. Although mixed metal oxide-based systems have been thoroughly investigated, the information available on the use of perovskite for H_2 generation is extremely meagre. In this study, Mn and Ti- based perovskite materials (as mixed phase) were synthesized using surfactant templating assisted sol-gel technique, which were analyzed using powder x-ray diffraction, BET, SEM/EDX, and utilized for thermochemical water-splitting process for H_2 generation. This material was loaded inside Inconel reactor and multiple thermochemical cycles were performed. H_2 generation was continuously monitored using online H_2 sensor and GC. As calcined perovskite materials in the form of mixed phase were utilized to fabricate supercapacitors, which were characterized for their gravimetric capacitance. In addition, supercapacitors were also fabricated using carbon electrode made from the HTL derived biochar. The cyclic voltammetry curves obtained at 5 V and 2 A were used to determine the specific capacitance and compared with commercial supercapacitors.

599875 Hydrogen and Ammonia for Renewable Energy Storage: The Economics of Location, Scale, and Demand Type

Abstract:

Generation of electrical power from intermittent sources of renewable energy such as the wind or the sun is a promising approach to improving energy sustainability. Energy storage is necessary to balance this intermittent renewable generation with electrical power demands. Batteries are the most commonly used energy storage technology, but their capital cost and relatively low energy density makes them unsuitable for longer-term, high-capacity energy storage. Hydrogen, produced via water electrolysis and later converted back to electrical energy using a fuel cell, is another potential energy vector. The main limitation of hydrogen-based energy storage is the storage infrastructure itself; high-pressure gaseous storage necessitates expensive types and amounts of material, and hydrogen liquefaction requires significant additional energy. Ammonia production using electrolysis-derived hydrogen and nitrogen obtained via electrically driven air separation has recently received worldwide attention [1-3] as a potential route to decrease these storage costs, as ammonia is liquid at ambient temperature and moderate pressures. Ammonia can be used to generate electrical power through traditionally fossil-fueled technologies (e.g. internal combustion gensets, fuel cells) with minor modifications. Potential limitations of ammonia-based energy storage are (i) the capital investment needed for additional process units (e.g. nitrogen production, ammonia production), (ii) the additional energy needed to transform the hydrogen to ammonia, and (iii) the potentially lower power generation efficiency of ammonia as compared to hydrogen which along with (ii) can lower overall power-to-power efficiency.

In this work, we elucidate this trade-off between storage cost, chemical production cost, and energy efficiency to determine the best chemical-based energy storage systems for a variety of locations, power demand scales, and power demand types. We perform this investigation through a mixed integer linear programming (MILP) optimal combined capacity planning and scheduling model, which minimizes the levelized cost of energy supply (LCOE) for islanded (not grid connected) renewable energy systems. The model optimally selects and sizes the units in the system, while simultaneously scheduling the operation of these units (i.e. on/off, chemical production rates, storage inventories, power generation rates) during each distinct period of a scheduling horizon which captures both diurnal and seasonal variation in intermittent renewable generation and power demand. The modeling framework generates variable length scheduling periods by clustering consecutive hours of full year time series data for wind capacity factor, solar capacity factor and power demand. This temporal aggregation makes the model computationally tractable, allowing for high throughput computational studies. In our previous work [4], we used our capacity planning and scheduling model to determine the minimum LCOE of using hydrogen and/or ammonia as energy storage for 1 MW residential renewable energy supply systems in 15 American cities which represent climate-demand regions throughout the United States. We found that ammonia is generally more economical than hydrogen as a single method of energy storage and that in every location, cost savings can be achieved by using both synergistically. In the present study, we expand the scope to optimally size and schedule hydrogen and ammonia renewable energy storage systems in each location for power demand scales up to 100 MW, with various combinations of commercial and residential demand profiles at each scale. The results provide a more comprehensive understanding of the economics of chemical-based energy storage and reveal the most promising applications for ammonia energy storage.

References

[1] Klerke, A., Christensen, C. H., Nørskov, J. K., & Vegge, T. (2008). Ammonia for hydrogen storage: challenges and opportunities. *J. Mat. Chem., 18(20),* 2304-2310.

[2] Zamfirescu, C., & Dincer, I. (2008). Using ammonia as a sustainable fuel. *J. Power Sources, 185(1)*, 459-465.

[3] Lan, R., Irvine, J. T., & Tao, S. (2012). Ammonia and related chemicals as potential indirect hydrogen storage materials. *Int. J. Hydrogen Energy, 37(2),* 1482-1494.

[4] Palys, M. J., & Daoutidis, P. (2020). Using hydrogen and ammonia for renewable energy storage: A geographically comprehensive techno-economic study. *Comput. Chem. Eng, 127*, 106785.

610513 Hydrothermal Liquefaction of Bagasse for H_2, Bio-Oil and High Value Products

Abstract:

Hydrothermal Liquefaction of Bagasse for H_2, Bio-oil and High Value Products

Vinod S Amar, Anuradha Shende, Bharath K Maddipudi and Rajesh V Shende* Department of Chemical and Biological Engineering, South Dakota School of Mines and Technology, Rapid City, South Dakota 57701, United States of America

Corresponding Author: *Rajesh Shende, Department of Chemical and Biological Engineering, South Dakota School of Mines and Technology, Rapid City, South Dakota 57701, United States of America (E-mail: *Rajesh.Shende@sdsmt.edu*)

Abstract

Beer bagasse (spent brewer grain), a by-product generated in breweries, is rich in fiber and proteins. Being a low-cost biomass substrate, in this research, we investigated hydrothermal liquefaction (HTL) of bagasse for the production of H_2, bio-oil and other value-added products. HTL of beer bagasse was performed at 250-300°C under non-catalytic and catalytic conditions with bagasse to solvent (B:S) ratios of 1:5, 1:10, and 1:30. While HTL reaction was in progress, the product gases were analyzed using GC. After HTL reaction, biocrude was separated from the biochar and analyzed using HPLC, GC-MS and TOC analyzer. The HTL derived light bio-oil (LBO) and heavy bio-oil (HBO) were blended with biodiesel prepared from the spent cooking oil. HHVs of biochar, LBO, HBO, and blended fuels were calculated based on their elemental analysis. The results obtained with H_2, bio-oils, blend fuels, and value-added products from the HTL of beer bagasse will be presented.

600663 CuO-Cu_2O Redox Thermochemical Storage for High-Temperature Solar Process Heat

Abstract:

The integration of concentrating solar thermal technologies with thermal energy storage enables dispatchable power generation and around-the-clock operation of high-temperature industrial processes such as cement manufacturing, mineral and metallurgical extraction, waste recycling, and thermochemical fuels production. In particular, thermochemical energy storage (TCS) utilizes endothermic/exothermic reversible chemical reactions to store and release solar process heat at temperatures exceeding 600°C and with relatively high gravimetric heat capacity. We report on the experimental demonstration of a lab-scale TCS reactor prototype for storing heat at 1000°C. It consists of a packed-bed reactor filled with CuO-based granules undergoing reduction-oxidation (redox) cycles with a gravimetric heat capacity of about 470 kJ/kg, as measured by thermogravimetry. Two series of 30 consecutive redox cycles were performed by swinging either the oxygen partial pressure of the gas inflow (e.g. with air or nitrogen) or the air inflow temperature, yielding stable conversion in both cases. During the heat release, a spontaneous stabilization of the outflow temperature of the system at around 1025°C (equilibrium temperature of the CuO/Cu_2O system in air) was observed, which is beneficial for the downstream process. The influence of the gas flow rate and gas inflow temperature on the TCS system performance was parametrically investigated.

599901 Responsible Innovation in Chemical Process Design: Stakeholder-Driven Multi-Objective Optimization of a Modular Food Waste Valorization Process

Abstract:

Food waste is produced at staggering rates (1.3 billion tons/year),[1] consumes vast amounts of energy (26 exajoules/year) and is a significant source of land, water, and air pollution.[2,3] Food waste is also a major component of the organic matter that flows into landfills, and it imposes substantial costs—through collection and disposal—on individuals, businesses, and local and state governments.[4] In addition to poor economics and environmental footprint, the odors, increased traffic, and public health impacts related to landfills are shown to disproportionately burden low-income and minority communities.[5] It is therefore imperative to not only develop technologies that divert food waste from landfills, but to do so through a systems approach that simultaneously considers economic, environmental, and societal objectives.

Since food waste is a geographically distributed feedstock, modular processing at the sources (e.g., restaurants, cafeterias, grocery stores, etc.) can minimize overall costs and environmental impact by eliminating the need to transport raw materials to a centralized processing facility. When designing distributed chemical or biological processes for implementation in such public areas, it is critical to work closely with all stakeholders involved; these include business owners and employees, municipal and state policymakers, and members of the general public. Each of these groups has their own set of priorities and constraints which must be met to ensure the process is practical, supported by policy, and safe to implement.

To address these challenges, we develop a *stakeholder-driven* multi-objective optimization framework for the design of modular processes that upgrade food waste to valuable chemical products. Our approach combines SWOT (Strengths, Weaknesses, Opportunities, Threats) analysis with a fuzzy analytic hierarchy process (FAHP) [6] to, for the first time, allow public and private-sector stakeholders to both propose and weight the quantitative metrics used as objectives in process optimization. For example, business owners will naturally want to maximize profitability metrics such as net present value, while policymakers may prioritize emissions reduction. By having stakeholders generate objectives via SWOT analysis and then systematically quantifying the priority of each objective via FAHP, we minimize the non-technological barriers to adoption of the feasible process alternatives resulting from the optimization. This novel approach to "responsible innovation" [7] in chemical process design increases the likelihood that the proposed technologies have beneficial real-world impact.

As a proof-of-concept, we apply this method to the optimization of a simplified model for a modular food waste valorization process involving the following basic steps: (1) fungal aerobic digestion of mixed food waste to produce liquid "digestate," (2) *de novo* synthesis of aromatics from digestate using engineered microbes, (3) separation (e.g., adsorption) of products from the reactor effluent for subsequent purification. Process optimization is carried out using standard routines, and techno-economic analysis and life-cycle assessment are implemented to evaluate economic and environmental objectives, respectively. We demonstrate the effect of different stakeholder priorities on the optimal process design and the resulting economic and sustainability metrics. Additionally, scenario analysis is conducted to assess the sensitivity of the optimal design to uncertain economic and physical parameters and other assumptions. Future work will include the development of more detailed empirical and semi-empirical models of each process unit, allowing the proposed multi-objective optimization framework to be used for process synthesis.

1. (2019) Key facts on food loss and waste you should know! http://www.fao.org/save-food/resources/keyfindings/en/
2. DSM Environmental Services (2017) Analysis of Organic Diversion Alternatives.
3. United States Environmental Protection Agency (2010) Greenhouse Gas Reporting Program (GHGRP). https://www.epa.gov/ghgreporting/ghg-reporting-program-data-sets
4. (2016) A Roadmap to Reduce U.S. Food Waste by 20 Percent. https://www.refed.com/downloads/ReFED_Report_2016.pdf
5. Mohai P, Saha R (2015) Which came first, people or pollution? Assessing the disparate siting and post-siting demographic change hypotheses of environmental injustice. *Environmental Research Letters*, 10:1–17. https://doi.org/10.1088/1748-9326/10/11/115008
6. Darshini D, Dwivedi P, Glenk K (2013) Capturing stakeholders' views on oil palm-based biofuel and biomass utilisation in Malaysia. *Energy Policy*, 62:1128–1137. https://doi.org/10.1016/j.enpol.2013.07.017
7. Stilgoe J, Owen R, Macnaghten P (2013) Developing a framework for responsible innovation. *Research Policy*, 42(9):1568–1580. https://doi.org/10.1016/j.respol.2013.05.008

601847 Solvent Optimization in Reactive Extraction for Process Intensification in Biomass Upgrade

Abstract:

In chemical processes when desired products are reactive intermediates, reactive separations (i.e. simultaneous reaction and product separation) are essential for process intensification toward high yields and selectivities. The focus on limited sets of common solvents for liquid-phase extraction of such products leads to non-optimal results and a systematic framework for solvent selection in biphasic systems is lacking at present. In this work, we have explored the reactive solvent extraction of 5-hydroxymethylfurfural (HMF), a biomass-derived platform chemical and a reactive intermediate in hexose hydrolysis.

A multifaceted study was conducted starting with in silico solvent screening using the multiscale COSMO-RS model that identified potential high-performing solvents to maximize HMF extraction from a water-organic biphasic system. This was followed by experimental determination of HMF partition coefficients (PHMF) of selected solvents both at room temperature and reaction temperature. To preserve phase compositions at elevated reaction temperatures, an in-situ sampling and subsequent sample handling procedure was established. For biphasic systems, substituted phenols predicted by COSMO-RS exhibit order-of-magnitude increase in PHMF compared to conventional solvents. However, predicted high-performing anilines were determined to be reactive with HMF, highlighting the importance of combining in silico screening with reactivity and stability experiments. This comprehensive study considers solvent-HMF reactivity, solvent-water miscibility, solvent stability at reaction temperature, partitioning of the substrate (fructose), acid (HCl) and other reaction products (levulinic acid (LA), and formic acid (FA)) in fructose dehydration and provides the first systematic framework of solvent selection for reactive extractions, which can be extended to other biomass derived products such as furfural.

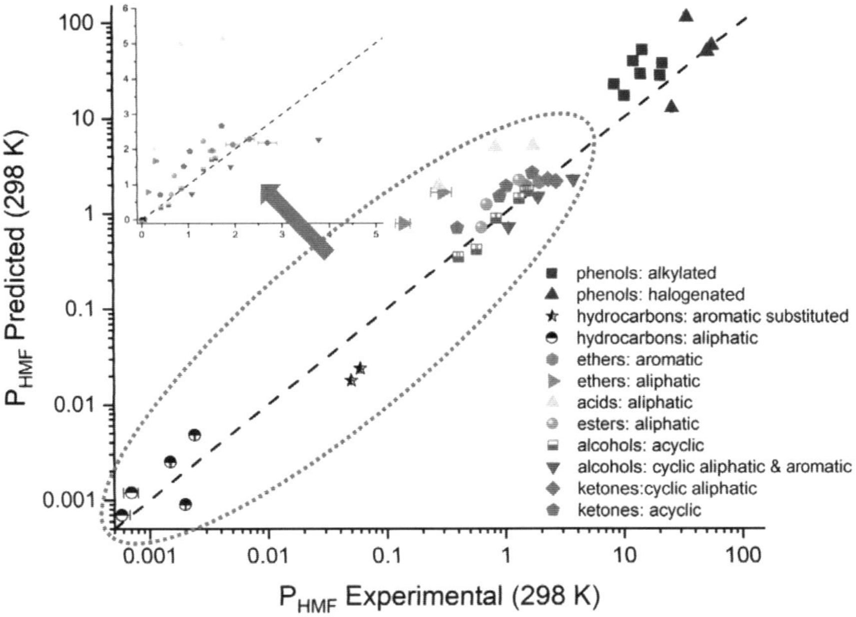

Parity plot showing comparison of experimental and predicted P_{HMF} at 298 K. Conventional solvents used in HMF extraction are highlighted (inset)

602342 Process Synthesis and Intensification for Upgrading Natural Gas Liquids in Shale Gas

Abstract:

Shale gas is a promising energy resource and chemical feedstock for the transition period towards a sustainable economy and has the potential to be a carbon source for the long term. However, huge amount of shale gas at remote shale gas basins is directly flared[1] due to the lack of infrastructure to transport the gas from the well head to the central processing plant. To avoid long distance transportation, it is essential to covert shale gas to value-added products on site at a distributed scale. The value-added products could be gasoline, diesel, aromatics, alcohols, or any other liquid chemicals which can be easily shipped. Unlike current shale gas process where large scales are preferred, simple and intensified processes with least processing steps and least pieces of equipment are favored for remote shale plays.

Natural gas liquids contained in shale gas are especially of our interest since they are comparably easier to activate and transform to liquid products. In the current work, novel processes are developed to achieve the upgrading process and more importantly, several key process intensification principles are summarized which can be potentially applied to any process that requires a modular and intensified design. While conventional shale gas processing usually follows a four-section hierarchy of "gas treatment - NGL recovery - NGL fractionation - NGL activation", our innovative configurations provide a paradigm shift and result in much simpler and more intensified process designs, including NGL co-processing[2], integrated NGL recovery and activation, switched NGL recovery and activation, and eliminated NGL recovery. A two-step conversion of NGLs to liquid hydrocarbons via dehydrogenation followed by oligomerization is used as an example to show how these innovative process designs evolve. Simulation results show that the loss of ethane, the NGL component with the highest concentration, could be largely reduced by the innovative process configurations. At the same time, higher yield of liquid products, fewer processing steps, reduced pieces of equipment and elimination of energy and capital-intensive units can be achieved. The intensification of process here would benefit the modularization of shale gas plants and make it possible for onsite distributed production of liquid hydrocarbons for remote shale locations.

Reference

1. Natural centers for environmental information. Global gas flaring observed from space. https://ngdc.noaa.gov/eog/viirs/download_global_flare.html (accessed 04/06/20)
2. Ridha, T., Li, Y., Gençer, E., Siirola, J. J., Miller, J. T., Ribeiro, F. H., & Agrawal, R. (2018). Valorization of Shale Gas Condensate to Liquid Hydrocarbons through Catalytic Dehydrogenation and Oligomerization. Processes, 6(9), 139

602631 Pilot-Scale Demonstration of Coal Waste and Torrefied Municipal Solid Mixtures with an Entrained Flow Gasifier

Abstract:

Conversion of waste coals and torrefied municipal solid waste (MSW) mixtures to syngas was demonstrated with a pilot-scale entrained flow gasifier (EFG). Coal waste and coal plus opportunistic fuel mixtures present an alternative feedstock for energy production. Semi-sorted MSW was torrefied at 350 °C using a large-scale torrefaction reactor to generate a coal-like feedstock for blending and feeding with coal waste. Bituminous waste coal mixed with torrefied MSW at weight ratios of 100:0, 75:25, 50:50, 25:75, and 0:100 were gasified in the EFG while H_2, CO, CH_4, CO_2, and H_2S concentrations were continuously measured in the syngas. Emissions defined by the EPA OSWI regulations (CO, NO_x, PM, SO_2, dioxins/furans, HCl, Cd, Pb, and Hg) were also determined. The optimal steam-to-O_2 ratio was determined for each waste coal and MSW blend tested and product yields were quantified, including the ash and slag. Analyses of slag, filter ash, and quench water for the Resource Conservation and Recovery Act trace metals were completed for each feedstock blend along with a toxicity characteristic leaching procedure (TCLP) on the slag and filter ash to verify that trace metals were vitrified in the melted ash stream. Steady-state pilot-scale EFG operation along with syngas and byproduct yield measurements experimentally validated the use of coal waste and MSW in an EFG. Based on the pilot-scale optimization data, a demonstrator-scale (100 kg/h) system was modeled and designed to estimate capital and operating costs, as well as the levelized cost of energy (LCOE) for small, modular gasifier systems.

613467 Machine Learning Reactivity in Networks of Coupled Gas Phase Reactions

Abstract:

The main computational cost of reaction rate constant calculations lies in the exploration of reactive pathways on potential or free energy surfaces. To overcome this challenge, recent efforts have employed machine learning to predict chemical reactivity at the level of single reactions.[1–3] In this context, we have investigated the use of machine learning to predict reactivity for networks of coupled gas phase reactions.[4] Single reaction rate constants were obtained by training deep neural networks on existing datasets. Chemical species were classified based on their properties and the kinetics of the reaction networks was evaluated using direct or stochastic integration of ODEs at various temperatures. This effort has enabled us to gain insight on how ML error on single rate constants influences the kinetics in a network of coupled of reactions.

1. Slakman, B. L. & West, R. H. Kinetic solvent effects in organic reactions. *J. Phys. Org. Chem.* **32**, 1–12 (2019).
2. Grambow, C. A., Pattanaik, L. & Green, W. H. Deep Learning of Activation Energies. *J. Phys. Chem. Lett.* **11**, 2992–2997 (2020).
3. Komp, E. & Valleau, S. Machine learning quantum tunneling in the kinetics of chemical reactions, *in preparation*. (2020).
4. Komp, E. & Valleau, S., *in preparation*. (2020).

613415 Progress in Applying AI/ML to Reaction Engineering and Chemical Synthesis

Abstract:

Machine learning, artificial intelligence, and other forms of computer assistance have the potential to augment human intuition about chemical reactivity to accelerate scientific discovery. There are many applications of computer-aided chemistry relevant to catalysis and reaction engineering, from developing quantitative structure-activity/property relationships for model-guided optimization to building statistical models able to describe chemical reactions at the molecular or atomistic level. In this talk, I will highlight recent work applying AI/ML to reaction engineering and share some thoughts on future directions for the field.

607357 An Additive Manufacturing Process for Producing Integrated Optoelectronic Flexible Electronics

Abstract:

The integration of flexible photonic components into flexible electronics has created new applications in imaging, sensing and optical interconnection between devices. While the majority of photonic devices are printed on rigid substrates, the photonic integration of devices on flexible polymer substrates has been demonstrated in recent years. However, current technology of contact and non-contact processes limits the type and performance of optical devices on these flexible substrates. Here we show that by utilizing recent advances in additive manufacturing, we can print photonic interconnects directly on flexible substrates. For our proof of concept devices, we have shown negligible changes in transmission losses through substrate bending at moderate bend radii down to about 20 mm. The results demonstrate how flexible optoelectronic devices can be produced using additive manufacturing tools. Furthermore, our composite direct write additive manufacturing system of micro-dispensing, fused deposition modeling, and laser subtraction has the unique ability to produce flexible optoelectronics in a single automated process. We anticipate that our flexible integrated optoelectronics devices to be a starting point in the development of more sophisticated devices.

608403 Lightweight Polymer Lattices with Unconventional Mechanical Properties

Abstract:

Advances in 3D printing enable the fabrication of complex lattice structures with regular repeat units. These lattices use geometry to reduce weight and impart unconventional mechanical properties to the structure. The use of polymers introduces both time and temperature dependence to the response of the structure to mechanical loads. However, despite the prevalence of polymer base additive manufacturing techniques, limited research has been performed on polymer-based lattices. The investigations that have been performed are typically conducted under quasi-static, compressive loading conditions and at room temperature. Therefore, we investigate polymer-based lattices subjected to both thermal and mechanical loads. We use fused deposition modeling to fabricate planar lattice structures. The structures are tested in a custom test fixture mounted in an Instron tensile tester. Tension and compression tests are performed under both uniform and spatially varying thermal loads. In addition, we compare experimental results to predictions from finite element analysis. The understanding gained by this work provides insight into the mechanical behavior of polymer lattices in critical temperature regimes, around the glass transition temperature. Our work leads to the further development and application of these lattices to structures which require light weight and tunable mechanical properties.

606803 Numerical Simulation of the Extrusion Process in Fused Deposition Modeling: The Effects of Process Variables on the Pressure Drop in the Nozzle

Abstract:

Fused deposition modeling (FDM) has been used and developed as one of the most popular techniques in additive manufacturing. One of the important aspects to be considered in process conditions of FDM is energy requirement to push filaments into the nozzle, which is directly related to pressure drop. In this work, we performed CFD simulation of the extrusion process in FDM using OpenFOAM, an open source simulation tool based on the finite volume method. A computational module for solving the equations of non-isothermal multiphase flows, which we developed, is used to systematically investigate the effects of process variables, such as, nozzle temperature, substrate velocity, feed stock thermal conductivity, gap distance, feed velocity inside the nozzle, and the nozzle geometry, on the pressure drop inside the nozzle. The simulation results are then compared with analytical and experimental results available in the literature.

606454 Quantitative Analysis of Hydrogel-Based Cement Pastes for 3D Printing Applications

Abstract:

Robust cement-based printing pastes for extrusion-based 3D printing have been developed using hydrogel-forming polymers. Not all gel-forming polymers, however, produce good outcomes. Gel rheology, pore structure and the nature of confined gel-water are being studied in an attempt to discover physical reasons for printing efficacy. In addition, despite growing interest in 3D printing by the construction industry, there is still a lack of quantitative metrics to assess the quality of 3D printed objects. Three printability metrics were developed to assess deviations from target objectives. Regression analysis were performed to obtain the most influential mix formulation effects. Conjectures based on objective measures of printability and gel and paste rheological characteristics are emerging that could help target better printing aid selection and design. The class of hydrogels explored exhibit time-independent shear stability, a characteristic that may be a pre-requisite for printing aid applications. These particular gels are also highly shear thinning and produce workable, but cohesive pastes when mixed with cement. While notable differences in gel-pore structure and the amount of confined water are quantifiable, it is not so clear how or if these factors are directly related to printability.

609588 3D Printing of SiC Rocket Nozzles Using Vibration Assisted Printing

Abstract:

The study and use of the space domain, including the recent reinvigoration of manned space exploration to the moon and beyond, drives the search for higher performance materials for high temperature spacecraft components such as thermal protection systems (TPS) and rocket nozzles. Ceramics and high-performance carbon both exhibit material properties that are suitable for TPS applications, but their performance can be maximized using additive manufacturing (AM) methods. Vibration Assisted Printing (VAP) is a newly developed AM system that can process extremely viscous mixtures, such as those consisting of ceramic forming polymers with solid ceramic particles, to yield net-shape parts with minimized porosity. This work explores the formulation development and the AM of ceramic structures utilizing VAP. A highly viscous mixture was prepared from a polycarbosilane polymer and pure SiC powder, which could be 3D printed using VAP. SiC components were cured up to 248.8°C and pyrolyzed at up to 1,600°C and were characterized via SEM, EDS, XRD.

597841 Surpassing the Detection Limit and Accuracy of Electrochemical DNA Sensor through the Application of CRISPR-Cas Systems

Abstract:

We present a CRISPR Cas system enhanced electrochemical DNA (E-DNA) sensor with unprecedented sensitivity and accuracy. The principle of the E-DNA sensor is the target induced conformational change of the surface signaling probe (containing an electrochemical tag), leading to the variation of the electron transfer rate of the electrochemical tag. With the introduction of CRISPR cleavage activity into the E-DNA sensor, a more apparent signal change between with and without the presence of the target can be achieved. We compared the performance of Cas9 and Cas12a enhanced E-DNA sensor and optimized the chemical environment of CRISPR, achieving a femto-molar detection limit without enzymatic amplification. Moreover, we correlated the CRISPR cleavage signal with the original E-DNA signal as a strategy to indicate potential mismatches in the target sequence. Comparing with classic DNA electrochemistry based mutation detection strategy, CRISPR enhanced E-DNA sensor can determine the presence of a single mutation at an unknown concentration condition. Overall, we believe that the CRIPSR enhanced E-DNA sensing strategy will be of especially high utility for point-of-care systems owing to the programmability, high-sensitivity and high-accuracy.

592702 In Planta Detection of Synthetic Auxin Plant Hormones Using Single-Walled Carbon Nanotubes (SWNT) Based Nano-Sensors

Abstract:

1-Naphthalene acetic acid (NAA) is one of the most commonly-used synthetic auxin plant hormones in agriculture as rooting agent or as sprays to prevent premature flowering and fruit drop. NAA is subject in some countries to regulatory control, requiring monitoring of NAA levels in vegetables and fruits. Unfortunately, state-of-the-art NAA sensing methods are destructive and laborious. In this work, we show the development of a novel NAA nano-sensor based on the concept of corona phase molecular recognition, using cationic polymer wrapped single-walled carbon nanotubes (SWNTs). This approach allows real-time non-destructive *in planta* detection of NAA. Using a semi-rational approach to polymer design, we developed a sensor that is selective for auxins, with particularly large responses to NAA as well as indole-3-butyric acid (IBA), both of which are used as rooting agents. An unresponsive control sensor was also developed to enable a ratiometric sensing platform. The sensor is demonstrated *in planta* for living spinach plants. We further probed the corona phases of both the active and control sensors, using the molecular probe adsorption (MPA) method to measure their accessible surface area for analyte adsorption. In both sensors, the Gibbs free energy of the riboflavin molecular probe binding to SWNT scales inversely with cube root of accessible surface area, which is in agreement with the structure-property relationship established for corona structures across different systems.

Since its initial discovery and commercialization in the 1940s, 2,4-Dichlorophenoxyacetic acid (2,4-D) has been an selective auxin herbicide commonly used for agricultural purposes. With increased usage of 2,4-D, herbicide resistance in weeds is of concern. To date, there are 28 reported species of weeds resistant to 2,4-D. Application of ineffective herbicides with poor levels of control causes wastage of money to farmers and detriment to the environment. In this work, we have developed a selective SWNT-based nano-sensor for 2,4-D based on the same corona phase molecular recognition approach, which also allows real-time non-destructive *in planta* detection of 2,4-D. This helps prevent wasteful spraying and ineffective deployment of herbicides onto resistant weeds. It also allows us to calibrate the herbicide dosage to minimize its usage. Further, it provides a rapid screen for herbicide susceptibility without need to monitor plant growth over days. The sensor is demonstrated *in planta* for living non-model plant species such as spinach, pak choy (*Brassica rapa subsp. chinensis*) and choy sum (*Brassica chinensis var. parachinensis*).

610521 *Electrochemical Sensors for Microbial Activities in Benthic Sediments: A Sentry for Lacustrine P Biogeochemistry*

Abstract:

Internal loading (i.e. release from benthic sediments) of PO_4^{3-} to the lacustrine water column could induce harmful algal blooms (HABs). The extents of internal PO_4^{3-} loading are difficult to constrain and detect in situ and in real time because fluxes are measured in laboratory incubations, monitoring approaches are difficult to deploy in the field, and the scales of microbially-induced redox gradients can occur on sub-centimeter scales. Release or entrained of PO_4^{3-} from benthic sediments is controlled by the activities of microorganisms in those sediments. In this work, we develop an electrochemical approach to detect, identify, and monitor benthic sediment-associated microbiological activities that could be deployed in situ and in real-time.

A series of laboratory and field experiments was conducted to identify electrochemical signatures associated with a variety of microbiological activities occurring in Lake Erie sediments. Our technique makes use of zero-resistance ammetry (ZRA) measurements, which we have previously used to diagnose microbiological activities. The approach uses inexpensive materials, demands little power, and can interrogate microbial activities at small spatial scales. The ZRA approach detects internal PO_4^{3-} loading (the release of PO_4^{3-}, as "dissolved reactive phosphorous" from sediments), which can contribute to harmful algal blooms (HABs). Microbial degradation of algal biomass and microbially-induced mineralogical transformations can profoundly influence the PO_4^{3-} concentration in the water column, and therefore, the potential for HABs despite adequate management of external PO_4^{3-} loading. Thus, microbial metabolism in PO_4^{3-}-loaded benthic sediments represents a remnant threat to induce or contribute to HABs well into the future. Microbiological activities induce redox gradients in sediments, depending on the predominant terminal electron accepting process. Most notable from the standpoint of PO_4^{3-} flux is the anaerobic reductive solubilization of Fe(III) (hydr)oxide phases by Fe(III) reducing bacteria (FeRB), which releases PO_4^{3-} that was previously adsorbed to those phases. However, microbial activity can induce a variety of other chemical conditions that enhance or limit PO_4^{3-} solubility (e.g. Fe(II) oxidation, organic carbon mineralization, acid production, primary productivity). *ZRA-based monitoring can determine extents and types of microbial metabolism in the sediments, be exploited for long-term stewardship of Lake Erie sediments, and serve as an "early-warning system" for internal PO_4^{3-} loading events.*

This presentation describes our newly-developed, ZRA-based technique for monitoring sediment activity and internal PO_4^{3-} loads. We believe that this will eventually be an "early warning system" for internal PO_4^{3-}-supported HABs.

608727 Iot Based Fuel Adulteration Detection Using Dispersion Model and Tail Pipe Emissions

Abstract:

Over the years up-gradation to cleaner fuel aims to reduce the rising environmental pollution. Evidence from the past states that adulteration of fuel exacerbates the problem. To cut the cost of highly taxed improved hydrocarbon, it is usually blended up to 80% with cheaper hydrocarbon. The detection of adulteration is to be achieved by implanting sensors within the fuel tank and the exhaust tailpipe. This whole process will be brained by the Engine Control Unit (ECU) and the data will be transmitted to the pollution monitoring agencies via Internet of Things (IoT) over an interval of time. The data received from the vehicle will be modeled and analyzed on the basis of the dispersion model. The molecular diffusion of the adulterate in the quality hydrocarbon will result in the intermixing of molecules, by extending this study to reverse engineering, we will determine the amount of adulterate injected. The data will be assimilated with the fuel and emission standards that will estimate the quantity and the injection time period of the adulterate. Looking closer to the data, it can be easily estimated the adulteration cases over a zone by analyzing the chronicle of refueling. This ensures the environmental agencies to keep track of the adulterate hotspots. Presently various testing parameter are available such as ASTM D6730 (Gas Chromatography), ASTM D2425(Mass Spectroscopy), ASTM D2425 (Distillation). The common limitation of the tests available is that analysis must be performed outside the vehicle body, that will no longer provide the chronology behind the addition of adulterant. It is estimated that when this model will be implemented to the vehicles, it will uplift the aspect of the cleaner fuel to deliver better engine performance with reduced emission levels. This will empower every citizen to monitor the quality of fuel easily by accessing the data from the vehicle.

605985 In-Situ Chemical Composition Sensing Using IR Adsorption in Micro-Structured Flow Reactors

Abstract:

Flow reactors are attractive options for moving beyond small-scale batch synthesis for some high-value chemicals, particularly in the fine chemical industry. One major limitation at the present is the lack of online sensors, for example it is challenging to design a reactor to carry out a complex organic chemical reaction, since most available analytics require sampling and relatively slow, offline analysis, making the results of limited use for process control.

In our DFG-funded Research Unit called Promise (https://promise.kit.edu) we explore multiple options for the in-situ sensing and control of a range of different physical and chemical processes. One project in particular explores the possibility for using relatively low-cost, solid-state IR sensors embedded into a flow reactor along the flow path, to give real-time, spatially resolved measurements of chemical compositions, which can be correlated to the extent of the chemical reaction (and even more complex measures like product selectivity, by-products and so on). In this presentation we outline our latest reactor design, which features three distinct and individually temperature-controlled reaction zones, two IR sensing zones (with specific materials of construction that allow for IR-transmission over a very wide range of wavelengths, much greater than can be achieved using conventional glass materials), and unique flow paths allowing for full system sealing at the top and bottom of the reactor. This apparatus was constructed in the micro process engineering workshop and combined with unique and custom-built IR sensors, resulting in a system that can make real-time IR analysis and use the results for process monitoring and eventually process control.

The Figure shows exploded views (from the top and bottom), indicating the complexity of the device and the careful design decisions necessary to achieve optical transparency in all required regions, while meeting flow and sealing requirements. Initial results for several photochemical reactions will be presented.

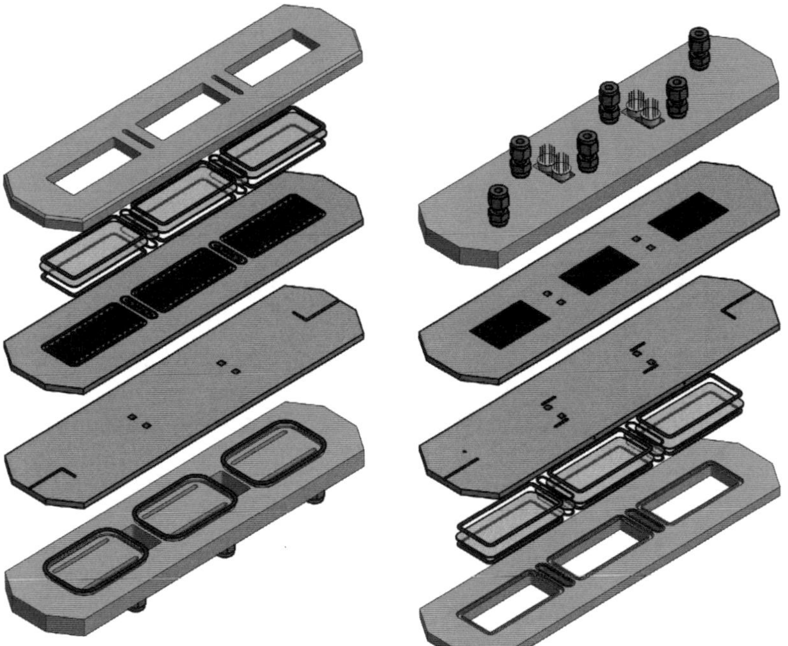

611466 Rapid Detection of Tuberculosis Breath Biomarkers Using Engineered Electroactive Solutions (EAS)

Abstract:

The World Health Organization has designated the development of a rapid, inexpensive, point-of-care screening test for Tuberculosis (TB) as high priority in their annual Global Tuberculosis report, reporting nearly 1.5 million TB related deaths in 2019 [1]. Screening via associated breath biomarkers has been identified as a promising emerging technology faster and easier to use at the point of care than standard sputum-based tests. Traditional GC-MS testing has identified methyl nicotinate (MN) and methyl p-anisate (MPA) as specific biomarkers of mycobacterium tuberculosis that are not readily present in ambient air or the breath of healthy patients, however this mode of testing is prone to confounding and is often prohibitive at the point of care due to its high cost and need for highly skilled technicians. The relatively high solubility of these semi-volatile compounds in aqueous solution provides a unique ability to isolate the biomarkers from more volatile breath components by a diffusive gas to liquid transfer process. This step serves to both purify and effectively concentrate the sample for electrochemical analysis.

Previously, our lab developed a sensing platform utilizing Engineered Electroactive Solutions (EAS) on inexpensive, commercially available screen-printed electrodes, in which the change in a transition metal of known electrochemical activity is observed when exposed to an analyte of interest [2]. The original study was successful in using a divalent copper electroactive solution (Cu2-EAS) to detect MN down to 1 mM using cyclic and square wave voltammetry techniques (CV and SWV, respectively). However, the relatively high reactivity of copper may limit specificity. The sensing platform has been improved to use a divalent cobalt electroactive solution (Co2-EAS), with a limit of detection of about 0.01 mM of MN, about 100x more sensitive than Cu2-EAS. When Co2-EAS is mixed with a sample solution of MN, a peak specific to the metal-biomarker complex is present in both the forward and reverse scans, the height of which corresponds to the amount of biomarker present (see figure 1). The addition of this peak leads to a more sensitive and specific test for MN than the Cu2-EAS, The Cu2-EAS utilized a “fingerprint” method in which the relative heights of copper’s existing redox peaks were compared. Unlike the Cu2-EAS, the Co2-EAS is also able to identify MPA in solution by a significant change in the existing cobalt peaks when it is added to solution, as shown in figure 2.

The simple sample collection and testing methods for an EAS based sensor is ideal for resource limited communities where TB is prevalent. The ability to perform sample collection and testing remotely as well as store collected breath samples for later testing provides flexibility to both patients and technicians. Disposable testing kits can be assembled for less than $5 per test, meeting the goal price point for TB triage testing put forth by Van’t Hoog et al, estimating a 42% reduction in cost compared to “Xpert for all” testing in Uganda [3]. Field testing of this novel sensing platform will be conducted in collaboration with Makerere University in Kampala, Uganda to determine sensor performance and to better understand patient needs at the point of care.

References:

[1] World Health Organization, “Status of the TB epidemic,” 2019.

[2] Willis C, Misra M, Mohanty S, 2019, “Engineered Electroactive Solutions for Electrochemical

Detection of Volatile Organic Biomarkers Associated with Disease," 235th *ECS Meeting*, Dallas TX, 26-30 May

[3] A. H. van't Hoog et al., "Optimal Triage Test Characteristics to Improve the Cost-Effectiveness of the Xpert MTB/RIF Assay for TB Diagnosis: A Decision Analysis," PLoS One, vol. 8, no. 12, p. e82786, Dec. 2013.

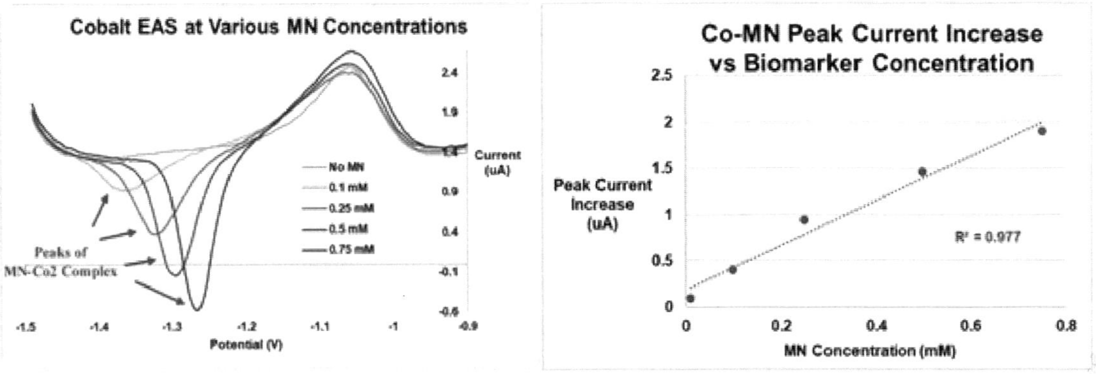

Figure 1. Size of peak corresponding to MN-Co2 complex is proportional to concentration of MN in solution

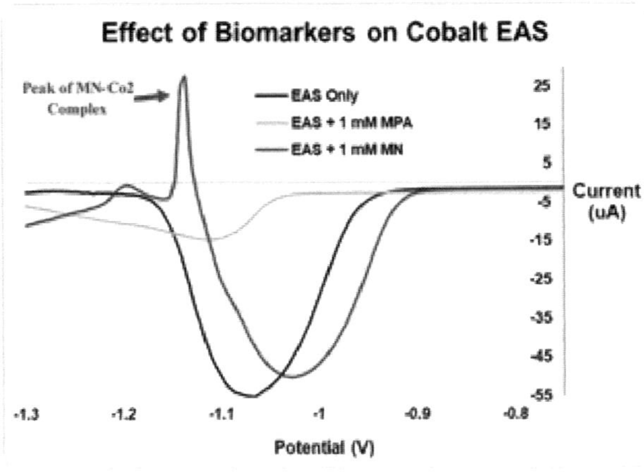

Figure 2. Comparison of square wave voltammograms of Co2-EAS with and without biomarkers present

606167 Deposition of Oxide Coatings Using Non-Thermal Atmospheric Pressure Plasmas

Abstract:

Atmospheric pressure plasmas (APPs) have shown great promise in a variety of manufacturing applications. APPs offer unique opportunities for the industrial application of plasmas, due to the versatility of the technique, the ease of scalability, and the opportunity to perform the technique point of manufacturing. This work focuses on depositing zirconia-silica bilayer coatings as a replacement for hexavalent chromium-based coatings. A new non-thermal process to deposit zirconia was developed using APPs enabling processing on temperature sensitive substrates. Electrochemical corrosion performance will be compared to other passivation and barrier coatings. Additionally, we investigate APPs to enable lightweight manufacturing for the automotive industry by depositing a thin film of silica that acts as a water intrusion barrier on bonds joints between dissimilar materials. The adhesion tests performed after water soak meets the industry benchmarks. These processes aim to replace wet chemical processes often used by the automotive and aviation industries. The mechanical properties of the films are tested using lap shear, both before and after a water soak test, which successfully demonstrates the films capabilities as a water barrier. The chemical composition is confirmed using both XPS and RBS, and the films morphology is examined using cross sectional SEM.

608084 Treatment and Extraction of Copper from Electronic Waste Via Induced Morphological Changes Utilizing Supercritical CO_2

Abstract:

The management of electronic waste (e-waste) has become a significant challenge faced by humanity. Globally, the number of discarded computers, phones, and other appliances, often defined as Waste Electrical and Electronic Equipment (WEEE), doubled between 2009 and 2014 to 41.8 million tons per year. Currently, e-waste disposal factories tend to recycle the non-toxic, valuable components of e-waste (metals) and release the toxic components into the surroundings via pyrometallurgy (burning). Chemical treatment, known as hydrometallurgy, is also used and involves large amounts of corrosive acids. As the amount of e-waste grows, the availability of easily mineable elements, particularly metals, decreases, disturbing the overall material cycle. Thus, new, environmentally-friendly approaches to recover valuable components from e-waste are urgently needed. This study focuses on the use of supercritical CO_2 ($scCO_2$) and sulfuric acid, 2 M H_2SO_4, as a novel solvent system for the treatment of e-waste, specifically in the extraction of copper. Printed circuit board (PCB) was used as the e-waste of study and is generally composed of 40% metals, 30% plastics, and 30% refractory material. Through the synthesis of metal and polymer composites, which reduced the complexity of PCB, the effects of $scCO_2$ in the solvent system were investigated. The results showed that $scCO_2$ and acid induced a permanent morphological change in the polymer component, creating voids and causing delamination. This finding was translated to the actual waste PCB system and a novel two-step process was developed. This two-step process involved the pre-treatment of small pieces of waste PCB with $scCO_2$ and 2 M sulfuric acid at 120 °C and 148 atm for 30 minutes followed by leaching of the pre-treated PCB in a solvent containing 2 M H_2SO_4 and 0.2 M H_2O_2 at ambient conditions.Experimental results showed that 82% of the copper contained in the PCB was extracted in under 4 hours. The characterization of the PCB demonstrated that the pre-treatment with $scCO_2$ and acid induced the crystallization of the plastics (polymer component), creating pores and weakening the structure of the PCB, thus enhancing transport of the solvent to the buried metal interfaces. This novel process utilizing $scCO_2$ minimizes intensive physical processing (e.g. grinding of the PCB) and reduces acid usage during the extraction of Cu from e-waste, presenting a greener alternative to the current methods of recycling of metals from e-waste, which tend to be energy-intensive and hazardous.

596648 Annealing Carbon By Pulsed Laser Light

Abstract:

Laser processing of materials is not new, but, using lasers to anneal, bond and otherwise transform carbons is. Fundamental understanding of the dependence upon carbon structure, morphology and chemistry is critical to implementing this technology into manufacturing and processing applications. In this work a Q-switched Nd:YAG laser and a continuous wave CO_2 laser are used to anneal carbon materials. Lasers provide rapid heating and cooling with high temporal control. The extent of transformation is kinetically controlled by time above the threshold temperature for transformation. Enabling the production of carbon structures not possible via traditional furnace annealing. The potential technological importance of laser annealing carbon is demonstrated. Examples of material processing and synthesis not possible via traditional furnace annealing are provided. To resolve the nanostructure changes HRTEM is employed.

Perhaps the most over looked application of laser annealing carbon is the ability to do so continuously and rapidly. Graphitization furnaces operate as a batch processing system. Time from start to completion is on the order of a day due to the slow (~25 °C/min) heating rates, long hold times, and slow cooling of the heavily insulated furnace. Additionally, these furnaces require routine maintenance and replacement of expensive specialty graphite heating elements. CO_2 laser annealing provides equivalent material transformation on the order of seconds and modern CO_2 lasers are designed for years of maintenance free use [1]. Although laser annealing is limited to thinner material due to limited heating depth, materials can be annealed continuously and with potentially high throughput.

The material was heated to 2,600 °C in 1.4 ms Under the action CO_2 laser radiation carbon can be heated to 2,600 °C in 1.4 ms. After 20 s, the structure is equivalent to that obtained from furnace annealing at 2,600 °C for 1 hr. as has been shown elsewhere [2]. In contrast a Q-switched Nd:YAG laser with pulse width ~ 8 ns heats carbon materials to graphitization temperatures with a heating rate on the order of a few 10^{11} °C/s. The peak temperature is controlled by the laser pulse energy; typical temperatures range between 2,500 °C to the C_2 sublimation temperature of 4,184 °C [2,3]. The time at elevated temperature is limited (time above 2,000 °C is 1.5 μs). On this time scale, the long-range material motions are kinetically restricted. Consequently the Nd:YAG laser annealing trajectory deviates from traditional furnace pathways. The limited time at elevated temperature can be used for the purpose of surface modification. Surface modification via kinetically limited oxidation is one potential application. Another application is making connections between carbons without additional material. Laser sintering a mixture of carbon materials could result in a wide range of potential applications.

Heat-Treatment-Temperature (HTT) and resulting carbon structural transformation has been extensively studied. A detailed quantification of HTT and resulting material annealing was provided by Oberlin in 1984 [4]. The four stages outlined in Oberlin's HTT diagram (temperature dependent) are believed to be separated by "very rapid" transitions [8]. So rapid, in fact, that intermediate structures are not observable on furnace heating timescales (several minutes minimum) due to slow ramp rates and have thus been ignored. These intermediate structures remain unexplored and represent an entirely new class of carbon materials. It has been demonstrated that CO_2 laser heating results in equivalent end structure as compared to furnace annealing [1]. The high temporal control of

CO_2 laser annealing allows for exploration of such intermediate structures that can be captured by the rapid heating and cooling timescales. In summary lasers are poised to be instrumental in the advancement of carbon science and technology.

References

[1] J.P. Abrahamson, Pulsed laser annealing of carbon, Ph.D. Thesis, Penn State, 2017.

[2] J.P. Abrahamson, M. Singh, J.P. Mathews, R.L. Vander Wal, Pulsed laser annealing of carbon black, Carbon 124 (2017) 380–390. doi:10.1016/j.carbon.2017.08.080.

[3] H.A. Michelsen, C. Schulz, G.J. Smallwood, S. Will, Laser-induced incandescence: Particulate diagnostics for combustion, atmospheric, and industrial applications, Prog. Energy Combust. Sci. 51 (2015) 2–48. doi:10.1016/j.pecs.2015.07.001.

[4] A. Oberlin, Carbonization and graphitization, Carbon 22 (1984) 521–541. doi:10.1016/0008-6223(84)90086-1.

604429 An All-in-One Approach for Training Deep Learning-Based Control Laws

Abstract:

Model predictive control (MPC) provides a general framework for controlling nonlinear systems. However, the high online computational latency limits the scope of its application for systems with fast dynamics. One way to enable the real-time application of MPC is to approximate the implicit MPC control law with a neural network [1,2,3]. The hypothesis of this approach is that a deep learning neural network can sufficiently approximate the nonlinear behavior of MPC, but at a negligible online computational cost that is significantly better than that of a full-fledged MPC. For linear systems with quadratic cost function, a neural network with rectified linear units as activation functions can exactly represent piecewise affine optimal control laws [4].

An essential aspect of the above-mentioned approaches is the training of large dimensional neural networks. Many researchers follow the "optimize then train" protocol. The idea in these algorithms is to simulate the optimal controller, generate corresponding state and optimal control action data pairs and then use these data to train a neural network via supervised learning to approximate the original optimal controller. Unfortunately, this conventional "optimize then train" protocol is not guaranteed to work on nonlinear systems. For nonlinear systems, it is possible that multiple optimal control actions exit for the same initial state. In this case, the MPC control law is a set-valued function. The non-uniqueness of the state-action data pairs makes it impossible to learn the underlying MPC control law through this two-step approach, as illustrated in Figure 1. Even if the uniqueness of optimal control actions is guaranteed, the optimal control problem might have multiple local optimal control actions, which are typically identified using local solvers. Moreover, for general nonlinear systems, no neural network structure with finite neurons can guarantee an exact representation of the MPC control laws. The training error of the neural network can lead to sub-optimal control actions and state constraint violations. The resulting errors could potentially accumulate over time and eventually making the neural network deviate significantly from the optimal control.

We propose a new "optimize and train" method that combines the steps of data generation and neural network training into one single large-scale stochastic optimization problem. This approach directly optimizes the closed loop performance of the DNN controller over a number of possible initial states. Directly optimizing the control law instead of control actions avoids the difficulty caused by the nonuniqueness of optimal control actions, as shown in Figure 2. Moreover, all constraints are guaranteed to be satisfied for all initial state scenarios considered in the optimization, while this guarantee does not hold for the conventional "optimize then train" method. We also propose methods to ensure constraints satisfaction for all possible initial states. The critical challenge is the solution a large-scale nonlinear stochastic optimization problem. However, tremendous progress has been made recently in the field of stochastic optimization that allows us to design and implement efficient local and global algorithms [5,6]. Besides directly using stochastic optimization solvers, we also proposed a recurrent neural network formulation for nonlinear systems with input constraints and soft state constraints. This approach can take advantage of existing packages (e.g., TensorFlow) in the machine learning community and can potentially tackle extreme-scale problems efficiently with GPUs. The benefits of our all-in-one approach over the conventional ``optimize then train" protocol is illustrated through numerical results.

References:

[1] Parisini, T., & Zoppoli, R. (1995). A receding-horizon regulator for nonlinear systems and a neural approximation. Automatica, 31(10), 1443-1451.

[2] Csekő, L. H., Kvasnica, M., & Lantos, B. (2015). Explicit MPC-based RBF neural network controller design with discrete-time actual Kalman filter for semiactive suspension. IEEE Transactions on Control Systems Technology, 23(5), 1736-1753.

[3] Chen, Y., Shi, Y., & Zhang, B. (2018). Optimal control via neural networks: A convex approach. arXiv preprint arXiv:1805.11835.

[4]Karg, B., & Lucia, S. (2018). Efficient representation and approximation of model predictive control laws via deep learning. arXiv preprint arXiv:1806.10644.

[5] Cao, Y., Zavala, V. M., & D'Amato, F. (2018). Using stochastic programming and statistical extrapolation to mitigate long-term extreme loads in wind turbines. Applied energy, 230, 1230-1241.

[6] Cao, Y., & Zavala, V. M. (2019). A scalable global optimization algorithm for stochastic nonlinear programs. Journal of Global Optimization, 75(2), 393-416.

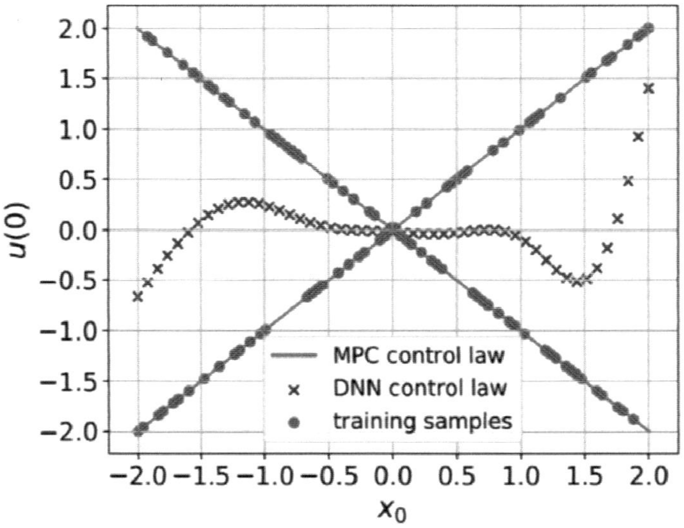

Fig. 1. DNN control law trained with "optimize then train" method

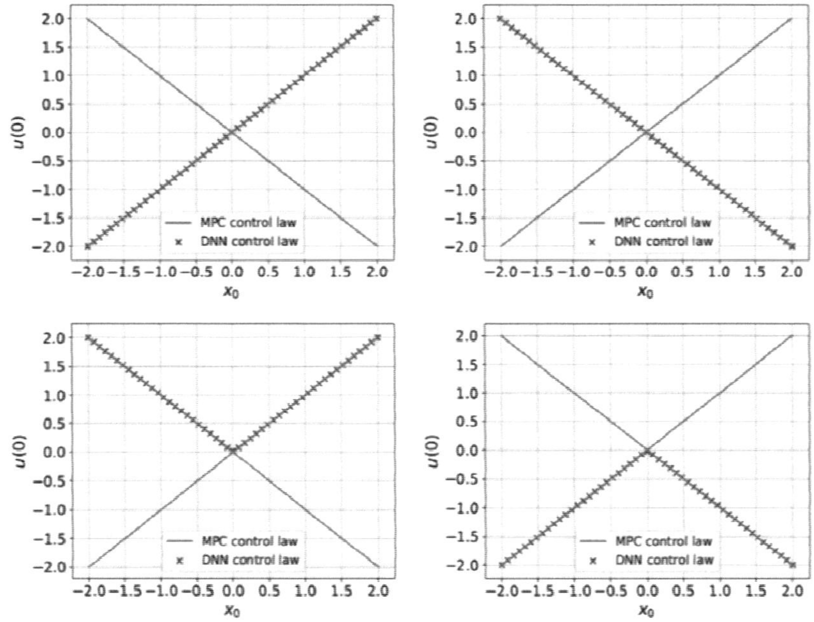

Fig. 2. DNN control law trained with "optimize and train" method

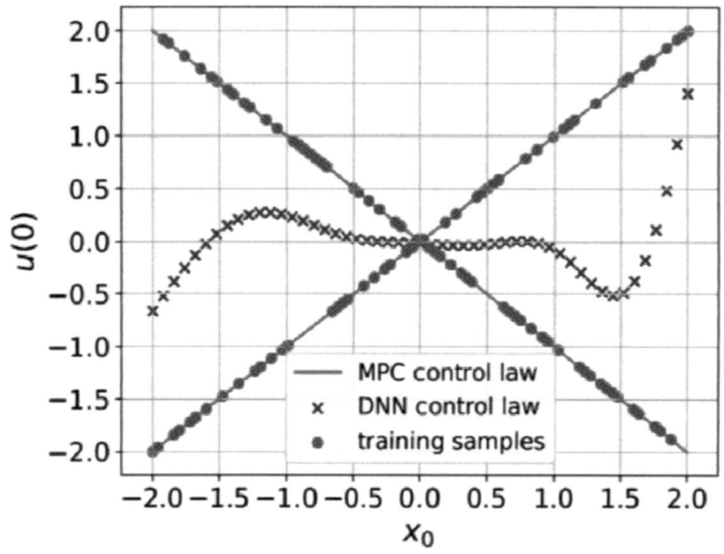

Fig. 1. DNN control law trained with "optimize then train" method

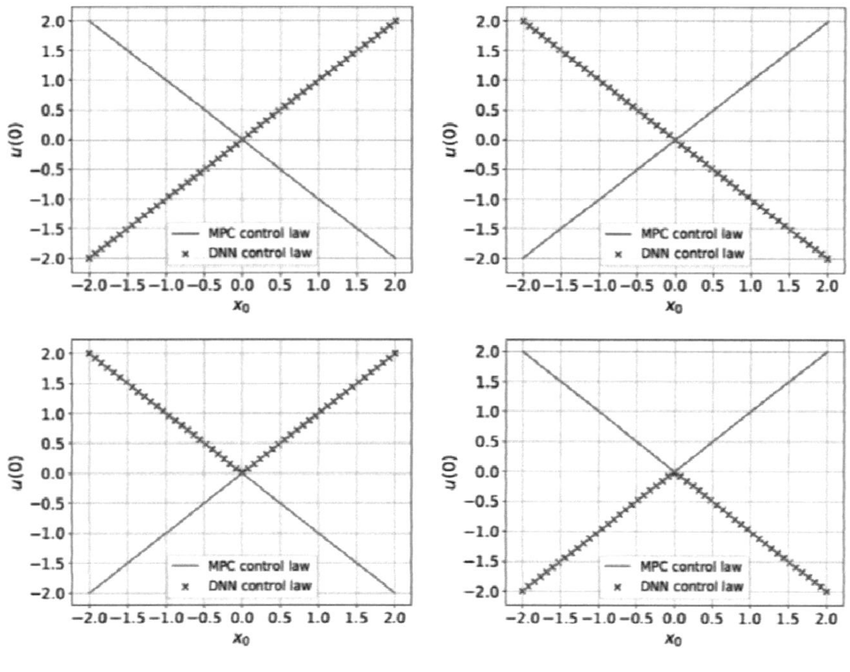

Fig. 2. DNN control law trained with "optimize and train" method

601990 Cloud-Based Control of a Robotic Manufacturing Process for Personalized Medicines

Abstract:

This abstract focuses on an automated production process enabling the manufacturing of personalized drug products. Particularly, we developed a system able to fulfill the needs of small-batch, high-mix, personalized, or even on-demand, production. Recently, the concept of precision medicine has received great attention. Finding ways to manufacture personalized therapies, in a consistent and controlled way is becoming the holy grail in the pharmaceutical field. However, existing manufacturing processes are not suitable for this mass-customization approach. In fact, while traditional systems can repeatably manufacture hundreds of thousands of identical dosage forms, having being designed for one specific product at a time, personalized therapies (*i.e.* extemporaneous formulations) are produced by technicians on a small scale only. Additionally, most of these production methods require labor to monitor, regulate and complete material handling for the equipment in use.

Based on these considerations, the availability of a fully automated architecture (*i.e.* hardware and software) capable of manufacturing personalized pharmaceutical products at scale would be of utmost importance. In this respect, we have developed a modular manufacturing system and a supporting cloud-based software architecture for its control. The modular manufacturing system, referred to as "robotic cluster", consists of *i)* different manufacturing modules, each of them performing specific fabrication steps, and *ii)* a transfer system that moves work-in-progress units (WIPUs) from module to module in a distributed and parallelized way. During the manufacturing process, WIPUs are shuttled into a given module, which performs a specific step of the process. Then, the WIPUs are moved into another module, and this process is repeated for all the following operations, including quality control. The sequence of manufacturing operations for a small batch of WIPUs is performed by moving it through the modules until all needed operations are completed. This sequence can be performed in a fully-planned, fixed-scheduled, a flexible-scheduled, or dynamic manner.

Each manufacturing module is connected to a dedicated control system and to other equipment *via* its control box, enabling the input of commands for the next fabrication steps while transmitting out the data pertaining to the work that has already been performed. All critical process parameters are monitored locally and in the cloud. The strategy pursued here not only ensures the security of the manufacturing system, but also increases its versatility and flexibility. It also makes this architecture suitable for scaling, by deploying many robotic clusters at the same time. Moreover, unlike flexible manufacturing systems described in literature, ours is comprised of a set of swappable units that are readily changed, updated, repaired and maintained. This way, complex operations are divided into simpler steps and for each of them there is a corresponding custom-built module. Therefore, the principal application of this technology is in mass customization, where products can be fully personalized to each order and large quantities of products can be fabricated in parallel to achieve the same scale and throughput of traditional serial manufacturing.

As a proof of concept, we evaluated the suitability of the discussed architecture for the manufacturing of multi-compartment capsules for personalized therapy. The capsules here proposed have a modular structure that consists of multiple inner compartments, which are independent of one another. Each compartment can be filled with different active ingredients and with a range of doses, as well as formulations, for any given drug. Multi-compartment capsular devices are especially advantageous for therapies involving multiple drugs, because they simplify the dosing schedule and have a positive potential impact on overall patient compliance.

Personalized filling of each compartment of the capsule shells was performed thanks to a custom-built robotic cluster, composed of a set of automated filling modules working simultaneously, in parallel. A robotic arm (designed to safely collaborate with human operators) moves the capsular devices form one filling station to the next, and quality control machines check the deposition of the desired amount of active ingredients after each step.

Within this proof of concept, we performed an optimization on the manufacturing steps as a fixed-scheduled sequence. We created a list of variable orders *i.e.* virtual patient prescriptions of varied precise dosages of multiple drugs, and input them into a schedule optimization algorithm. Said algorithm outputs the sequence of robot arm and module actions which maximizes throughput of the cluster. Generally, this algorithm maximizes the amount of work being done in parallel at once (many batches worked on in separate modules), and schedules the order in which the cluster outputs patient-scale batches to minimize the time taken to complete all batches. To ensure quality control we bound the time between manufacturing step and quality control: in-process controls should be measured more-or-less contemporaneously with the action. No WIPU can wait for longer than 10 min - say, sitting in a buffer or idle in a module - before being processed by the corresponding quality control module. For example, we bound the time between a filling module dosing powdered formulations and a weighing module recording the dosed weight. This scheduling optimization algorithm enables the parallelism of the robotic cluster. It is through this novel planning algorithm that we achieve the same scale and throughput of traditional serial manufacturing.

The automated deposition stage allows the rapid and accurate adjustment of the combination and dosage strength of the active ingredients. Cross-contamination is avoided by enclosing each filling station in a dedicated clean room. The capsule filling systems were validated with preliminarily filling trials, which used powders with different particle size distribution and bulk density. We considered 25 batches of 50 capsule each for each powder formulation. In all cases, the amount of active ingredient deposited by the robots had an average coefficient of weight variation (*i.e.* the ratio between the standard deviation and the average value of the weight of the deposited powder) below 3%. These results demonstrate the accuracy and the flexibility of the powder deposition systems. Additionally, we use spectroscopic techniques to check the weight uniformity and identity of the pharmaceutical formulations. Every measurement is collected and saved in the cloud based quality control system, allowing the creation, in real time, of a digital master batch record with all the information relevant to each manufacturing step (*e.g.* time, date, type of machine, weight of drug deposited), which could be accessed only by verified personnel and regulatory entities.

611934 Emerging Cognitive Engineering Approaches to Enhance Control Room Operator Capabilities in Chemical Industries

Abstract:

Control room operator's actions can have significant effect on the throughput, quality and safety of process plants (Baron, 1998). Typically, control room operators monitor the process via a Human Machine Interface (HMI) on which all the necessary information about the process state is presented. The complexity of operating modern chemical plants have increased significantly due to tight mass and energy integrations, deployment of sophisticated control automation strategies, and the need to operate the process in agile fashion in order to capture market and supply chain dynamics. However, there has been no concomitant increase in operations personnel's cognitive ability. Any mismatch between the operators' capability and the process control demands leads to high cognitive workload in human operators, often a precursor for poor performance which in-turn leads to sub-optimal operator actions or at worst accidents. This paper addresses the critical need to understand and enhance control room operator's cognitive capabilities.

Traditional operator performance assessment methods ignore cognitive aspects of performance. Operator's performance is assessed primarily in terms of number of successes and failures, response times, deviation of operator actions from predefined sequence (Lee et al., 2000), etc. However, in addition to overt performance (as done traditionally) it is imperative to consider covert aspects of performance i.e., cognitive workload for a holistic assessment of human performance. Various emerging cognitive engineering based sensors such as eye gaze tracking, Electroencephalography can reveal insights into the operators' cognitive state that determines these covert aspects. Specifically, in this paper we propose an Electroencephalography (EEG) based methodology to assess the operators' cognitive workload, especially during abnormal situation tasks where their active involvement is essential to safe process operations. The potential of EEG to understand cognitive workload for dynamic tasks which evolve with time has received some attention in aviation (Liang et al., 2018) and driving (Morales et al., (2017). However, there are hardly any studies relating to chemical control room operators.

We propose a single dry electrode EEG based methodology to dynamically assess the cognitive workload of operators while they perform various actions to control the process during abnormalities. The proposed methodology is illustrated on a simulated ethanol production plant testbed. A set of 10 different operators were considered in this work. Each operator's performance during the course of experiments involving monitoring and control of the process was studied. In addition to the process data and operator actions, we recorded the EEG data of the operators. Each experiment involved 6 tasks (comprising 5 different scenarios, out of which was repeated) of different abnormal situation. Each operator repeated the same experiment 7-8 times. Overall, 73 experiments were conducted by the 10 operators for a total of 438 tasks. Due to data quality issues, we evaluate results for 8 participants only.

We analyze the EEG signal in frequency domain by applying Fourier transform. Our results reveal that the power spectral density (PSD) of brain waves (4-30 Hz) is a biomarker that can identify the workload of operators. The PSD data of each operator was partitioned into two clustered using k-means clustering. Cluster 1 represents low power spectral density with low magnitude and Cluster 2 showed high power spectra density. We studied the workload variation with the help of cluster fraction over the course of the task as the operators' engagement with the process increased as well as with repetition. To evaluate the cluster variation within each task, a task is divided in three phases. In Phase 0 the operator monitors the process and ensures all the process variables are within the desired limits. During an abnormal condition an alarm is triggered, which leads to Phase 1 where the operator uses his mental model of the process to decide the control action(s) in order to clear the alarm. The portion of the task corresponding to all the control actions taken by the operator is called as Phase2. We observe distinct differences in the cluster fractions across the 3 phases. For a typical operator who successfully completes a task, during the monitoring Phase 0 we found the fraction of Cluster 1 to dominate (about 60%) indicating a correspondence between low workload during monitoring. During abnormality when alarms are triggered, in phase 1 the Cluster 1 fraction decreases (to about 40%) indicating that the operator uses his mental resources to decide the control action, i.e., increased workload. Further in phase 2, as the operator takes various control actions the Cluster 1 fraction further drops significantly (to about 0.20%) until the process becomes normal at which juncture it increases back to the 60% observed in Phase 0. The above patterns of change in the cluster pattern was observed for 81% (243 out of 300) of the tasks conducted by the 8 participants. This brings forth the capacity of the high accuracy of the EEG based cognitive sensor and the proposed machine learning algorithm to accurately estimate the workload of the control operator during normal and abnormal conditions. These distinct patterns were also highly correlated various process metrics (extent of variable deviation beyond alarm limits, time taken to clear alarm, number of standing alarms etc). In this paper, we will describe the proposed experimental and computational methodology and reported detailed results from our human subject experiments.

References:

Baron, RG. (1988). Human Factors in the Process Industries. Human Factors and Decision Making: Their Influence on Safety and Reliability, Symposium for the Safety and Reliability Society. ed. Sayers, BA. (pp.1-9)

Heard, J., Harriott, C. E., & Adams, J. A. (2018). A survey of workload assessment algorithms. IEEE Transactions on Human-Machine Systems, (99), 1-18.

Srinivasan, R., Srinivasan, B., Iqbal, M. U., Nemet, A., & Kravanja, Z. (2019). Recent developments towards enhancing process safety: Inherent safety and cognitive engineering. Computers & Chemical Engineering, 128, 364-383.

Neurosky. (2015). Greek Alphabet Soup – Making Sense of EEG Bands (Last accessed: Nov 10, 2018) http://neurosky.com/2015/05/greek-alphabet-soup-making-sense-of-eeg-bands/

Pfurtscheller, G., Stancak Jr, A., & Neuper, C. (1996). Event-related synchronization (ERS) in the α band—an electrophysiological correlate of cortical idling: a review. International journal of psychophysiology, 24(1-2), 39-46

Liang, Yongqiang, Wei Liang, Jue Qu, and Jie Yang. "Experimental Study on EEG with Different Cognitive Load." In 2018 IEEE International Conference on Systems, Man, and Cybernetics (SMC), pp. 4351-4356. IEEE, 2018.

Morales, J. M., Diaz-Piedra, C., Rieiro, H., Roca-González, J., Romero, S., Catena, A., Fuentes, L.J. & Di Stasi, L. L. (2017). Monitoring driver fatigue using a single-channel electroencephalographic device: A validation study by gaze-based, driving performance, and subjective data. Accident Analysis & Prevention, 109, 62-69.

Sharma, C., Bhavsar, P., Srinivasan, B., & Srinivasan, R. (2016). Eye gaze movement studies of control room operators: A novel approach to improve process safety. Computers & Chemical Engineering, 85, 43-57.

Bhavsar, P., Srinivasan, B., & Srinivasan, R. (2017). Quantifying situation awareness of control room operators using eye-gaze behavior. Computers & Chemical Engineering, 106,191-201.

Iqbal, M. U., Srinivasan, B., & Srinivasan, R. (2019). Electroencephalogram based Biomarkers for Tracking the Cognitive Workload of Operators in Process Industries. In Computer Aided Chemical Engineering (Vol. 46). https://doi.org/10.1016/B978-0-12-818634-3.50233-2

601227 The Flory-Huggins Parameters of Binary Ionic Liquid Mixtures with Ammonia

Abstract:

Phase equilibrium measurements and modeling are essential for the design of chemical processes such as separations, reactions, fluids flow, to name a few. One of the industrial processes that received increased attention in the last decade involves ionic liquids (ILs), which are low melting point salts composed entirely of ions. The field of ILs requires a more in-depth understanding of the interaction between ILs and other substances so that novel applications can be developed. In this work, our goal is to investigate the non-ideality of a binary mixture of ammonia and four imidazolium-based ionic liquids using the extended Flory-Huggins model, where both entropic (combinatorial) and enthalpic (residual) contributions to non-ideality are included. We previously reported the solubility of NH_3 in four ILs using the gravimetric method (1-butyl-3-methylimidazolium hexafluorophosphate, 1-butyl-3-methylimidazolium tetrafluoroborate, 1-ethyl-3-methylimidazolium bis(trifluoromethylsulfonyl)imide, and 1-ethyl-3-methylimidazolium 1,1,2,2-tetrafluoroethanesulfonate). Here, the Flory parameters (χ) of these four ILs and ammonia systems were regressed with the assumption that the χ parameter is dependent on temperature and independent of the concentration. The Flory-Huggins model with the combinatorial term only, and the extended Flory-Huggins model with both combinatorial and residual terms are compared. The results showed that the NH_3+IL systems have a strong negative deviation from the ideal solubility that cannot be solely described with entropic impacts, specifically at high ammonia concentrations in the ILs. This suggests an increase in the size of the IL molecule might improve the ammonia sorption capacity of an IL at low pressures. However, the ammonia sorption capacity could be significantly improved by considering the molecular interaction between ammonia and ILs (i.e., hydrogen bonding) in addition to the variation in size of the molecule.

601430 Predicting Hydrogen Bond Donating Ability between Nitroaniline Dyes and Type III DES Using Tzvp Based COSMO Models

Abstract:

Type III deep eutectic solvents (DES) are formed by hydrogen bond donors and acceptors (HBD and HBA respectively). These HBA's and HBD's form a liquid that has a lower melting point then either of the individual compounds. The main interaction that drives this depression in melting point is the hydrogen bonding between the HBD and the halogen cation of the HBA. In addition to this, the HBA itself may have hydrogen's able to participate in bonding; such as the primary alcohol in the common HBA choline chloride. These complex hydrogen bonding systems have been qualitatively described with methods such as pulse-field gradient nuclear magnetic resonance spectroscopy. This method does not necessarily describe the changes in hydrogen bonding potential difference between the individual HBA and HBD's and the resulting DES. Solvatochromic measurements are a quick experimental tool that can utilize dyes with UV-Vis spectra that change based on their environment. Dye pairs such as 4-nitroaniline and N,N-diethyl 4-nitroaniline can be used with Type III DES to measure hydrogen bonding acceptance capacity; so how saturated the DES is with hydrogen bonds.

In addition to experimental methods, computational modeling can attempt to predict the ability of a DES to form additional hydrogen bonds. COSMO (Conductor-like Screening Model) uses electron distribution calculations to determine thermophysical properties and phase properties such as liquid-liquid equilibrium of binary and ternary systems. One output of these electron density calculations for Type III DES is the surface potential of the molecules. This potential is directly related to the molecules ability to interact with potential bonding sources, such as hydrogens available for bonding. Knowing how different HBA's and HBD's interact will be useful for designing DES for target specific applications (such as liquid-liquid extraction or gas absorption) where hydrogen bonding may be desirable.

This study evaluates Type III DES's ability to form hydrogen bonds through both experimental measurement with solvatochromic dyes and computational prediction through COSMO calculations with a TZVP basis set. DES made from HBA (quaternary ammonium and polyphenol phosphonium salts) and HBD (glycol, glycerol, and carboxylic acids) were used with dyes of 4-nitroaniline, N,N-dimethyl 4-nitroaniline, and N,N-diethyl 4-nitroaniline to examine hydrogen bonding ability of the DES with dyes of different polarity and basicity. Electron distribution based modeling in COSMO was used to predict the relative effect of each component used to form DES in the solvent's ability to form hydrogen bonds with the selected dyes.

610647 Thermally Stable Ionic Liquid As Media for Separating Aliphatic and Aromatic Compounds

Abstract:

The process of separating aromatic and aliphatic hydrocarbons is energy intensive due to similar boiling points and formation of azeotropes. The use of ionic liquids (ILs) for these separations has the potential to lower these energy costs. We show that thermally-robust peraryl ILs especially, triphenyl-*p*-phenoxylphenylphosphonium bistriflimide [Ph_3P-*p*-$POPNTf_2$] have a high affinity towards aromatic compounds while remaining immiscible with aliphatic compounds. Furthermore, when mixed with excess aromatic the IL forms two distinct phases: an IL-phase saturated with aromatic and a nearly pure aromatic phase. The molecular-level interactions leading to this behavior, however, still remain unclear. Here, we use molecular dynamics (MD) simulations to provide a detailed molecular-level understanding of these systems. The fundamental interactions between ILs and model compounds such as *ortho-*, *meta-*, *para-* xylene, toluene, hexafluorobenzene, cyclohexane, naphthalene, ethylbenzene, pentylbenzene, bromobenzene, cyanobenzene and *n*-heptane are assessed. MD correctly predicts the formation of two phases and there is close quantitative agreement with the saturation limit of aromatics in the IL phase, while *n*-heptane, cyclohexane and pentylbenzene were completely immiscible with the ILs. The presence of aromatics in the ILs disrupt the IL network by forming cation-aromatic, anion-aromatic while weakening ion-ion repulsive interactions, meanwhile counter ions network remained intact. The change in IL-aromatic structure is further analyzed through radial/spatial distribution functions, nearest neighbors, pair-interaction energies between several atom pairs. The results showed formation of cage like structures (liquid clathrate) which are dynamic in nature. These efforts provide an improved insights into molecule-level understanding of the solvent environment provided by the cation and anion during separation and further help in tuning the ILs for efficient separation applications.

604062 Effects of Physical Structure of Alkylammonium Hydrogen Sulfate Ionic Liquids on Acidity

Abstract:

Alkylammonium [HSO_4] protic ionic liquids (ILs) are a promising class of ILs, shown to be effective in a range of applications such as biomass pretreatment, catalysis, chemical synthesis, and protein stabilisation [1]–[5]. Their main advantage is their extremely low cost ($0.78/kg [6]), due to a simple one-step synthesis of two cheap reagents. Previous studies on ILs have found that the chemical activity of the IL is dominated by the anion. Therefore, while a number of physical properties of hydrogen sulfate ILs have been investigated, they have tended to involve a small number of cations, with relatively little investigation into the effect of cation structure on chemical behaviour. Additionally, most applications have focused on the use of tertiary alkylammonium cations. Changing the physical structure of the cation could potentially further reduce costs, as well as widen the range of potential applications by modifying chemical behaviour. Four alkylammonium cations were screened in this study, with three different degrees of substitution (shown in Fig. 1). The acidity of the resulting IL solutions (containing 20% wt. water) was quantified using the Hammett acidity function.

Decreasing degree of alkylammonium substitution was found to significantly enhance solution acidity. The acidity of [MBA][HSO_4] was found to be equivalent to that of [DMBA][HSO_4] with 11% excess sulfuric acid, while that of [HA][HSO_4] and [BA][HSO_4] was found to be equivalent to that of [DMBA][HSO_4] with 31% and 32% excess acid respectively. Addition of up to 20% excess base to [BA][HSO_4] expectedly made the IL solution less acidic, although its Hammett acidity was still substantially lower than that of [DMBA][HSO_4], as well as lower than that of [MBA][HSO_4]. Such a strong trend in acidity was unexpected from aqueous acid-base theory, due to the narrow range of constituent amine pKa's and small differences in pH of heavily diluted IL solutions. This was attributed to increased solvation of primary and secondary alkylammonium cations, decreasing the solvation of the [HSO_4^-] anion and thus increasing its proton activity. The effect of solution density was found not to play a significant role. When adding excess sulfuric acid to [DMBA][HSO_4], the IL solution displayed a buffering effect at low excesses (1-5% wt.), showing a higher Hammett acidity than purely aqueous sulfuric acid.

Biomass pretreatment was identified as an application which would benefit from enhancement of alkylammonium [HSO_4] solution acidity. Pretreatments using primary and secondary alkylammonium cations were carried out and compared to those using [DMBA][HSO_4], a current gold-standard protic IL. Pretreatment with primary alkylammonium [HSO_4] ILs led to overly severe pretreatments with very accelerated kinetics, while pretreatment with [MBA] [HSO_4] ILs showed equally high performance to [DMBA][HSO_4] but with accelerated kinetics. Addition of up to 20% excess base to primary alkylammonium ILs had relatively little effect on pretreatment severity. Additional pretreatments were performed using [DMBA][HSO_4] with excess acid, matching the Hammett acidities of [HA][HSO_4] and [MBA][HSO_4]. Despite equal Hammett acidities, pretreatments using excess acid were significantly more severe than those using the primary and secondary alkylammonium HSO_4 ILs. The observed differences in pretreatment severity suggest different sources of acidity between the two types of solutions, which cannot be distinguished using the

Hammett acidity parameter. Pretreatment findings add credence to the theory that the acidity of the primary and secondary alkylammonium ILs is due to the increased activity of protons from lower solvation of the [HSO_4^-] anion, rather than increased proton dissociation in solution. In addition, the development of [MBA][HSO_4] as a promising pretreatment solvent highlights the possible improvements from screening of ion structure in ILs.

[1] F. J. V. Gschwend, C. L. Chambon, M. Biedka, A. Brandt-Talbot, P. S. Fennell, and J. P. Hallett, "Quantitative glucose release from softwood after pretreatment with low-cost ionic liquids," *Green Chem.*, vol. 21, no. 3, pp. 692–703, 2019.

[2] F. J. V. Gschwend, F. Malaret, S. Shinde, A. Brandt-Talbot, and J. P. Hallett, "Rapid pretreatment of Miscanthus using the low-cost ionic liquid triethylammonium hydrogen sulfate at elevated temperatures," *Green Chem.*, pp. 1–3, 2018.

[3] P. Attri and P. Venkatesu, "Exploring the thermal stability of α-chymotrypsin in protic ionic liquids," *Process Biochem.*, vol. 48, no. 3, pp. 462–470, Mar. 2013.

[4] E. T. Kermani, H. Khabazzadeh, and T. Jazinizadeh, "Friedländer synthesis of poly-substituted quinolines in the presence of triethylammonium hydrogen sulfate [Et3NH][HSO4] as a highly efficient, and cost effective acidic ionic liquid catalyst," *J. Heterocycl. Chem.*, vol. 48, no. 5, pp. 1192–1196, Sep. 2011.

[5] M. R. M. Shafiee, B. H. Najafabadi, and M. Ghashang, "Triethylammonium Hydrogen Sulfate as a Catalyst for the Preparation of 1,5-diaryl-3-(arylamino)-1 *H* -pyrrol-2(5 *H*)-one derivatives," *J. Chem. Res.*, vol. 35, no. 11, pp. 634–636, Nov. 2011.

[6] Baaqel et al, *Green Chemistry*, 2020 (in press)

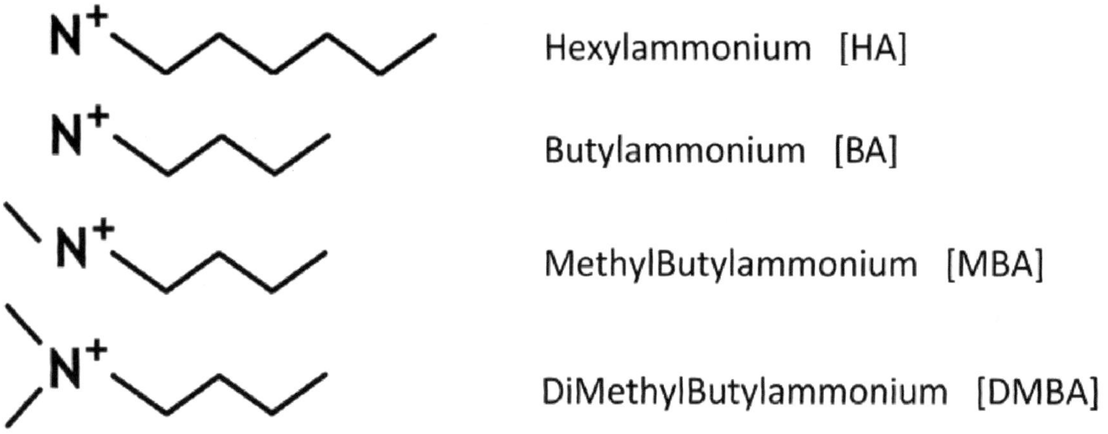

Figure 1: Cations screened, all using the [HSO_4^-] anion.

608640 Highly Selective Electrochemical Reduction of CO_2 to CO Via a Novel Catalyst with Mn-N_3 Site

Abstract:

Electrochemical CO_2 reduction reaction (CO_2RR) is one of the most promising ways to deal with the excess CO_2 and resolve the energy crisis [1]. However, due to the stability of CO_2 molecule and reaction involves multiple electrons and protons, poor selectivity and low current density are the main challenges of this filed [2]. So, efficient catalysts are in great need.

As the third abundance of transition metals in Earth's crust, manganese (Mn) has shown great potentials on many electrocatalytic fields. Nevertheless, the performance of Mn based catalysts is largely limited for CO_2RR. Herein, we firstly prepared Mn single-atom catalyst (SAC) with Mn-N_3 site supported by g-C_3N_4. The prepared catalyst exhibited a maximum CO FE of 98.8% with 14.0 mA cm^{-2} CO current density (j_{CO}) at overpotential of 0.44 V in $KHCO_3$ electrolyte, outperforming all reported Mn SACs and being one of the best reported electrocatalysts for CO_2 conversion CO. Moreover, a higher j_{CO} value of 29.7 mA cm^{-2} was obtained at overpotential of 0.62 V, when ionic liquid was used as electrolyte. In-situ experiment and DFT results confirmed that Mn-N_3 site is the active site, on which the free energy barrier of key intermediate formation is greatly reduced. It can be anticipated that Mn SACs may also exhibit excellent performance in other electrochemical reactions through changing supporting materials to form exclusive active site.

References

[1] Feng JQ, Gao HS, Feng JP, Liu L, Zeng SJ, Dong HF, Bai YG, Liu LC, Zhang XP. Morphology modulation-engineered flowerlike In_2S_3 via ionothermal method for efficient CO_2 Electroreduction. *ChemCat Chem*, 2020, 12: 926-931.

[2] Feng JP, Zeng SJ, Liu HZ, Feng JQ, Gao HS, Bai L, Dong HF, Zhang SJ, Zhang XP. Insights into carbon dioxide electroreduction in ionic liquids: Carbon dioxide activation and selectivity tailored by ionic microhabitat. *ChemSusChem*, 2018, 11: 3191-3197.

608885 Thermodynamic Studies/Analysis of NH_3 Absorption in Pyridinium-Based Protic Ionic Liquids

Abstract:

Ammonia (NH_3) is undoubtedly another major threat to environment and human health since it is considered as a main precursor of PM2.5 (Particulate matter ≤ 2.5 μm)[1]. Meanwhile, NH_3 is an important chemical raw material and widely applied in some fields, such as fertilizers, nitric acid, freezing medium and so on. Therefore, the recovery of ammonia from tail gas is of great significance for pollution control and resource utilization. Although the water scrubbing has been widely used in industry, it possesses some inherent drawbacks, such as high energy consumption, huge water consumption. Hence, it is extremely essential to develop new absorbents with high efficiency, reversibility and low energy consumption for NH_3 recovery.

In our previous research, we developed the hydroxyl-functionalized ionic liquids[2] and protic ionic liquids[3] for absorption NH_3, respectively. They all displayed effective absorption performance for NH_3 through the hydrogen bonds. In this work, a novel type of pyridinium-based ionic liquids (ILs) with acidic protons and hydroxyl groups were designed and synthesized for NH_3 absorption. The NH_3 solubility in five pyridinium-based protic ILs at temperatures from 303.15 to 343.15 K and pressures up to 600 kPa were measured using gas-liquid equilibrium method, and the thermodynamic properties including reaction enthalpy, entropy, Gibbs free energy were further obtained by a Reaction Equilibrium Thermodynamic Model (RETM)[4] to better understand the NH_3 absorption process. The results indicated that the PIL [4-MeOHPy][NTf_2] exhibits the maximum NH_3 solubility of 3.43 mol NH_3/mol IL at 313.15 K and atmospheric pressure, surpassing any nonmetallic ILs previously reported. Furthermore, the characteristics of isotherms under low pressures behaved an obvious chemical reaction between ILs and NH_3. This work provides an important reference for the thermodynamic properties of NH_3 absorption process in practical application.

612216 Aqueous Ionic Amines for CO2 Capture in Air Revitalization

Abstract:

Air revitalization in closed environments is a mission critical task for NASA, with the chief technical challenge being the removal of CO_2 from the recirculating air supply. Specifically, efficient removal of CO_2 from closed space craft environments is a NASA research and development priority as stated in the NASA Technology Roadmap Technical Area (TA): Human Health, Life Support, and Habitation Systems document. The practical significance of this requirement was recently illustrated in a NASA study that showed that 7% of the genes changed their expression during spaceflight, which was attributed to metabolic stress on the body due to elevated CO_2 exposure in the spacecraft.[1-2] Additionally, NASA is concerned with cognitive and behavior effects on humans exposed to increased levels of CO_2.[3]

On the international space station, CO_2 removal is currently accomplished using a thermally regenerable solid zeolite, Grace 13X. While effective, the solid zeolite has a number of logistical problems including the production of dust from the absorbent bed and sensitivity to ambient humidity. A similar separations challenge exists on submarines, where closed environment CO_2 removal is a crucial part of the life support system. On submarines, CO_2 is removed via chemical capture with an aqueous ethanolamine solution. This is also an effective process, but with significant disadvantages; chiefly, the volatilization of the amine, which creates a foul odor, can damage electronics and presents long term heath effects. Furthermore, regeneration of the solution (desorption of the CO_2) is somewhat energy intensive as the large thermal mass of the solution (often 70 mass % water) must be heated to promote desorption. While the energy requirement is moot on a nuclear submarine, power consumption is a key concern for spacecraft.

A related area of research is the removal of CO_2 from post-combustion flue gas streams, such as those from coal and natural gas-powered electrical generation plants. Although the goal in this application is also the removal from of CO_2 from ambient/near ambient pressure gas streams, the system temperature is somewhat higher, and the partial pressure of CO_2 is much higher than for air revitalization. However, the same technology, non-volatile aqueous amines, may present a solution for both challenges. The authors have worked with industrial partners to develop concentrated aqueous solutions of non-volatile (ionic) amines which have shown promise as viable post-combustion CO_2 capture agents. From this work, three joint process patents focused on CO_2-capture efficiency and minimizing regeneration energy requirements have been issued[4-6] in addition to the composition of matter patents already owned by South Alabama.[7-8] While the previous work was focused on optimizing the solutions for the industrial setting at higher temperatures and higher CO_2 partial pressure, the work described below focuses on the development and evaluation of these systems for used in closed environment air revitalization. Furthermore, as these are aqueous systems, they are inherently compatible with humid air and as hygroscopic salts, the potential exists for them to be used as part of an overall humidity control system.

The work presented here focuses on the thermophysical and thermodynamic properties of aqueous solutions of ionic amines for use in CO_2 capture for air revitalization, thermodynamic of cycling and thermal stability of the compounds. These properties are probed using a combination of DSC and SDT techniques along with data collected from a lab-scale absorber.

1. Drake, N., No, Scott Kelly's Year in Space Didn't Mutate His DNA. National Geographic 2018.
2. Witze, A., Astronaut twin study hints at stress of space travel. Nature 2017.
3. Stankovic, A.; Alexander, D.; Oman, C. M.; Schneiderman, J. A review of cognitive and behavioral effects of increased carbon dioxide exposure in humans; 0499-9320; Massachusetts Institute of Technology: 2016; pp 219277/1-219277/22.
4. Chinn, D.; Cooper, R. E.; He, Z.; Davis, J. H., Jr.; West, K. N.; Timken, H. K.; Driver, M. S. Aqueous solutions of amine functionalized ionic compounds for carbon capture processes. US20120171094A1, 2012.
5. Chinn, D.; Cooper, R. E.; He, Z.; Timken, H. H.; Driver, M. S.; Davis, J. H., Jr.; West, K. N. Process for the separation of carbon dioxide from flue gas. WO2012092176A2, 2012.
6. Cooper, R. E.; Chinn, D.; He, Z.; Davis, J. H.; West, K. N.; Timken, H. H.; Driver, M. S. Method for improving total energy demand in a post-combustion carbon capture process with ionic absorbent. WO2012092204A2, 2012.
7. Davis, J. H., Jr. Functionalized ionic liquids for removal of acid components from sour natural gas. WO2003086605A2, 2003.
8. Davis, J. H., Jr. Carbon dioxide scrubbing using ionic materials. WO2008122030A2, 2008.

602555 Separation of Hydrofluorocarbon Mixtures Using Ionic Liquids

Abstract:

Hydrofluorocarbons (HFCs) are organic compounds that are commonly used as refrigerants. HFCs were introduced as a replacement to chlorofluorocarbons (CFCs) in response to the Montreal Protocol due to their low Ozone Depletion Potential (ODP). Although HFCs have low ODP, many have a high Global Warming Potential (GWP) and are now being phased out as a result of the Kigali agreement and replaced with hydrofluoroolefins (HFOs). The phasing out of HFCs and HFC mixtures alone is leading to thousands of tons of refrigerants being unusable with the only options for disposal being to vent or incinerate them. It is important that new techniques are developed to sustainably separate and recycle HFC mixtures so they can be disposed of properly or repurposed in new, more environmentally friendly refrigerants such as HFO mixtures.

Project EARTH (Environmentally Applied Research Toward Hydrofluorocarbons) is a collaborative project between multiple universities and industry companies that is focused on developing efficient and selective separation processes and recycling methods for HFC mixtures. The separation of HFC mixtures is difficult due to the fact that many of the mixtures are azeotropic or near azeotropic. For this reason, Project EARTH is looking into multiple approaches to this challenge, one of which being the use of ionic liquids (ILs). We believe that IL's properties (no measurable vapor pressure, dissolution of many organic and inorganic compounds, variable solubility of gases and liquids, and high thermal, chemical, and electrochemical stability) make them uniquely suited to creating selective separation processes for various HFC mixtures.

This presentation will cover results on the use of various ILs for the separation of refrigerant mixture R-410a which is a 50/50 mixture by weight of R-125 (CHF_2CH_3) and R-32 (CH_2F_2).

608807 Functionalized Ionic Liquids Based on Vegetable Oils for Rare Earth Elements Recovery

Abstract:

Rare earth elements (La-Lu, Y and Sc) and their compounds are critical materials in computers, audio systems, electric and hybrid vehicles, cell phones, wind turbines, magnetic resonance imaging machines, et al. To extract rare earth metals from aqueous solution is an extremely important topic in the world today and is anticipated to be of even greater future importance due to the increasing demand for these critical metals. Unfortunately, producing rare earths is a problem of economical extraction and always requires using large volumes of environmentally harmful chemicals. In this work, functionalized ionic liquids (FILs) based on vegetable oils have been directly synthesized and used to effectively extract rare earth elements. Carbon dioxide gas is introduced to successfully strip rare-earth-loaded FILs in the presence of water. The novel extraction process reveals great advantages of accessibility, biocompatibility and sustainability as well as cost efficiency.

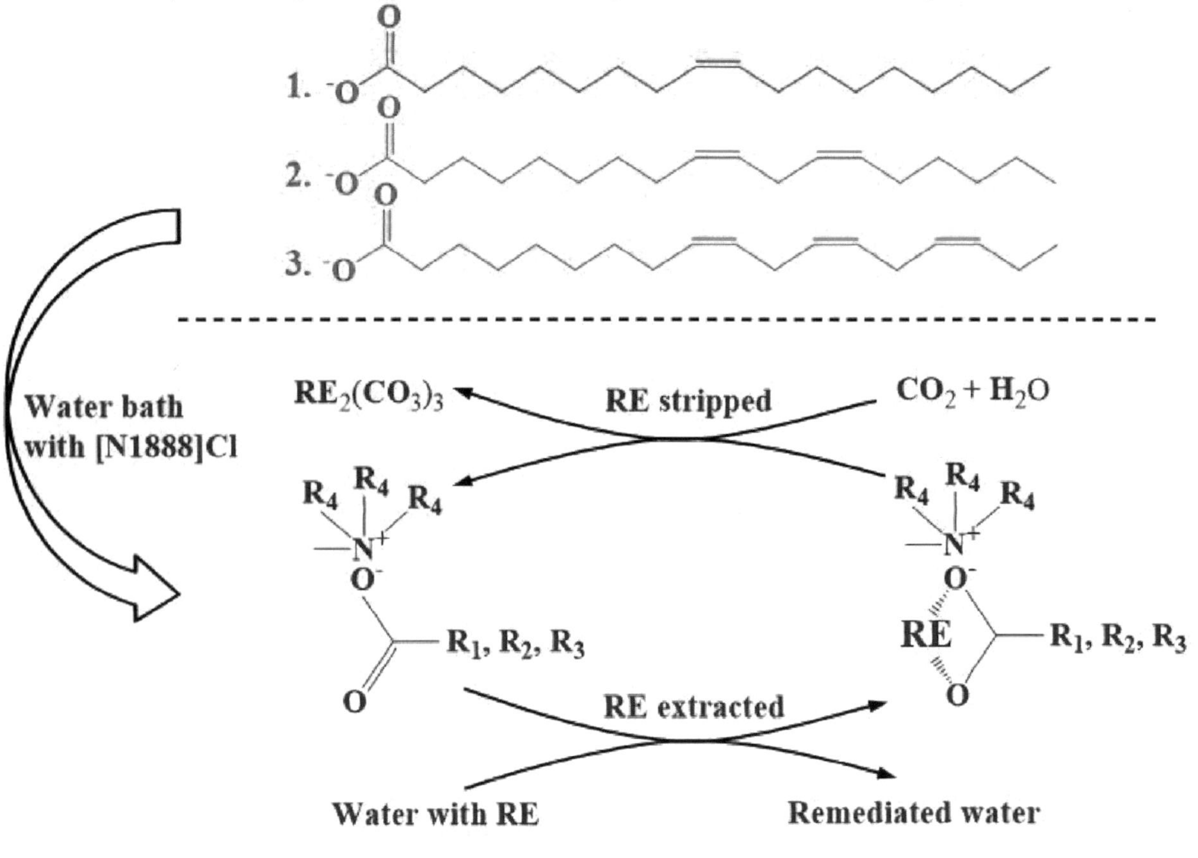

618370 Sponsored Technology Workshop - Digital Transformation for a Sustainable Future - AVEVA

Abstract:

AVEVA is proud to host a live workshop discussing how the current digital transformation companies in the Energy sector are undergoing today can embrace and directly influence a sustainable future. This will be an interactive session with Ian Willetts, VP Process and Simulation and Julien de Beer, VP Engineering Portfolio who will share examples of how today's digital twin technologies are helping Energy companies drive for a lower carbon future. This interactive session will give all attendees the ability to share and discuss different ways that companies are working toward new environmental goals while adapting to the current business climate where the majority of employees are working remotely due to the on-going global pandemic.

599348 Scalable Modular Manufacturing of Amino Acid Crystals with Improved Uniformity

Abstract:

The pharmaceutical and chemical industry has increased demands for reproducible and efficient green manufacturing of crystalline products. For many key ingredients of these products (e.g., amino acids), uniformity and consistency in quality across different scales/stages is a must, and a huge challenge for existing batch processes.[1–8] This talk focuses on the slug-flow (or segmented flow) crystallizer design, which is one of the effective continuous processes demonstrated to allow enhanced control of organic crystal properties such as size and shape[2–8]. The slurry flow is combined with an air flow and fed to a tube to induce a multiphase hydrodynamic instability that spontaneously generates well-mixed slugs where the crystals continue to grow. Experimental validation confirms that the proposed designs reduce development time by orders of magnitude while suppressing secondary nucleation, attrition, and clogging/fouling—dominant but undesired phenomena that worsen the ability to quality control.

References

(1) Sun, Z.; Ya, N.; Adams, R. C.; Fang, F. S. Particle Size Specifications for Solid Oral Dosage Forms: A Regulatory Perspective. *Am. Pharm. Rev.* **2010**, *13* (4).

(2) Jiang, M.; Zhu, Z.; Jimenez, E.; Papageorgiou, C. D.; Waetzig, J.; Hardy, A.; Langston, M.; Braatz, R. D. Continuous-Flow Tubular Crystallization in Slugs Spontaneously Induced by Hydrodynamics. *Cryst. Growth Des.* **2014**, *14* (2), 851–860.

(3) Eder, R. J. P.; Schrank, S.; Besenhard, M. O.; Roblegg, E.; Gruber-Woelfler, H.; Khinast, J. G. Continuous Sonocrystallization of Acetylsalicylic Acid (ASA): Control of Crystal Size. *Cryst. Growth Des.* **2012**, *12* (10), 4733–4738.

(4) Jiang, M.; Papageorgiou, C. D.; Waetzig, J.; Hardy, A.; Langston, M.; Braatz, R. D. Indirect Ultrasonication in Continuous Slug-Flow Crystallization. *Cryst. Growth Des.* **2015**, *15* (5), 2486–2492.

(5) Su, M.; Gao, Y. Air-Liquid Segmented Continuous Crystallization Process Optimization of the Flow Field, Growth Rate, and Size Distribution of Crystals. *Ind. Eng. Chem. Res.* **2018**, *57* (10), 3781–3791.

(6) Daniel Scott, C.; Labes, R.; Depardieu, M.; Battilocchio, C.; Davidson, M. G.; Ley, S. V.; Wilson, C. C.; Robertson, K. Integrated Plug Flow Synthesis and Crystallisation of Pyrazinamide. *React. Chem. Eng.* **2018**, *3* (5), 631–634.

(7) Lu, J.; Litster, J. D.; Nagy, Z. K. Nucleation Studies of Active Pharmaceutical Ingredients in an Air-Segmented Microfluidic Drop-Based Crystallizer. *Cryst. Growth Des.* **2015**, *15* (8), 3645–3651.

(8) Mou, M.; Jiang, M. Fast Continuous Non-Seeded Cooling Crystallization of Glycine in Slug Flow: Pure α-Form Crystals with Narrow Size Distribution. *J. Pharm. Innov.* **2020**. https://doi.org/10.1007/s12247-020-09438-0.

605142 On Demand Treatment of Wastewater Using 3D-Printed Membranes

Abstract:

Wastewater treatment is critical to the chemical industry. While conventional membranes have been utilized in oil-water separation for some time, demonstration of a 3D-printed membrane with well-controlled local structures, which renders it to have multi-selectivity, is still lacking to-date. Moreover, state-of-the-art wastewater treatments often involve multiple steps. Consequently, a more intensified process, which is enabled by a single multi-selective membrane, is highly desirable. In the current study, we present the first-of-its-kind demonstration of multicomponent and multiphase oil-water separation using 3D-printed membranes and an intensified process design. The driving force here is surface selectivity and topography rather than pressure. First, two-component oil-water separation with high efficiency, high flux, and excellent durability is achieved using 3D-printed polymethylmethacrylate (PMMA) membranes with zwitterionic hydrogel coating. The membranes are simultaneously hydrophilic/oleophobic and mechanically strong. Second, robust on-demand separations of immiscible liquid mixtures with high flux are enabled by a novel separation membrane with microscale repeating re-entrant structures. The re-entrant membrane is fabricated by two-photon polymerization (2PP) 3D-printing, and the re-entrant angle can be precisely manipulated to achieve on-demand separations. Third, novel supported ionic liquid membranes (SILM) have been developed to separate miscible liquid mixtures, which has been a key challenge in wastewater treatment using membranes. Ionic liquid (IL), which serves as the extraction solvent, is impregnated into PVDF membranes. The SILM is optimized with curvatures to have larger interfacial area and higher separation rate. Lastly, the successful separation of the three-component/two-phase liquid mixture (benzene/heptane/water) has been demonstrated with an integrated 3D-printed box. Efficient separation of water from benzene/heptane is achieved using the 3D-printed PMMA membranes with zwitterionic hydrogel coating at the bottom of the box, while the separation between benzene and heptane is achieved using the SILM on the sides of the box.

606466 Three Dimensional Printing (3dP) Fixed-Bed Unit Operations: Future Manufacturing Techniques

Abstract:

Unit operations are single, basic steps in a chemical process. Typically, these are units are physical separations, such as distillation, or chemical transformation, such as isomerization. Recently, there has been a large focus on 3d-printing (3dP) unit operations. Despite the advances and recent growth in adoption of this technique, most research has focused on manufacturing reaction containers and microflow devices, such as the 'lab on a chip.' While these research avenues take advantage of the on-demand manufacturing and custom design capabilities of 3dP techniques, they fail to utilize one of the most important factors setting apart additive manufacturing from other methods: complex internal geometries. Leveraging the ability to 3d-print complex internal geometries, a variety of new unit operations can be tailored according to need. Here, we present a modifiable platform for use in multi-phase processing as a fixed-bed unit operation. By changing the manufacturing parameters, the platform can be tailored on-demand for specific use. We describe a customizeable column that can act as a fixed bed catalyst, phase dispersion device in continuous emulsification or foaming, and as a separator in filter or de-emulsification. In reacting systems, the surface area is, maximized and the internal geometry is chosen to facilitate solid-liquid/gas contact. The surface can either be decorated with catalyst or fabricated using catalyst-impregnated filament. For use in continuous phase dispersion there is a balance between capillary extrusion and wall shear, therefore the porosity is balanced with an internal geometry to promote both avenues of emulsification and foaming. In phase separation applications, careful choice of filament material, porosity, and internal geometry allow the column to be transformed from a continuous emulsification unit to a continuous de-emulsification unit. The adaptability and on-demand nature of 3dP is shown through modification of a base design to encompass continuous heterogeneous reactions, phase dispersion, and phase separation.

609334 3D Printed Polymer/Nanoparticle Layered Structures

Abstract:

3D printing has been applied in automobile, aerospace, biomedical, pharmaceutical, and plastics recycling field. 3D printing has an advantage over traditional manufacturing in processing composite materials because of their capability in managing multi-materials at multi-scales. More importantly, 3D printing can process solids (e.g., filaments and powders), liquids (e.g., inks and solutions), and gels (e.g., polymer or blend networks). One challenge in additive manufacturing that remains to be solved so far is the printing resolution at nanoscale. Our research is seeking innovation to add one layer of nanoparticles on polymer surfaces for layered structures with unique properties. In this project, the layer-by-layer assembly was merged on the 3D printing platform. The first layer of polymers was controlled with precise dimensions and morphologies, depending on which the nanoparticles were organized with long-range orders and controlled orientations. Simple materials of UV-light-sensitive monomers and carbon nanofibers (CNFs) were used for demonstrating the feasibility of composite processing. The CNFs were found to anchor on specific surface features (i.e., rough surface with comparable nanoparticle sizes). The prepared polymer/nanoparticle composites were used as environmental and health-monitoring sensors for liquid and gas chemicals. It was found that the prepared nanocomposites can serve as chemiresistors that function precisely as responses to chemical solvents, volatile organic compounds, and mechanical strains. The high sensitivity and selectivity of these sensors from our 3D printing shed light on rapid prototyping of electrical, mechanical, and other devices with complex structures.

611251 Electron Tunneling and Mechanical Compression of Aligned 3D Printed Structures of Graphene and MoS2/BN Gel

Abstract:

Designing 3D printed micro-architectures using electronic materials with well-designed electronic structures will potentially lead to accessible device fabrication for 'on-demand' applications. Here we show gelation of six different concentration of graphene/MoS2 (100:0, 10:90,30:70, 50:50, 70:30, 90:10) and Graphene/BN (50:50) alloys. Rheology studies show the viscoelastic properties of the gels, in addition to providing shield yield stress. The shear-thinning behavior provides the ability of fabrication of 3D printed devices in an aligned controlled nozzle-extrusion based modified 3D printer, enabling the fabrication of electrodes for analyzing electron-tunneling barrier width between conductive graphene-centers and studying the resultant mechanical properties. The 3D printed devices were solidified by removing water through lyophilization. The ordered graphitic region with sp^2 hybridized carbon atoms in the graphene sheets is in the order of 8.28 ±0.23 nm. The temperature-dependent electronic transport 3D printed electrodes structures exhibited a transport-barrier of 16 meV and a tunneling width of 0.56 nm (Fowler Nordheim electron tunneling) with graphene centers having a carrier concentration of $2.63457*10^{12}/cm^2$. By modifying the graphene/MoS2 concentration, we were able to control the Young's Modulus of the structure between 0.6 and 1.6 MPa. We envision that the proposed 3D-printing of gels of nanomaterial will lead to an evolution in the design of next-generation of 'on-demand' printed electronic and electromechanical devices.

610408 Xpcs Studies of a Cellulose Nanocrystal Thermoset Ink during Extrusion Printing

Abstract:

Cellulose nanocrystal (CNC)-epoxy composites have received a great deal of attention as light-weight, high strength materials due to the lower cost, biodegradability, and high aspect ratio of the filler particles. The particles have low density (~1.6 g/cm^3) and are quickly processed; however, the utilization of CNCs in more common non-polar polymers is still limited to their hydrophilicity and poor dispersibility. Therefore, to utilize such particles as mechanical reinforcements in, for example, epoxy resins, the hydrophilicity of the particle would need to be adjusted. The bulk properties of such composites depend on (i) the properties and shape of the nanofiller, (ii) the dispersion of the nanofiller in the matrix (microstructure), (iii) the orientation of the nanofillers inside the matrix, and (iv) the manufacturing process of the composite. Higher alignment has resulted in enhanced strength and conductivity in several instances. However, a key component that is required and is currently missing in the literature is a clear understanding of the underlying structure-dynamics of the particles, its recovery after processing, and how this influences macroscopic properties, such as tensile strength. This would help in developing novel design rules for processing.

In this work, using state of the art scattering techniques at the National Synchrotron Light Source at the Brookhaven National Lab (BNL), we address the dispersion of surface-modified CNC in a commonly used epoxy resin (EPON 828) and explore processing such composites through direct-ink write (DIW) printing with a DIW printer installed on the beamline. X-ray photon correlation spectroscopy (XPCS), coupled with shear rheology studies, is utilized to study the morphology and dynamics before, during, and after printing of a naturally hydrophilic CNC particle in a mostly hydrophobic thermosetting epoxy resin. The hydrophobicity of the particles was varied over three different concentrations, allowing for a systematic increase of grafting on the surface of the particles. Each level of surface modification was dispersed into the EPON resin at loadings up to 40 wt%. The rheological measurements revealed a clear distinction in macroscopic properties (viscosity, viscoelasticity, recovery) due to changes in the particle-particle and particle-matrix interactions amongst the different levels of modification. Additionally, results from the XPCS studies show drastic changes in the dynamics of the particles as a function of surface modification during the recovery of the material after printing. Results from the scattering studies are then correlated with the macroscopic properties (tensile strength) after printing to delineate the underlying structure-function relationships. Our in-operando results provide insight into the importance of understanding the microstructure properties of such materials to allow for the development of enhanced polymeric nanocomposites.

610454 2D-Stationary Computational Printing of Cement-Based Pastes.

Abstract:

The 3D printing of cement-based materials requires two primary paste characteristics: (1) extrudability, (2) resistance to deformation under layering load. To study the range of properties that effect printing outcomes, a computational strategy has been developed that captures the dynamics of 2D slices at a stationary location along the print path. It is well known that the rheology of cement paste is time-dependent and coupled to both the physical and chemical (hydration) behavior of cement particles. Thus, 3D printing of cement pastes is related to rheological (flow) characteristics of the material, e.g. yield stress, plastic viscosity, and time-dependent effects. To target optimal paste design parameters and correlate rheological properties to printability, this work introduces an efficient new computational printing strategy for predicting the flow behavior of cement pastes. A rheometer and a mini conical slum flow test were used to experimentally establish fluid properties and to calibrate rheological models prior to conducting computational experiments. The relative importance of rheological properties such as yield stress, and plastic viscosity were quantified and correlated with printed hollow cylinder benchmark objects. The layer height of printed objects was compared to 2D-stationary computational printing outcomes. Finally, based on the relationship of experimentally determined rheological measures of printing pastes to computational printing constructs, optimal rheological properties were suggested for scaling purposes.

607634 Liquid Metal Microparticle-Based Polymer Nanocomposites and Their 3D Printing

Abstract:

Introduction: Polymer nanocomposites (PNCs) offer exciting opportunities by synergistically combining the favorable properties of their polymer matrices and nanoparticle fillers. Particularly conductive PNCs consisting of mechanically compliant polymers and electrically conductive nanoparticles are emerging material systems in applications such as flexible, wearable electronics [1] and solid-state energy storage [2]. Additive manufacturing (3D printing) methods such as direct-ink-writing, involving layer-by-layer deposition of "inks" derived from these materials in the form of micro-scale filaments is recently emerging as a promising processing method for these materials systems. This approach carries the potential to control constituent morphology in each filament [3] which drives the functional properties of the fabricated structures [4], provided that the ink rheology and associated deposition mechanisms are well understood. This presentation focuses on the rheological characteristics and 3D printing of PNC systems featuring graphene flakes and/or novel liquid metal micro/nano particles in a polyethylene oxide (PEO) matrix.

Materials and Methods: The graphene flakes are of 8 nm thickness and have lateral size of approximately 5 mm. The PEO matrix is a blend of two different molecular weight polymer: primarily 100kDa with trace amounts of 5MDa to tune/vary the viscoelastic properties of the composite [5], [6]. The liquid metal microparticles are obtained by sonication of eutectic Gallium Indium (EGaIn) inside the ink solvent (water or acetonitrile) to obtain an average spherical particle diameter of 1.5 μm [6]. These particles consists of a liquid core encapsulated in an elastic solid skin of gallium oxides [7] and thus overall exhibit a deformable behavior unlike the conventionally rigid nanofillers used in PNCs. Several different compositions of PNCs including both or only one type of fillers were prepared and rheologically characterized to understand their (1) shear thinning behavior through cone and plate rheometry, (2) transient extensional viscosity variation and relaxation profiles under constant rate uniaxial deformation. Same inks are 3D printed using a custom direct-ink-writing system using positive pressure dispensing through cylindrical and conical nozzles with 200 μm diameter opening. In these experiments, lines were printed using various inks at several flow rates, printing speeds and nozzle-to-substrate distance levels. Experiments were imaged using high magnification microscopes to examine the filament geometry for various inks under same process parameters. Printed lines were measured using white light interferometry to characterize their geometry. Finally electrical conductivity of the lines were characterized to study the combined influence of PNC composition and printing process parameters on the electrical conductivity.

Results: The rheological characterization showed that increasing EGaIn microparticle content in the PNC inks led to reduction in the shear viscosity while increasing viscoelasticity, unlike the graphene flakes, where increased content led to higher shear viscosity and reduced viscoelasticity [6]. These rheological changes led to liquid metal microparticle loaded PNCs to exhibit continuous filamentary deposition at lower flowrates, higher printing speeds and higher nozzle-to-substrate distances. Under the increased extensional flow, these filaments experience visco-capillary thinning [8] to form line diameters smaller then nozzle diameters . Studies on the electrical conductivity of the printed patterns showed that PNCs including EGaIn nanoparticles only did not for conductive patterns, however, when introduced as a secondary filler in graphene-based PNCs they improved electrical conductivity. Increased extensional flow in the deposited filaments generally reduced the electrical conductivity. Printing through cylindrical nozzles also led to increased electrical conductivity as compared to conical nozzles likely as a result of the increased shear stresses aligning the graphene flakes in the printed structures to form conductive pathways.

Discussion: The influence of liquid metal particles on the PNC rheology contradict that of the rigid nanoparticles [9], due to the unique deformable nature of these particles. Particularly, these particles can stretch and deform under shear and extensional stresses during flow, manifesting in increased viscoelasticity during extensional flow and reduced viscosity during shear flow. The capability of liquid metal particle loaded filaments to experience higher strains under extensional flow allowed thinner filaments to be deposited at higher printing speeds, effectively increasing the process resolution and throughput. However, it can be postulated that the under the uniaxial stresses, the average distance between the conductive particles reduce leading to reduced conductivity. Lack of conductivity in PNCs only including EGaIn particles is due to their non-conductive oxide skins [10]. These particles increasing electrical conductivity in graphene-based PNCs is possibly due to two reasons: (1) the EGaIn particles can form conductive interfaces with graphene flakes, thus acting as conformable anchors between these particles which can conduct electricity regardless of relative orientation of these particles, (2) the rheological changes induced by the liquid metal particles reduces the energy required for the graphene particles to align and form conductive pathways, particularly under shear flow. These results showed that EGaIn particles can be used as rheological modifiers in PNCs along with other conductive nanofillers as they can counter the effects induced by the rigid fillers. Furthermore, these particles can enhance electrical conductivity in such PNCs. Through 3D printing, unique properties of the liquid metal particle-based PNCs can be utilized to achieve high resolution printing and filament-by-filament control of the electrical conductivity.

References:

[1] J. T. Muth *et al.*, "Embedded 3D Printing of Strain Sensors within Highly Stretchable Elastomers," *Adv. Mater.*, vol. 26, no. 36, pp. 6307–631, Jun. 2014.

[2] M. Wei, F. Zhang, W. Wang, P. Alexandridis, C. Zhou, and G. Wu, "3D direct writing fabrication of electrodes for electrochemical storage devices," *J. Power Sources*, vol. 354, pp. 134–147, 2017.

[3] J. P. Lewicki *et al.*, "3D-Printing of Meso-structurally Ordered Carbon Fiber/Polymer Composites with Unprecedented Orthotropic Physical Properties," *Sci. Rep.*, vol. 7, no. January, p. 43401, 2017.

[4] S. Colonna, O. Monticelli, J. Gomez, C. Novara, G. Saracco, and A. Fina, "Effect of morphology and defectiveness of graphene-related materials on the electrical and thermal conductivity of their polymer nanocomposites," *Polym. (United Kingdom)*, vol. 102, pp. 292–300, 2016.

[5] S. Nesaei, M. Rock, Y. Wang, M. R. Kessler, and B. A. Gozen, "Additive Manufacturing With Conductive , Viscoelastic Polymer Composites : Direct-Ink- Writing of Electrolytic and Anodic Poly (Ethylene Oxide) Composites," *J. Manuf. Sci. E-T ASME*, vol. 139, no. November 2017, pp. 1–12, 2019.

[6] S. Nesaei, D. J. Cavanagh, and A. Gozen, "Rheology of liquid metal particle-based polymer composites : A comparative study," *J. Rheol.*, vol. 63, no. 4, pp. 559–568, 2019.

[7] T. R. Lear, S.-H. Hyun, J. W. Boley, E. L. White, D. H. Thompson, and R. K. Kramer, "Liquid metal particle popping: Macroscale to nanoscale," *Extrem. Mech. Lett.*, 2017.

[8] H. Yuk and X. Zhao, "A New 3D Printing Strategy by Harnessing Deformation, Instability, and Fracture of Viscoelastic Inks," *Adv. Mater.*, vol. 30, no. 6, pp. 1–8, 2018.

[9] J. Abraham, Rene Muller, S. Thomas, S. C. George, and N. Kalarikkal, *RHEOLOGY AND PROCESSING OF POLYMER NANOCOMPOSITES*. 2016.

[10] M. G. Mohammed and R. Kramer, "All-Printed Flexible and Stretchable Electronics," *Adv. Mater.*, p. 1604965, 2017.

608433 3D-Printing of Self-Healing Polymer Composites

- Download/View Extended Abstract

Abstract:

3D printed components typically show poor mechanical strength when compared to parts fabricated by traditional methods, leading to increased potential for material damage during fabrication and use. Additionally, many of the polymer material options for 3D printing are not intrinsically recyclable, requiring disposal if they suffer any damage or degradation which increases environmental waste through premature end of their useful life. Thus, we aim to improve the lifetimes of 3D printed polymeric objects through the incorporation of self-healing properties. Inspired by biological self-healing, in which a damage event triggers an autonomic healing response, microcapsules containing healing agents can be embedded into a host material. During a damage event these microcapsules rupture, release the healing agent and heal the surrounding material by polymerization, entanglement or cross-linking. 3D printable self-healing polymer composites will increase product shelf life, and thereby improve material sustainability and long-term cost. Double shell wall polyurethane/poly (urea-formaldehyde) microcapsules are synthesized by in-situ-interfacial polymerization. Microcapsules with both solvent and monomer core fluids are prepared to investigate both solvent-healing and monomer self-healing mechanisms. Microcapsules containing healing agents are either incorporated into the host polymer matrix or are coated onto 3D printing polymer filaments to create 3D printed objects capable of self-healing. Microcapsule survivability and self-healing properties of these composite materials after 3D printing are evaluated via examination of the healing efficiency and mechanical strength of the fabricated 3D printed objects.

609215 Dual-Layer Filament for Material Extrusion Additive Manufacturing

Abstract:

Presence of carbon based and/or inorganic fillers in 3D filaments has recently attracted considerable interest from academic and industry to enhance properties of 3D-printed parts. Although 3D-tailored composites have been developed, very little work has been done on the production of advanced 3D filament feedstock for FDM; particularly for highly filled 3D filaments that require (i) high filament strength, (ii) high dimensional accuracy and (iii) superior surface finish. Current FDM filaments rarely exceed filler concentration of 10%, for example, in case of calcium phosphate without sacrificing aforementioned qualities. In this work, a co-axial 3D feedstock filament up to 40% filler concentration has been fabricated via melt spinning; such filament met the desired properties with uniform filler dispersion. We gain further insights on rheological behaviors of such system by performing multiscale flow simulation. On a continuum scale, 2D FEM flow simulations have been conducted to investigate the interfacial instabilities in co-axial flows of dissimilar material. On a mesoscale, we utilize coarse-grained molecular dynamics simulations to predict and investigate the multilayer structure-properties by investigating the effects of i) core/shell layer thickness ratio, ii) filler geometry and concentration, and iii) strain rate. Experimentally, the dual-layer filament with high precision has been fabricated 1.75 +/- 0.02 mm) with an excellent surface finish; its printability and mechanical properties of the printed parts are compared to those of a single layer filament.

607093 Development of a Novel Bioprocess for CO_2 Conversion to Biochemicals Using Synthetic Designed Proteins and Non-Growing Cells

Abstract:

We are developing a new bioprocess for the production of biofuels from CO_2 using renewable energy. Autotrophic organisms frequently grow slowly, rely on the Calvin cycle for carbon fixation, and can be challenging to engineer for high yields and productivities. We are working to address these challenges by replacing the Calvin cycle with artificial photosynthetic proteins and designed metabolic pathways that will operate in non-growing bacteria cells suspended in photobioreactors. Renewable electricity will be used to power LEDs that specifically energize the artificial photosynthetic pathway in the stationary cells. This will allow for the efficient conversion of CO_2 to biofuels and biochemicals such as glycerol.

We have developed a novel designed light harvesting complex that can split water and regenerate the ubiquitous biological cofactor NADH. At the same time, we have developed a synthetic metabolic pathway that utilizes NADH to reduce CO_2 to formaldehyde and then the formaldehyde is ligated to create glycerol using another designed enzyme (formolase). Once installed in E. coli, the pathways will allow for light-powered NAD(H) regeneration which will power CO_2 reduction to glycerol that is independent of the cellular metabolism. Thus the system will be operated in photobioreactors which will specifically energize the designed photosynthetic pathway in cells that are not growing.

The most recent results of this promising new platform will be presented, including performance metrics for the newly designed photoprotein, pathway fluxes through the carbon fixation pathway, and protein engineering results on key enzymes in the carbon fixation pathway.

603157 Partial Oxidation of Methane to Value-Added Products over Iron Oxide Nanocatalysts

Abstract:

Methane, the primary component of natural gas, is a promising energy source, but billions of cubic meters are flared and wasted every year in the U.S. alone. Researchers have been working towards a process that can utilize the energy stored in methane and eliminate flaring by converting the methane on-site to a value-added product, such as methanol, for the past several decades with limited success. However, recent breakthroughs bring us closer to realizing a process fit for practical CH_4 conversion. Here, we report the development of a scalable, electrochemical device that has been designed to incorporate a gas-diffusion electrode (GDE) at the interface between the methane gas and the aqueous electrolyte. The GDE is necessary to address the transport limitations of the diffusion of nonpolar methane to the electrode surface and the low solubility of methane in an aqueous electrolyte. The device is set up with an *in-situ* Differential Electrochemical Mass Spectrometer (DEMS) to monitor product selectivity as a function of gas composition, applied potential, and temperature. This device utilizes an earth-abundant iron oxide nanosheet catalyst with active sites that resemble the biological enzyme methane monooxygenase (MMO), which converts methane to methanol in nature. We report Faradaic efficiency and product selectivity of methane partial oxidation while co-flowing methane and oxygen. Several strategies to desorb methoxy intermediates will be discussed, which prevents overoxidation to CO_2. Additionally, a basic electrolyte was chosen to sequester any produced CO_2 as carbonate, which mitigates the CO_2 emission. The use of a biomimetic catalyst on a GDE in our electrochemical device with *in-situ* DEMS has provided critical steps towards an efficient process of partial methane oxidation to value-added chemicals.

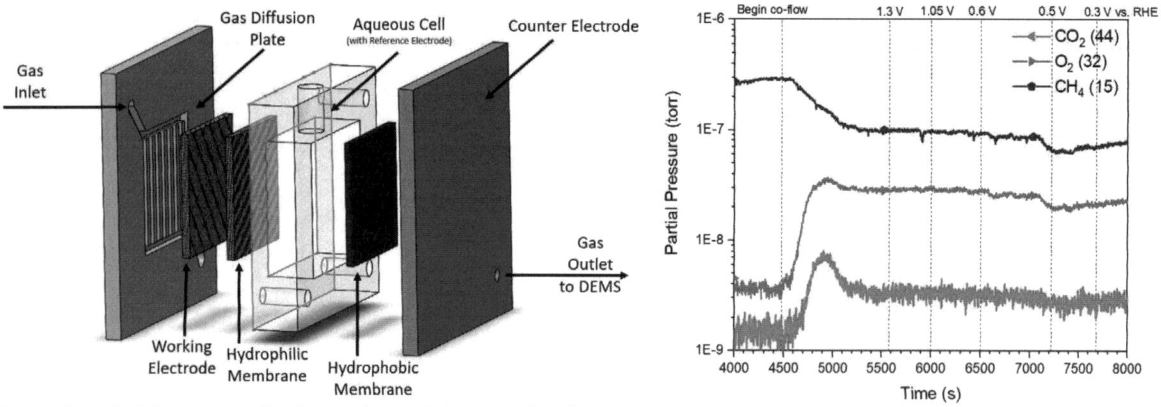

Figure 1: (left) Schematic of the electrochemical device with a GDE, (right) example DEMS data from the electrochemical device.

610896 Comprehensive Process Modeling for Converting Natural Gas to Lower Olefin Via Reforming and Fischer-Tropsch Process

Abstract:

Synthesis gas or syngas, mixture of hydrogen (H_2) and carbon monoxide (CO), generated from the upgrading of traditional chemical feedstocks such as natural gas, coal and biomass, is utilized industrially for the production of value-added chemicals which include chemical intermediates (olefins, alcohols), fuels and synthetic wax. Syngas conversion to products is commonly practiced via the Fischer-Tropsch synthesis (FTS) process, where metal catalysts (e.g., solid iron- and cobalt-based catalysts) are contacted with syngas at elevated temperatures and pressures to produce hydrocarbons (e.g., olefins and paraffins) and/or oxygenated organic products. However, these processes also coproduce C1 byproducts which include carbon dioxide (CO_2) and methane (CH_4). Furthermore, catalysts/processes for Fischer-Tropsch may pose disadvantages due to low conversion/activity resulting in high concentrations of unconverted syngas in the product mixture. In this presentation, we discuss scenarios to maximize the yield of olefins from the syngas conversion reaction and analyze consequent carbon efficiency and energy consumption.

A process model is developed in process flowsheeting software for lower olefin production based on natural gas as a feedstock for syngas production via methane reforming. Steam methane reforming and dry reforming are considered for this study. The heat balance for reformer is carried out using a fuel-fired furnace. Syngas is fed to an olefin synthesis reactor; excess H_2, if any, is separated and can be used as a fuel. The product stream from the syngas conversion/olefin synthesis reactor, composed of unreacted CO and H_2, CH_4, CO_2, and heavier hydrocarbon products is subjected to a series of separation steps including H_2 membrane separation, CO_2 separation units, and a sequence of distillation column including de-methanizer, de-ethanizer, C2-splitter, de-propanizer, C3 splitter, etc. to produce final olefin products. The unconverted reactants can be recycled to appropriate units in the system. A heat and mass balance is performed and several scenarios are evaluated based on catalyst performance and recycle options.

The results show that conversion in olefin synthesis reactor has a profound effect on recovery section. The low boiling-point components (CO + H_2) results in an aggressive cryogenic process and/or high pressure systems to recover lower olefins (C2–C4) from the product mixture where temperature as low as -200 ^{0}C have to be implemented for separation of olefins. Yield vectors are generated based on a series of catalyst/s and results of these sensitivity analysis suggest that separation can be achieved at higher temperature. Further, recycle options to the reformer were evaluated. Although the recycle results in higher overall yield of lower olefins, this also results in higher load on the furnace and results in lower energy efficiency and thus, similar overall carbon efficiency.

An integrated process model including reformer, reformer furnace, olefin synthesis reactor, recovery units is developed to evaluate carbon efficiency and energy consumption for natural gas to lower olefin production via syngas conversion. This work demonstrates process modeling capabilities that enables comparison between catalysts that have contrasting yields and selectivity, and provides direction for future development efforts.

601267 Simultaneous Activation of Methane (CH_4) and Nitrogen (N_2) in a Microwave-Enhanced Catalytic Reaction: A Combined DFT Modelling and Experimental Study

Abstract:

In this study, we demonstrate a microwave-assisted catalytic reaction that can activate CH_4 and N_2 simultaneously to produce ammonia (NH_3) and C_2 products in a single step process, represented schematically in Figure 1. This approach integrates microwave irradiation with novel heterogeneous catalysts that can activate CH_4 and N_2 molecules at atmospheric pressure and moderate surface temperatures. This new reaction scheme aims to utilize CH_4 as a direct H_2 source to produce NH_3 and valuable C_2 products (ethylene and acetylene) thus offering two major advantages: (1) potential to build small-scale ammonia production plants which meet the capacity of non-conventional energy sources, (2) direct utilization of stranded shale gas eliminating H_2 production cost in NH_3 synthesis.

DFT and MKM simulations can be a very useful tool in elucidating heterogeneous catalyst behavior, especially for a reaction as complex as the one demonstrated in this work. This report investigates the feasibility of such a complex reaction using plane wave density function theory (DFT) and microkinetic modelling (MKM). The production trends and the catalyst behaviors are rationalized using these computational models and conformed experimentally. Further, an in-situ catalyst regeneration method is proposed based on the nature of the coke deposited on the catalyst surface. Lastly, the coked catalyst is regenerated in the microwave reactor to examine catalyst stability and reproducibility.

The co-production of ammonia, ethane, ethylene and acetylene demonstrated the activation of two highly stable molecules, CH_4 and N_2. The results from density functional calculations and micro-kinetics modelling predicted the reaction behavior at temperatures ranging from 573 K to 973 K and atmospheric pressure. The significant influences of terrace (111) and step (211) sites on K promoted Ru crystals were established in product formation and catalyst coking. NH_3 formation occurs predominantly on the step (211) sites while C_2 product formation can occur on both the sites. It was found that the elementary reaction steps leading to NH_3 synthesis occurs due to the non-thermal effects of microwave irradiation while C_2 products formed simultaneously are due to the thermal effects only. A reaction mechanism is postulated on the catalyst coking and validated experimentally by in-situ regeneration using H_2 and O_2.

In conclusion, this first of a kind reaction demonstrated in this study establishes the possibility of building a single-step small scale reaction system that has the potential of utilizing stranded natural gas resources in direct manner to produce value-added chemicals. The small-scale atmospheric pressure ammonia synthesis can not only lower the high energy requirements of the conventional production plants but also promote the use of the low-capacity and intermittent, renewable energy resources

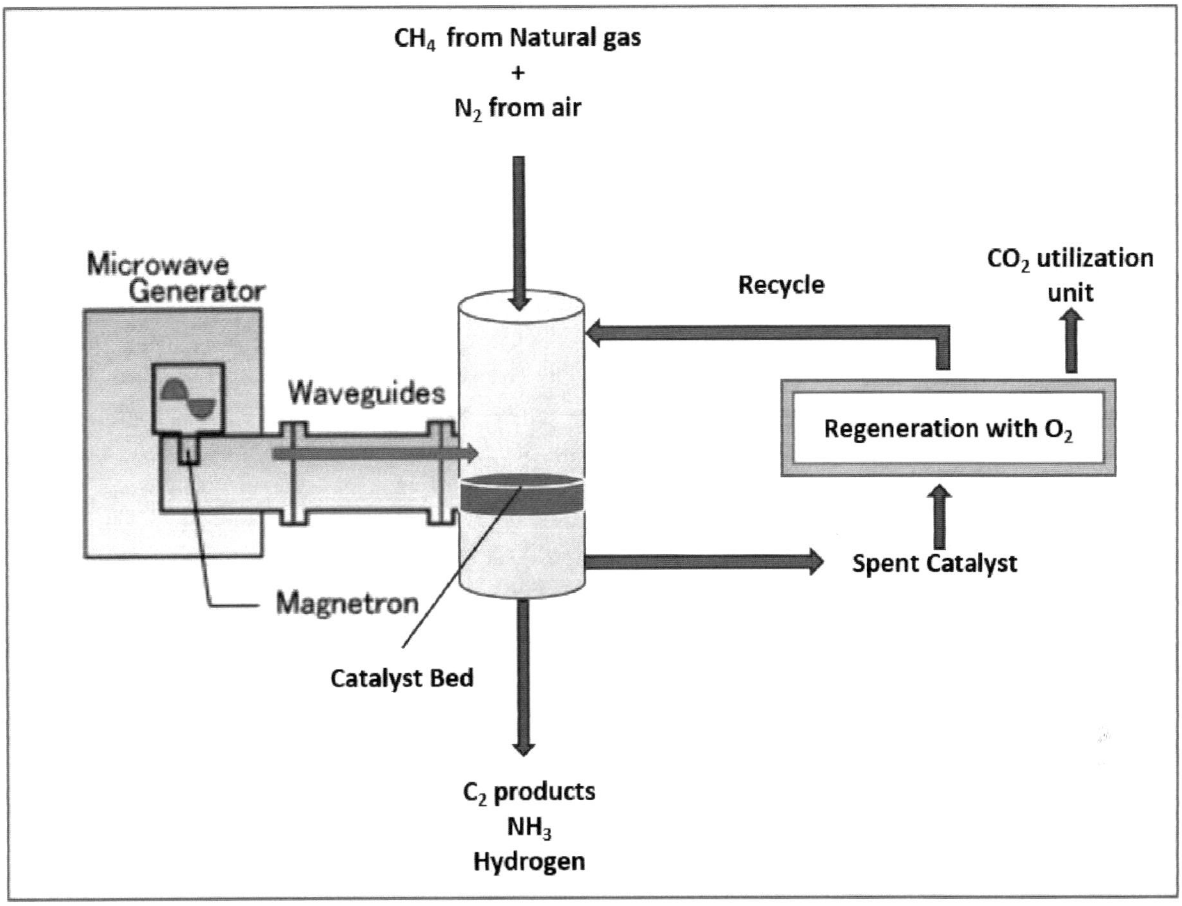

Figure 1: Proposed Process scheme for single step utilization of methane and nitrogen in a microwave catalytic reaction system

597349 CH_4 Transformation to Fuels By Photocatlysis: Rational Design of Catalysts

Abstract:

Methane hydrate/shale gas transformation is a potentially efficient pathway for both high value chemical production and renewable energy storage, thus has been attracting substantial interest over the last decades. However CH_4 activation is energy intensive and kinetically very challenging so that methane activation is regarded as the "holy grail" in the catalytically chemical process. Photocatalysis provides a cost efficient potential to activation of such small molecule at very mild conditions, while to achieve the potential is a huge challenge.[1]

Stimulated by our recent research outcomes on the charge dynamics in inorganic semiconductor photocatalysis, which reveal that the low reaction efficiency is due to both fast charge recombination and large bandgap of an inorganic semiconductor [1,2], together with the recent findings on atomic catalysis [3], we developed novel material strategies for photocatalytic methane conversion to methanol.

Highly dispersed atomic level iron species immobilised on a TiO_2 photocatalyst show an excellent activity for methane conversion, resulting into ~97% selectivity towards alcohols operated under ambient conditions by a one-step chemical process [4]. Such photocatalyst is also very stable, promising an attractive industrial process of methane upgrade. The dominating function of the iron species has also been investigated in detail.

1. Wang, H. Suzuki, J. Xie, O. Tomita, D. J. Martin, M. Higashi, D. Kong, R. Abe and J. Tang, *Chem. Rev.,* **2018**, 118, 5201-5241.
2. Tang, J. R. Durrant and D. R Klug, *J. Am. Chem. Soc.,* **2008,** *130*(42) 13885-13891.
3. Wang, J. Li and T. Zhang. *Nat. Rev. Chem.,* **2018,** 2, 65-81.
4. Xie, R. Jin, A. Li, Y. Bi,... G. Sankar, D. Ma and J. Tang, *Nature Catalysis,* **2018,** 1, 889-896.

612158 Big Data-Based Fault Detection with Advanced Analytics in the Pharma Industry

Abstract:

Motivated by market pressure to reduce product costs, regulatory incentive to modernise manufacturing and a growing environmental concern to reduce waste has led the pharmaceutical industry to improve its manufacturing process. The impetus to adapt and modernise the manufacturing process can largely be attributed to the promotion of the Quality by Design paradigm (QbD) and guidance on Process Analytical Technology (PAT) by the FDA in 2004. The promotion of these methods coincided with a time where the cost of computing and data storage had declined, but also with developments in non-invasive online sensor technology, such as Raman and Near-Infrared (NIR) spectroscopy. This culminated into an increase intro the application of multivariate analysis (MVA) and specifically Multivariate Statistical Process Control (MSPC).

MSPC exploits the Big Data generated from a complex chemical process such as pharmaceutical production and can be used for process understanding, troubleshooting and on-line fault detection. MSPC tools include multivariant process charts and soft sensors and can be combined with APC methods to improve overall plant efficiency. MSPC aims to use big data to better understand a process, to develop a definition of normal operation and then use that to detect faults on-line. There are three main categories to classify as SPC methods (1) model-based (2) knowledge-based (3) data-based. Data-based methods are often the preferred method for pharmaceutical manufacturing. The advantages of data-driven methods are that they do not require first principals understanding of the process or a unit as they construct a model using input and output data. The most popular data-driven methods are multivariate latent/MVA methods (which include PCA, PLS and their adapted versions). These methods project data into the latent space and work as a feature extraction method and reduce the dataset into a smaller or manageable set that can be used in further applications of fault detection.

This paper presents an industrial case study which focused on fault detection towards operational improvements for a key API production, pursued in collaboration with a major multinational UK pharmaceutical company (GSK). The aim of this project is to exploit data availability towards constructing and validating a data-driven model of a specific API production unit of the GSK plant (Montrose, Scotland, UK). The project involves using large data sets of process data (temperatures, partial pressures, flow rates, and stream compositions, from an on-site GCMS analysis station), and implementing advanced MVA methods in MATLAB, in order to diagnose operational patterns and recommend feasible improvements towards reliable API production intensification.

LITERATURE REFERENCES

[1] E. Tomba et al., "General procedure to aid the development of continuous pharmaceutical processes using multivariate statistical modeling-An industrial case study," Int. J. Pharm., vol. 444, no. 1–2, pp. 25–39, 2013.

[2] T. Kourti, "Application of latent variable methods to process control and multivariate statistical process control in industry," Int. J. Adapt. Control Signal Process., vol. 19, no. 4, pp. 213–246, 2005.

[3] J. F. MacGregor and T. Kourti, "Statistical process control of multivariate processes," Control Eng. Pract., vol. 3, no. 3, pp. 403–414, 1995.

[4] T. Kourti, “Multivariate dynamic data modeling for analysis and statistical process control of batch processes, start-ups and grade transitions,” J. Chemom., vol. 17, no. 1, pp. 93–109, 2003.

[5] J. F. MacGregor, P. Nomikos, and T. Kourti, “Multivariate Statistical Process Control of Batch Processes Using PCA and PLS,” IFAC Proc. Vol., vol. 27, no. 2, pp. 523–528, 1994.

[6] S. Garcia-Munoz and D. Settell, “Application of multivariate latent variable modeling to pilot-scale spray drying monitoring and fault detection: Monitoring with fundamental knowledge,” Comput. Chem. Eng., vol. 33, no. 12, pp. 2106–2110, 2009.

[7] M. Boiret, D. N. Rutledge, N. Gorretta, Y. M. Ginot, and J. M. Roger, “Application of independent component analysis on Raman images of a pharmaceutical drug product: Pure spectra determination and spatial distribution of constituents,” J. Pharm. Biomed. Anal., vol. 90, pp. 78–84, 2014.

598331 Recent Advances in Kinetic Parameter Estimation Toolkit (KIPET) with Spectra

Abstract:

Developed over the past four years, the Kinetic Parameter Estimation Toolkit (KIPET), is an open-source toolbox for the determination of kinetic parameters from a variety of experimental datasets including spectra and concentrations. *KIPET* seeks to overcome limitations of standard parameter estimation packages by applying a unified optimization framework based on maximum likelihood principles and large-scale nonlinear programming strategies for solving estimation problems with nonlinear differential algebraic equations (DAEs). The software package includes tools for data preprocessing, determination of parameter confidence levels for a variety of problem types, and informative wavelength selection to improve the lack of fit (Schenk et al., 2020). All these features have been implemented in Python with the algebraic modeling package *Pyomo*. *KIPET* exploits the flexibility of *Pyomo t*o formulate and discretize the dynamic optimization problems that arise in the parameter estimation algorithms. Solutions of these optimization problems are obtained with the nonlinear solver *IPOPT* and confidence intervals are obtained through the use of either *sIPOPT* or a newly developed tool, *k_aug.*

In this talk, we discuss a recently enhanced *KIPET* package (Short et al., 2020) that considers multiple experiments with potentially different reactants and kinetic models, different dataset sizes with shared or unshared individual species' spectra, leading to fast parameter estimation and confidence intervals based on the NLP sensitivities. In addition, we present a new variance estimation technique based on maximum likelihood derivations for unknown covariances from two sample populations. This approach leads to a straightforward deconvolution of variances between noise in model variables and in spectral measurements. Moreover, we discuss a new estimability analysis approach that systematically determines a ranked subset of kinetic parameters with well-defined confidence intervals (Chen and Biegler, 2020). With nonlinear kinetic models and limited measurements, it is often difficult to correctly estimate all the parameters, due to linear dependence and low correlation among the parameters. A common approach is to estimate a subset of the parameters by fixing the others at reasonable (often literature) values. However, it is often challenging to determine which parameters can be properly estimated. In this talk we present an efficient approach that ranks the estimable parameters, and discards those that cannot be estimated accurately. Based on reduced the reduced Hessian information with simple Gauss-Jordan elimination, the proposed approach leads to fast parameter selection and estimation within a simultaneous collocation framework. This approach becomes much more efficient for large problems than competing approaches based on multiple eigenvalue decompositions (Quaiser and Mönnigmann, 2009). Several case studies with increasing complexity are presented to demonstrate the performance of this proposed approach.

1. W. Chen, L. Biegler, "Reduced Hessian Based Parameter Selection and Estimation with Simultaneous Collocation Approach," submitted for publication (2020)
2. Schenk, M. Short, J. S. Rodriguez, D. Thierry, L. T. Biegler, S. Garcia-Munoz, W. Chen, "Introducing KIPET: A novel open-source software package for kinetic parameter estimation from experimental datasets including spectra," Computers and Chemical Engineering, 134(4), 106716 (2020)
3. Short, L. T. Biegler, S. Garcia-Munoz, W. Chen, "Estimating Variances and Kinetic Parameters from Spectra Across Multiple Datasets Using KIPET," Chemometrics and Intelligent Laboratory Systems, to appear (2020)
4. Quaiser, T., Mönnigmann, M. "Systematic identifiability testing for unambiguous mechanistic modeling-application to JAK-STAT, MAP kinase, and NF-κB signaling pathway models," BMC Systems Biology, 3,50 (2009).

601039 Teaching Computers to Interpret Mdsc Thermograms

Abstract:

We present a novel algorithm for the automatic data analysis of modulated Differential Scanning Calorimetry of thermograms designed to minimize subject matter expert input reduce the man reliability.

Differential Scanning Calorimetry (DSC) and mainly modulated DSC (mDSC) analysis are widely used to support the production of amorphous solid dispersions for the identification of kinetic and thermodynamic events such as glass transition, crystallization and melting. However, interpretation of thermograms require careful analysis of subject matter experts (SMEs) which can be time-consuming and thus can be a bottleneck to, e.g., formulation development where large numbers of samples are generated are require prompt analysis for reiteration. Furthermore, often there are elusive phenomena that can sometimes go unnoticed or cause dubious interpretation from different SMEs.

The presented Python-based software tool's computation of the derivatives and the maximums, minimums and inflections of each curve, is the base of the software. It then uses this information to search for events, based on their magnitude. One essential advantage of using an automated software is the ability to use the derivatives of the heat flow curves to improve the precision of the results. The derivatives are essential not only to the detect the exact temperature at which events occur but also to help define their onsets. In order to tackle the inherent noise of the DSC analysis and data acquisition, the proposed tool employs a low pass, Fourier transform-based filter, which enables the use of the thermogram derivatives.

We show case studies where the software analysis increases the detection rate of events that can easily be unnoticed by visual assessment due to their magnitude or shape, as they only cause very subtle variations. We also present case studies where smaller events are distorted or confounded with a nearby dominant events. However, following the digital detection of these events a more detailed visual assessment confirms their presence and often indicative of the onset of undesirable events such as phase separation or crystallization. Furthermore, the presented digital analysis tool promotes a faster and objective analysis of generated DSC thermogram data.

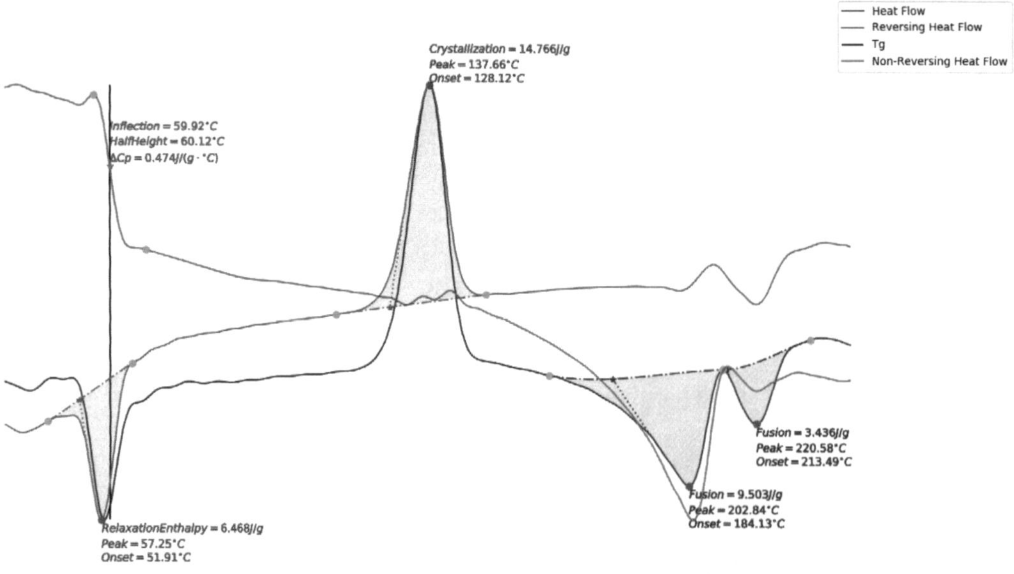

595673 Data-Smart Machine Learning Methods for Predicting Youngs Modulus of Directly Compressed Blends of Pharmaceutical Powders

Abstract:

The ability to predict the mechanical properties of compacted powder blends of Active Pharmaceutical Ingredients (API) and excipients solely from those of the individual components can reduce the amount of "trial-and-error" involved in formulation design. Machine Learning (ML) models can reduce model development time and effort with the imperative of adequate historical data. This work evaluates linear and non-linear ML for predicting the Youngs Modulus of directly compressed arbitrary blends of known excipients and API from readily available information about the API. The training data obtained from three BCS Class I APIs and four excipients demonstrate "data-smart" strategies to train ML models efficiently. The results indicate that even simple linear ML model provide reasonable prediction accuracy. The practical benefits of this method and how it compares with other mechanistic models are discussed. Finally, we demonstrate an application of the model to enable Quality-by-Design in drug product development.

593356 Advances and New Developments on Eiot for PAT Applications in Pharmaceutical Development and Manufacturing

Abstract:

The EIOT method[1] was proposed as an extension to the IOT[2] method as an alternative approach to extracting information from spectroscopic data into interpretable mass fractions of the resolved chemical species in the powder mixture.

The method in essence seeks to use Beer-Lambert's law as the deterministic functional relationship between spectra and mass fractions. The main difference with IOT is that EIOT does not assume the spectrum of the pure components is a measurable quantity; but a quantity that can only be estimated from measurements of known composition. The EIOT method proposes to decompose the measured spectra as the summation of the chemical signal plus the effect of the non-chemical interferences. The chemical signal is given by the summation of the apparent pure spectrum weighted by the mass fraction of each specie; the non-chemical interferences as the summation of the non-chemical interference signals weighed by their respective strength. The method relies on optimization techniques to estimate the apparent pure spectra and the signal of the non-chemical interferences; and when the method is used in real-time, to estimate the mass fractions and the strengths of the non-chemical interferences in each new sample.

Since the method was published, a number of enhancements have been done to the method, this talk presents such enhancements that i) the consideration of supervised signatures and ii) the consideration of exclusivity in a set of signatures (i.e. only one in a set is allowed to contribute). Details around the maintenance and long-term ownership of the EIOT method are discussed for its potential inclusion as a central component for a control strategy in drug product manufacture.

EIOT (as well as its predecessor IOT) bring a disruptive change in the practice of chemometrics; retiring the use of fixed regression models, to the use of real-time optimization and computer decision-making instead. As such, the requirements and expectations for use and long-term maintenance might have to be revised and adapted.

[1] Z. Shi, J. Hermiller, S.G.J.A.J. Muñoz, Estimation of mass-based composition in powder mixtures using Extended Iterative Optimization Technology (EIOT), (2018).

[2] K. Muteki, D.O. Blackwood, B. Maranzano, Y. Zhou, Y.A. Liu, K.R. Leeman, G.L. Reid, Mixture component prediction using iterative optimization technology (calibration-free/minimum approach), Industrial & Engineering Chemistry Research, 52 (2013) 12258-12268.

607697 New Carbon and Zeolite Structured Adsorbents Made from Expanded PTFE

Abstract:

The overall objective of this work is to develop novel structured adsorbents from expanded PTFE (e-PTFE) comprising activated carbon, carbon molecular sieve and zeolite adsorbents within the PTFE framework. So far, structured columns have been fabricated from 5A zeolite, Shirasagi molecular sieve carbon (MSC)-3R Type 162 and BPL activated carbon. A nearly identical column has also been fabricated with the corresponding commercial beaded, granular or pelleted counterparts for testing under identical conditions. Details about the fabrication technique will be discussed including converting sorbent-PTFE composite tapes into spiral wound adsorber columns with minimal bypass of the gas mixture input stream. Also, representative results from four different experimental tests will be presented. These experimental tests include single gas adsorption breakthrough curve experiments for CO_2 in N_2, volumetric equilibrium adsorption isotherm experiments for column adsorbent mass determination with N_2, 1-bed pressure swing adsorption (PSA) cycling experiments with CO_2 in N_2, single gas 10 Hz or 100 Hz volumetric frequency response experiments with CO_2, N_2 and possibly other gases, and column axial pressure drop experiments with air.

609404 3D-Printed Activated Carbon with Controlled Porosity for Removal of Toxic Chemicals from Contaminated Water

Abstract:

Removal of toxic chemicals from water is an attractive subject now, as the growing population and effects of climate change have rendered access to potable water more important than ever before.[1] Atrazine (ATZ) and carbamazepine (CBZ) are among the most frequently detected and top 10 contaminates in the water system.[1] In this study, we focused on removing ATZ and CBZ from water sources using a continuous flow system. To implement activated carbon into our continuous flow process, we developed 3D-printed monoliths which were optimized for aqueous stability. 3D printing was selected for monolith production because it allows for fine tuning of the monolith geometry, and physio/chemical/mechanical properties of monoliths.[2–4] To formulate these monoliths, we used bentonite clay as a binder and polyvinyl alcohol (PVA) as a co-binder. The 3D printing technology provides the ability to build activated carbon monoliths with various thicknesses and porosity in a fast and smooth operation. The 3D-prinetd carbon monolith precursors were pyrolyzed in nitrogen environment and then characterized. To study competitive adsorption, both dynamic and steady-state adsorption experiments were conducted by using different ratios of ATZ and CBZ. The effective diffusion coefficient of the 3D-prinetd carbon monolith adsorbates were increased with decease in monoliths internal patterns. The adsorption of both compounds onto powder and 3D-printed monoliths in dilute aqueous solutions (1-10 mg/L) were investigated for 6 h time-on-stream. All samples were analyzed by HPLC/ultraviolet equipped with a C-18 column. From these experiments, it was revealed that the monoliths were able to adsorb 27% of ATZ and 24% of CBZ from the system.

References

[1] M.J. Benotti, R.A. Trenholm, B.J. Vanderford, J.C. Holady, B.D. Stanford, S.A. Snyder, Environ. Sci. Technol. 43 (2009) 597–603.

[2] F. Magzoub, X. Li, J. Al-Darwish, F. Rezaei, A.A. Rownaghi, Appl. Catal. B Environ. 245 (2019) 486–495.

[3] S. Lawson, B. Adebayo, C. Robinson, Q. Al-Naddaf, A.A. Rownaghi, F. Rezaei, , Chem. Eng. Sci. 218 (2020) 115564.

[4] X. Li, F. Rezaei, A.A. Rownaghi, React. Chem. Eng. 3 (2018) 733–746.

597419 Novel Polymeric Fiber Sorbents Embedded with Porous Organic Cages

Abstract:

Focus on the synthesis and functionalization of porous organic cages (POCs) for separations have quickly grown over the last decade. However, to fully realize the potential of solid-phase POCs, they must be effectively packed into ordered structures for wide-scale use and implementation. Processes less destructive than pelletization are preferred and this necessitates alternate processing methods. This work shows the first fiber sorbent embedded with POCs, namely CC3R embedded throughout a cellulose acetate (CA) polymer matrix. CC3R was found to be stable after exposure to spinning solvents as confirmed by NMR, PXRD, and N_2 physisorption. CC3R-CA fibers were spun using the dry-jet wet-quench spinning method. Spun fibers retained the adsorptive properties of CC3R as confirmed by CO_2 and N_2 physisorption, reaching upwards of 60 wt% adsorbent loading. Addition of PVP led to significant leaching so it and other pore formers were excluded from the polymer dope. Fibers show slightly higher than expected CO_2 sorption at room temperature, believed to be due to a change in cage packing driven by non-solvent phase inversion. Fixed bed breakthrough experiments performed on both fibers and powder samples allowed for direct comparison of the sorption performance of the sorbent in different structures. The development of fiber sorbents embedded with POCs and displaying their adsorptive properties will help realize greater use of POCs as sorbents in other separation processes.

595319 Modeling and Optimization of Structured Contactor Adsorption Processes Based on Breakthrough Experiments

Abstract:

A structured adsorbent module for a gas separation process has been proposed with significant potential in terms of both its design diversity and operational flexibility. To characterize the module designs as for a given gas separation process, the pre-existing process model is modified with an additional parameter representing the structure designs. Based on our breakthrough experiments, the parameter, mass, and heat transfer coefficients are estimated in two different designs and varying gas flowrates. The model and parameters from this characterization are employed to search the best gas separation performances in different contactors and operations of a cyclic adsorption process using the structured contactors.

609013 CO_2 Capture Using 3D Printed PEI Adsorbents Supported By Carbon Nanostructures

Abstract:

Processes that utilizes solid adsorbents to capture CO_2 are promising alternatives to state-of-art technologies using water-based amine solutions as absorbents for capturing CO_2 from large point sources such as power plants. Although the energy needs of solid sorbent-based processes are low, the process footprint and consequently the capital cost connected to its implementation can be large due to the relatively long cycle times needed to get the required capture rate and purity of the CO_2 product. To overcome this challenge, processes having structured adsorbents like laminates, monoliths etc. are needed owing to their low pressure drop and better mass transfer characteristics[1]. In recent times, 3D printing of structured adsorbents has been gaining significant attention due the versatility in shapes of the adsorbents and good control over the channel dimensions[2].

Even though there are several published studies in the context of structured adsorbents, they are mostly restricted to characterization of the adsorbents with respect to equilibrium and mass transfer. Very few published studies exist on a process level demonstrating an improvement in the process performance in comparison with pellets[3-4]. It has been shown that, the true potential of any adsorbent only be obtained through rigorous optimization of a cyclic adsorption process. Secondly, most of the published studies in the context of adsorption-based carbon capture have neglected the presence of water in the flue gas. Supported amine sorbents are currently being studied as alternatives to zeolites due their moisture tolerating abilities but like 3D printed adsorbents, their performance in an actual adsorption process is not well understood[5].

Therefore, the aim of this work is evaluating the performance of a 3D printed adsorbent in the context of a vacuum swing adsorption process for CO_2 capture from a wet flue gas. In this work a series of novel adsorbents were prepared combining polyethylene-imine (PEI) with various nanomaterials such as multi-walled carbon nanotubes (CNT) in different ratios (Figure 1).These adsorbents were then printed in the form of monoliths by the 3D printing technique called robocasting. Breakthrough experiments were then carried out with a dry mixture of CO_2 and N_2 and a humid mixture containing 5% H_2O. Pressure drop tests were also carried out estimate the pressure drop across the 3D printed structure. Further, the breakthrough experiments were also used to estimate the adsorption kinetics. The information from the adsorbent characterization was then used to simulate and optimize a vacuum swing adsorption process in order to identify the operating conditions corresponding to 95% CO_2 purity and 90% CO_2 capture rate targets to capture CO_2 from a coal fired power plant containing 15% CO_2, 5% H_2O in N_2.

Keywords: Structured adsorbents, 3D printing, Vacuum Swing adsorption

References

1. Rezaei et al., *Eng. Chem. Res.* **2010,** *49,* 4832–4841
2. Couck et al., *Micropor Mesopor Mater 255 (2018) 185-191*
3. Nina Mohammadi, PhD thesis, 2017

4. Sharma et al., *Eng. Chem. Res.* **2020,** https://doi.org/10.1021/acs.iecr.9b05337
5. Gelles et al., *Adsorption* **2020,** 26:5–50

600726 Heat Intensification of TSA Processes with Packed Metal Foams: An Experimental and Modeling Study Applied to Post-Combustion CO_2 Capture

Abstract:

When performing a gas separation by Temperature Swing Adsorption (TSA), two key factors to increase process efficiency are the acceleration of the temperature swing kinetics, and the reduction of thermal gradients along the reactor cross section. To both aims, it is necessary to overcome the intrinsic heat transfer limitations associated with the operation of a packed bed, both in static and in flow conditions. Following an approach already suggested in the context of catalytic processes [1], it is possible to enhance heat transfer by introducing highly conductive open-cell metal foams filled with sorbent beads. Namely, the faster heat conduction through the metal adds onto the conventional heat conduction of the bed and the convection of the gas phase. This speeds up the reactor response to a change in temperature of the heating/cooling medium, and flattens temperature profiles transversally to the flow direction.

In the proposed configuration, the metal foam is placed in the reactor first. The void cells of the foam are then filled with the sorbent pellets. Keeping the average diameter of the particles less than half of the average diameters of the foam cells, filling can be simply performed by gravity. According to experimental measurements, this configuration also allows to house solid inventories (per unit reactor volume) much larger than those obtained with typical washcoated structured supports, and almost comparable to the ones of packed bed reactors. With this contribution, we propose the – to the best of our knowledge – first application of the packed metal foams as an alternative to the standard fixed bed configuration in adsorption based separations.

In particular, in this study, a post-combustion CO_2 capture process application has been investigated. In the context of CO_2 capture from flue gas, TSA processes are a promising alternative to the more established amine-wash[2]. Differently from absorption-based separations, a TSA cycle is not operated as a steady state process, but rather as a sequence of steps undergone by the gas-solid contactors. The simplest possible cycle configuration consists of: 1. an adsorption step during which the flue gas is fed, performed at low temperature to maximize CO_2 intake; 2. a desorption step, i.e. a heating ramp with the consequent release of the adsorbate, where the high-purity CO_2 stream is produced; 3 a final cooling step, which brings the system back to its initial state. The cooling step is performed in static conditions (i.e., both column ends are closed), whereas the other steps are typically performed at low velocities, well under the fluidization threshold, to contain the pressure decay.

In order to observe temperature gradients in both the axial and the radial direction within the reactor, a jacketed large cross-section (80 mm diameter and 200 mm length) cylindrical adsorption column has been adopted as experimental setup. It has been filled packing spherical zeolite 13X pellets (average diameter 1.8 mm) into an aluminum foam with a void fraction of 0.95 and a cell size of 5 mm. For this arrangement a total sorbent volumetric fraction of 0.50 has been measured, which represents a 25% reduction of the solid inventory compared to a standard packed column (at fixed reactor volume). A first series of steady-state and dynamic heat transfer tests has been completed by flowing inert gas (Argon) at different velocities through the column, and analyzing axial and radial temperature profiles across the packed foam. A detailed mathematical model for the description of the system dynamics in two spatial dimensions (axial and radial) has been validated by means of the experimental results.

Moreover, a second series of experiments has reproduced the operation of a basic TSA cycle (adsorption, heated regeneration, closed cooling) for the separation of a cry binary 12/88%v. CO_2/N_2 mixture. The cycle have been performed both in a packed bed and in a packed foam configuration. Pressure and temperature profiles within the column, as well as the composition of the outlet gas flow, have been monitored. The aforementioned reactor model has been extended to account for adsorption, also in view of its implementation into a process simulation tool, which could enable a systematic optimization of the process design. The final model has also been validated by means of the experimental results, for both packed bed and packed foam configurations.

Thus, in conclusion, experimental and modeling results demonstrate that heat transfer rates are increased by up to a factor two, as conjectured at the beginning of the study. This translates into a flatter temperature distribution along the reactor radius, and a faster transition toward the steady state when switching between heated and cooled steps.

As regards the specific application for CO_2 capture from flue gas, simulations have allowed, on the one hand, to quantify the productivity benefits obtained by cutting off the duration of the regeneration and cooling steps by a factor of three, and, on the other hand, to assess the disadvantages related to the lower packing densities, which limit the total adsorption capacity per unit volume.

References

[1] Visconti C.G., Groppi G. and Tronconi E., *Catal. Today.*, 2016, 273,178-186

[2] Joss L., Gazzani M., and Mazzotti M., *Chem. Eng. Sci.*, 2017, 158, 381-394

Acknowledgements

The authors thank Casale SA for the financial and technical support granted to the project.

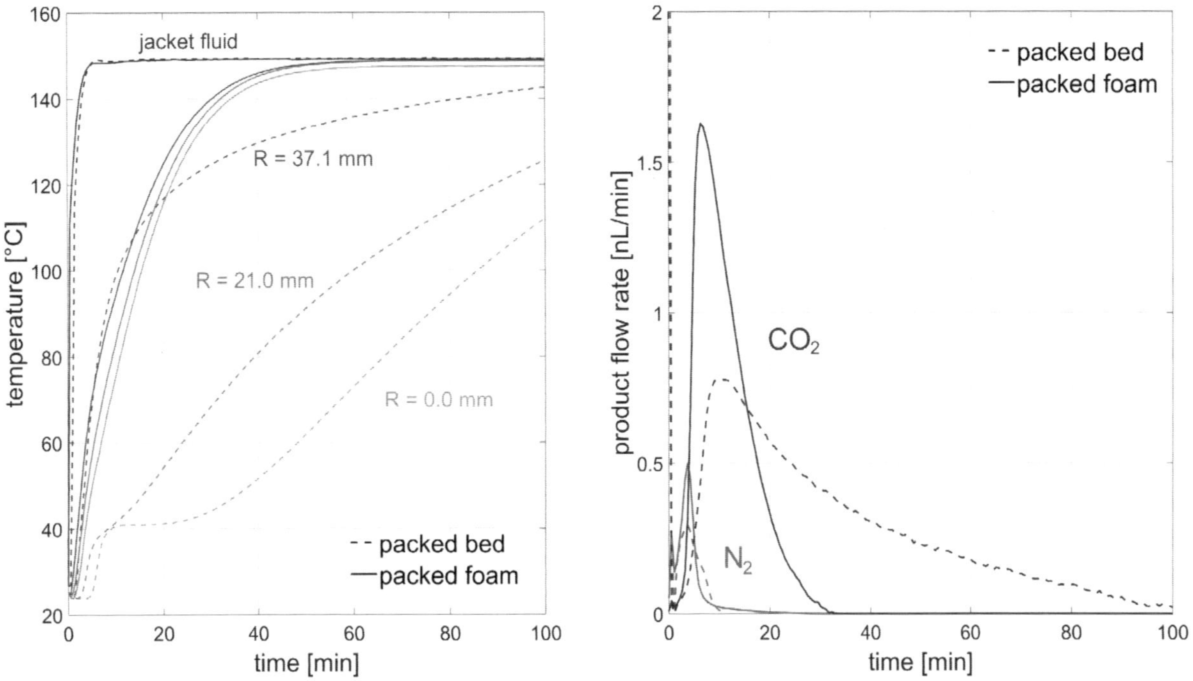

Figure 1: Experimental profiles in time for temperatures (at different radial positions) and outlet flow rates. Results for packed bed (dashed) and packed foam (continue) configurations during the regeneration step of a TSA cycle are shown. A hot fluid flowing in the jacket indirectly heats the reactor. The advantage of speeding up the transitory phases during the adsorption cycle is evident from the sharper CO2 flow peak at column outlet during the bed regeneration.

610654 4D Printing of Moisture Responsive Cellulose Nanocrystal Polymer Composites

Abstract:

Improving material selection for additive manufacturing (AM) has gained significant attention in recent years. Several researchers and companies have been pushed by consumers and industry to stretch the capabilities of the most common form of AM, known as fused filament fabrication (FFF), which involves extrusion of a polymer filament in a layer-by-layer fashion.

We have built on the development of a smart mechanically dynamic thermoplastic urethane/cellulose nanocrystal (TPU/CNC) composite, one which can respond to changes in environment simply by moisture exposure, to potentially increase design freedom and realize opportunities for expansion of FFF into new functional products. TPU/CNC composites have been documented to change modulus significantly upon exposure to moisture through decoupling of CNC hydrogen bonding and disruption of the mechanically percolating network. However, the effects of processing on the mechanical response of the composites is not well documented or understood.

Here we investigate the interplay between processing and structure-property relationships for 3-D printed TPU/CNC composites. We evaluate the impact of melt processing parameters on moisture induced mechanical modulus change following 1) TPU/CNC filament production from a masterbatched, solvent cast composite and 2) subsequent FFF of TPU/CNC filament into functional parts. The interplay between thermal history and shear induced particle orientation are investigated to determine effects on the dynamic modulus change upon moisture exposure. Further, we analyze the impact of geometry on moisture diffusion kinetics, and reversibility of the mechanical modulus change. Finally, we present the impact of FFF process parameters on the mechanical response of 3D printed parts, including the ability for the printed composite parts to change and/or hold their shape upon exposure to moisture and/or drying. These analyses help us understand how processing influences the structure and behavior of the composite, notably on its ability to mechanically adapt upon moisture exposure. This information is critical to successfully realize the ability to design and 3D print complex, mechanically moisture responsive composite structures.

597588 Evaluating Residual Stress in 3D Printed Parts

Abstract:

Residual stresses play an important part in layer-by-layer fabrication process of additively manufactured parts adversely affecting the mechanical properties and dimensional accuracy of the finished parts. In this work an experimentally validated simulation model for Digital Light Processing (DLP) 3D printing was developed to investigate residual stresses and shape distortion. Hole-drilling method was used in combination with two-dimensional subset-based digital image correlation software to quantify residual stresses in the 3D printed samples. An experimentally validated Finite Element Analysis (FEA) simulation model was developed to predict residual stresses formed during the layer-by-layer deposition. A nonlinear constitutive model involving the frontal photopolymerization was established. A parametric study was carried out using the FEA based simulation model.

607037 Tunable Structural Color of Bottlebrush Block Copolymers through Direct-Write 3D Printing from Solution

Abstract:

Additive manufacturing of functional materials exhibiting tunable structural color remains highly challenging, due to the need for on-the-fly control of printed structures at the nanoscale. We overcome this challenge by integrating nonequilibrium self-assembly of bottle brush block copolymers with a highly versatile direct-write 3D printing approach. We demonstrate that by varying printing conditions during deposition, peak reflected wavelength for bottlebrush photonic crystals prepared from a single ink can be controlled across a range of 403 to 626 nm (blue to red), corresponding to a predicted change in d-spacing of >70 nm (Bragg-Snell equation). We further demonstrate spatial and functional patterning of photonic crystals to achieve colorful prints of chameleon patterns through on-the-fly tuning of assembly kinetics during printing. Via scanning electron microscopy and synchrotron small-angle X-ray scattering analysis of printed films, we then confirm the microstructural underpinning of this phenomena as tuning of lamellar domain d-spacing (total thickness of A+B blocks) which we attribute to modulation of polymer conformation during printing. Finally, we establish the role of kinetic trapping of metastable microstructures as the mechanism for domain size control based on *in situ* optical microscopy and solvent-vapor annealing experiments. Thus, we demonstrate a new method for tunable block copolymer photonics without synthetic variation. Furthermore, we present a hardware and software integrated 3D printing scheme that can serve as a test bed for functional property modulation for a variety of material systems.

595727 3D Micro-Extrusion of Hierarchical Porous Structures for Sorption and Catalysis

Abstract:

This contribution presents the potential of 3D printed materials in the area of Carbon Capture and Utilisation (CCU). Carbon Capture, Utilisation and Storage (CCUS) technologies could enable large (90-95%) reductions in CO_2 emissions from power generation using fossil fuels and energy-intensive industrial processes. In CCU CO_2 is captured and the CO_2, as source of carbon, is utilized as a raw material for other processes like the manufacture of fuels, carbonates, polymers and chemicals. Capturing CO_2 using solid adsorbents is a promising alternative to state-of-art technologies using amine solutions as absorbents. In this study, solid adsorbents for CO_2 capture and heterogeneous catalysts for chemical conversion of CO_2 have been manufactured by 3D printing. 3D printing allows to tailor the porous architecture in view of optimal heat and mass transfer properties combined with a low pressure drop leading to an improved efficiency of sorption and catalytic processes.

3D printing was performed by a technology known as Direct Ink Writing or robocasting, which is based on the direct extrusion of viscous paste through a nozzle. By the controlled stacking of the extruded fibers layer by layer, structured porous architectures can be produced in a wide variety of materials that find application in sorption and catalysis ranging from metals supports, zeolites, oxide ceramics, activated carbon, (functionalised) polymers,... and their combinations. The paste formulation (active powder material, (in)organic binder, rheology modifier, dispersants,...), extrusion conditions and post-processing conditions have an impact on the porosity, structural properties, mechanical strength and performance (selectivity and kinetics) of the materials. Different strategies to control the porosity within the fibers by the use of different (in)organic binders have been developed.

Figure 1: 3D printed structures for CO_2 capture: (a) activated carbon, (b) zeolite.

(a) (b)

602424 Rapid Diffraction Limited Photo-Lithography of Organic Semiconductors

Abstract:

For any inorganic semiconductor device, photolithography is used to micro/nano pattern the areas over which metal electrodes are deposited, semiconductor is etched, ions are implanted, or oxide is grown. For organic semiconductors, academic articles only focus on materials property measurements because there is no easy and universal method to pattern the organic semiconductor or the doping density. Here we correct this gap in processing capabilities. We present a method to rapidly pattern any semiconducting polymer with diffraction limited resolution using a conventional maskless photolithography instrument. We demonstrate maximum write-speeds of several cm/min for patterning with <1 μm features. We demonstrate the writing and doping of semiconducting nanowires made from poly-3-hexylthiophene (P3HT) and Poly[2,5-bis(2-octyldodecyl)pyrrolo[3,4-c]pyrrole-1,4(2H,5H)-dione -3,6-diyl)-alt-(2,2';5',2'';5'',2'''-quaterthiophen-5,5'''-diyl)] (DPP-4T). These nanowires can be written at any location on the substrate and can include bends or T's. Typical wire width of 400 nm and length of >100 μm demonstrate the ability to pattern conjucated polymers with unprecedented resolution and speed. All equipment and processes are consistent with modern photo-lithography instruments in facilities and can be directly adopted.

604744 Kinetic Analysis of Degradable Resins in Digital Light Processing

Abstract:

Photocurable resins, typically consisting of a combination of multi-functional monomers and oligomers, a photoinitiator, and an optical absorber, are formulated for a wide array of applications in 3D printing. The complex reactions that occur during the formation of often highly cross-linked networks in 3D printing has seen minimal investigation, as the available experimental methods to analyze the final 3D printed specimen are limited when they will not readily go into solvent. By utilizing a methacrylated anhydride oligomer in a resin system that was engineered for 3D printing, a unique opportunity exists to provide insight into the initiation, propagation, and termination steps in free-radical polymerization by analyzing the linear-degradation products of the 3D printed photopolymer resin upon exposure to water. Ongoing efforts include the utilization of a specific synthesis protocol for the production of surface-eroding methacrylated polyanhydride oligomers that readily degrade in the presence of water. Addition of particular constituents in the reaction mechanism at different stages influences the physiochemical behavior of the oligomer in unique environments by incorporating different degrees of hydrophobicity into the the oligomer backbone while maintaining functionalized end groups that will later react to form cross-linked networks during the 3D-printing process.

Fabrication of a frugal microfluidic apparatus has been developed to access the kinetics of surface-eroding polymer degradation using a series of tunable, simple, and low cost processing steps. Theoretical models are being developed to aid in quantifying the surface-erosion data that is gathered by image analysis of the degradation device. In-depth analysis of the degradation rates and mechanisms with a novel and adaptable flow apparatus are being coupled with previously explored methods of analyzing linear-degradation products of multi-functional monomers to give insight into the complex chemical mechanisms that occur while printing resin formulations in specific environments as well as the effect of specific elements within the underlying molecular architecture. Preliminary data provides reproducible, material-saving, and time-efficient methods to measure the effects of the molecular architecture, printing parameters, and post-curing procedures on the mechanical properties of 3D printed acrylic-based degradable resins.

610412 Development of a Cost-Effective and Modular Closed Loop Process for Methanol Production in Wastewater Treatment Plants

Abstract:

Wastewater treatment plants (WWTPs) are under increasingly strict regulations to limit nitrogen and phosphorous in effluent discharge since they can have deleterious effects on aquatic species. In WWTPs, ammonia in wastewater is converted to nitrites, then nitrates and finally nitrogen in biological nitrification-denitrification processes. Methanol is commonly used as an external organic carbon source for denitrification since it is readily available, relatively cheap, and residual methanol can be removed from the treated effluent by aeration. Denitrification using methanol exhibits high potential removal efficiency, high process stability and reliability, relatively easy process control, low land area requirement, and moderate cost [1]. However, high competing demand for methanol from utility chemicals manufacturing can result in ever increasing cost of methanol for denitrification. Additionally, safety concerns related to methanol storage and handling facilities might preclude the purchase and transport of commercial methanol to WWTPs, highlighted by the ban of methanol use for denitrification in New York City in the late 2000s.

The presence of anaerobic digesters in WWTPs offers an opportunity to use biogas as a feedstock for on-site methanol production. Biogas, a mixture of 50-70% methane, 30-40% carbon dioxide, along with sulfur and trace elements, is used only in about half of WWTPs in the United States with digesters, for digester heating and combined heat and power [2]; in the rest of the cases, biogas is flared. On-site conversion of biogas to methanol has the potential to mitigate risks from cost and supply volatility by reducing the dependence on external carbon sources, address safety concerns and reduce the overall carbon footprint of the WWTPs [3]. However, traditional methanol synthesis processes from steam-methane reforming-based syngas ($CO+H_2$) are not viable since they cannot be scaled-down economically for use at WWTPs. This provides an opportunity for the development of a cost-effective and modular closed-loop process for on-site production of methanol at WWTPs.

RTI has developed and demonstrated a cost-competitive, internal combustion engine-based, impurity-tolerant syngas generator at near-deployment scales. RTI's pilot plant processes 50,000-60,000 standard cubic feet per day (SCFD) natural gas to produce syngas in a partial oxidation process. We have also integrated methanol production, with a 15% exhaust slipstream producing 1 barrel/day (BPD) of crude methanol. Instead of conventional process equipment that rely on economies of scale, the engine-based process uses economies of mass manufacturing to lower costs. The production of crude (75vol%) methanol alleviates safety concerns of handling, storage, and use. This paper will present technoeconomic evaluation of integrating the engine-based modular methanol synthesis technology in WWTPs for cost-effective on-site production.

A detailed biogas-to-methanol conversion process, including biogas cleanup, syngas production, conditioning and compression, and methanol synthesis sections, was simulated using Aspen HYSYS chemical process simulation software. Data from the operation of the engine and methanol synthesis pilot plant was used for reconciliation and validation of model performance. Several plant configurations were studied to optimize the biogas to methanol conversion process. Heat and mass balances, utility requirements, auxiliary power loads and equipment specifications were generated to calculate operating cost estimates. ASPEN Process Economic Analyzer (APEA) and vendor quotes were used to generate capital cost estimates. WWTP flow capacities determine the amount of external carbon needed and the biogas potential. Our analysis shows the engine-based methanol production process can meet methanol demand (~9 BPD) for WWTPs as small as 5 million gallons per day (MGD) capacities. Methanol costs are expected to be at par with commercial prices [4] from large-scale production at a 20 MGD scale, and much lower for larger WWTPs, up to 20% lower for a 30 MGD WWTP as an example. Our analysis indicates that the engine-based methanol production system is ideal for medium-sized WWTPs from 10-50 MGDs, with short payback periods and high returns on investment.

References

[1] H. H. Kung, MethanolProduction and Use Chemical Industries ; V. 57, 1994.

[2] United States Environmental Protection Agency, "Digesters at Water Resource Recovery Facilities," [Online]. Available:https://www.epa.gov/anaerobic-digestion/types-anaerobic-digesters#WRRFdigesters. [Accessed April 2020].

[3] S. K. C. Yu-Chen Su, "Bioaugmented methanol production using ammonia oxidizing bacteria in a continuous flow process," *Bioresource Technology,* pp. 101-107, 2019.

[4] Methanol Institute, "Methanol Price and Supply/Demand," March 2020. [Online]. Available: https://www.methanol.org/methanol-price-supply-demand/.

599047 Determining Materials Specifications Based on Technoeconomic and Life Cycle Assessment: A Case on Oxygen Career Material for Chemical Looping Steam Methane Reforming

Abstract:

Development of new process is one of the keys to achieve low-carbon production in the chemicals industry. Many new processes being studied rely on use of novel materials such as catalysts and chemicals careers. Since there are numerous possibilities of materials that can be used, it is important that a target-based development of these materials and processes are practiced for efficient process development with short lead time.

In this study, to delineate the feasible materials and their performance under operating conditions and a reactor of choice, process systems modeling, followed by assessment of economic and life cycle greenhouse gases emission is performed in an integrated manner. A novel chemical looping process for steam methane reforming (CL-SMR) as a part of methanol production is used as an example to demonstrate this approach. SMR is an endothermal reaction to produce CO and H2, the raw materials for commodity chemicals (ex. methanol). Conventional SMR is controlled at 700 - 900 °C to achieve a desired conversion, which not only leads to high energy consumption for supplying heat but also incurs expensive reformer equipment. CL- SMR is an attractive alternative, dividing the reaction system into two reactors where metal and its oxides is circulated among the reactors to carry oxygen atoms using two fluidized bed reactors connected with each other. In the steam reactor, H2O is reduced to yield hydrogen while oxidizing the metal into an oxide form, while in the methane reactor, CH4 is oxidized to produce carbon oxides (CO, CO2) and H2. In this way, the overall conversion is no longer bound by their original equilibria, allowing the operation at a significantly lower temperature while maintaining a sufficient conversion.

Using the process simulation combined with economic calculation and assessment of life cycle GHG emission, we present the target region for the career materials on a space defined by oxygen storage capacity and oxygen release rate in the methane reactor.

By presenting the simulation results, target temperatures and conversion, as well as the oxygen storage ccapacity and oxygen release rate were discussed.

610626 Minimization of Capital and Operating Costs in Cooling Water Distribution Systems

Abstract:

Cooling water is the fluid most commonly used in heat exchangers for cooling processes. The most widely used concept for cooling water semi closed loops is the one that a pumping system supplies water from the cooling tower basin through a header to branches where heat exchangers are located, and after heat absorption in these exchangers, the water returns from these branches through another header to the cooling tower top to reject the absorbed heat. The use of this kind of concept results in a reasonably simple design procedure for the cooling water system, but it can yield high electrical energy demands. As good engineering practice, the pressure drops in heat exchangers must not exceed 0.5 to 1.0 bar, but there are cases that this pressure drop may be higher than the recommended values. Cooling water consumers distant, above 500 m, from the cooling tower also impose high pressure drops in the control valves of the branches closest to the main pumping system. To avoid a short circuit, branches with heat exchangers with low pressure drops or near the cooling tower may have control valves that impose additional pressure drops and, therefore, mechanical energy losses. As a result, the pumping system must compensate these energy losses by increasing the consumed electrical energy.

Alternatives are analyzed to reduce the system total electrical energy consumed. Among these alternatives are the use of auxiliary (booster) pumps in the branches with the highest head losses, the installation of secondary headers, and the use of separate cooling towers. On the other hand, these alternatives result in an increase of capital costs. The mathematical modeling of the pumping system allows the estimate of system mechanical energy demands as a function of the presence of booster bumps and secondary headers. The synthesis of the cooling water distribution system must minimize the overall costs.

The optimal synthesis of the cooling water system shall determine the values for the decision variables of the system model that minimizes the capital and operating costs. Among the model decision variables are: capacity of the cooling water main pumping system; number of pumps in the cooling water main pumping system; capacity and location of the cooling water booster pumps; layout of the cooling water supply and return headers; required minimum pressure drop through control valves. An existing cooling water distribution system is selected as study case, so the optimal synthesis solutions can be compared to the existing one.

A sensitivity analysis is performed varying the design parameters such as cooling water total and individuals demands, cooling water consumers' locations, heat exchangers pressure drop and cooling tower elevation. Evidently, these project parameters have a direct impact in the cooling water distribution system and the resulting cost. As a result, a methodology is established so it can be applied to the detailed design of the cooling water distribution system.

609711 Economic Evaluation of Wastewater Treatment Plant Construction in Small Communities

Abstract:

Wastewater collection and treatment systems are an integral part of communities. In the life of a town, its investment in wastewater collection and treatment systems is hugely important for attracting industry, protecting the environment, and conforming to state law. Through this work we seek to understand the optimal size of a town for construction of a wastewater collection and treatment system.

We utilize data covering wastewater treatment plants in small towns across Oregon to estimate parameters utilized in a series of costing models. Using several capital cost estimating techniques, we identify the optimal size of town for capital investment in wastewater collection and plant systems.

We then compare early, late, and optimal construction timing with status quo (septic) setups to gain a better understanding of the consequences of ill-timed investment. Through the cost models, we analyze a number of different configurations for wastewater treatment plants including utilizing different machinery and perform sensitivity analysis on major parameters to gain an understanding of the uncertainties in our model.

Lastly, we utilize our model setup to observe several real cities amid wastewater collection and plant investment controversy to see the consequences in practice of regulatory and investment decisions. We quantify some of the projected costs associated with those communities and compare those to costs under ideal timeliness of investment.

613116 In Situ Nanometrology of Digital Light Processing Additive Manufacturing

Abstract:

Digital light processing (DLP) additive manufacturing (AM) is rapidly realizing its potential as a transformative manufacturing technology that affords unprecedented design flexibility and customization of final parts. With increased interest for deployment in critical applications such as healthcare, the need for fundamental insight into mechanical performance and feature resolution is increasing. In DLP, part formation occurs at millisecond to second times scales and micrometer length scales. Due to species diffusion, local depletion, and light absorption, the resultant parts exhibit inherent variation in conversion that corresponds with anisotropic chemical, thermal, and mechanical properties. This undesired heterogeneity compromises part performance compared to bulk counterparts. With high resolution and the ability to sense material properties, the atomic force microscope (AFM) is well suited to study the resin printing process in various in-situ and ex-situ modalities.[1] However, to date, integrated instruments combining 3D printing and nanometrology have not existed. Here, we introduce a hybrid instrument combining an AFM, an inverted optical microscope, and a 405 nm digitally programmable projection source.[2] We show that the instrument can be operated in two core modalities. In the first, the nanomechanical properties of patterned regions are spatially mapped as a function of print conditions. Notably, these properties can be compared in the just-printed, monomer-swollen state, and the ethanol-washed, dry state. Voxel modulus is found to vary depending on feature size and the absence/presence of neighboring voxels. In the second modality, the SPM acts as a stationary, local, high speed probe of cure rheology. It is uniquely capable of measuring polymerization at the small length scales and fast time scales of 3D printing. We find that for a thiol-ene resin, conversion is detected 10's of microns away from the light source. Furthermore, increasing light intensity to speed reaction rate does not fundamentally increase spatial control of cure by a significant amount. Overall, the hybrid instrument provides unique insight into future pathways to improve resolution and homogeneity of DLP parts.

1) Fiedler-Higgins, C. I., Cox, L. M., DelRio, F. W., Killgore, J. P. Monitoring Fast, Voxel-Scale Cure Kinetics via Sample-Coupled-Resonance Photorheology, *Small Methods* 2019, 3, 1800275.
2) Higgins, C.I., Brown, T.E., Killgore, J.P. Digital Light Processing in a Hybrid Atomic Force Microscope: *In situ*, Nanoscale Characterization of the Printing Process (submitted)

Figure 1: In-situ photopolymerization in the hybrid SLA/AFM. a) Optical view of AFM cantilever and projected 405 nm pattern, with composite map of resultant surface stiffness on right. b) Checkerboard 3D print nanomechanically mapped in monomer-swollen environment. Significant effects of inter and intra voxel interaction are observed as stiffness variations.

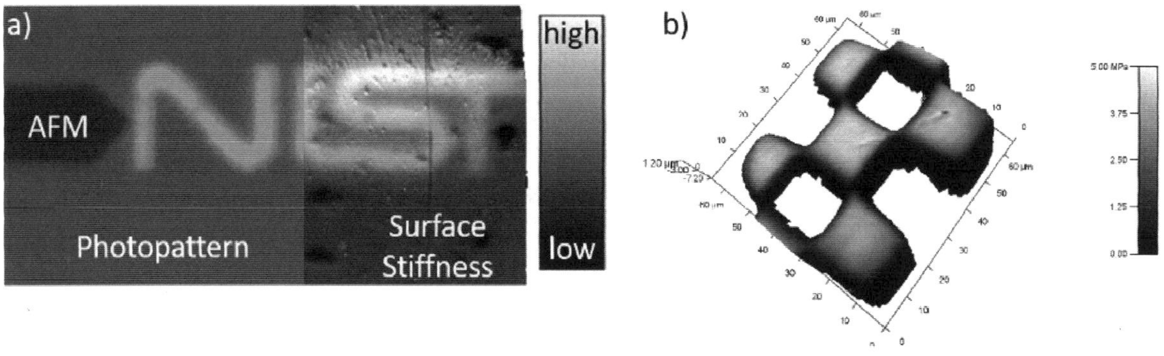

610006 Thermally Aged Poly(phenylene sulfide) Powder As a Feedstock for Powder Bed Fusion

Abstract:

Many new polymers are becoming commercially available for powder bed fusion (PBF) additive manufacturing (AM), including several regarded as high-performance polymers, like poly(phenylene sulfide) (PPS). PPS suits the needs of AM early-adopter industries, like aerospace, due to its high wear resistance, high specific strength, radiation resistance, and chemical resistance. Recent work has highlighted the process parameter selection considerations needed for processing PPS using PBF[1]. While there are many benefits to polymer PBF (e.g., no additional support material needed and batch-style production), one drawback is thermal ageing of powder that is not incorporated into a printed part. Ideally, the unfused powder can be reused as part of future builds; however, it is unlikely that printability properties of the thermally aged powder will match those of fresh powder. This change in properties leads to either limited reuse through a practice of mixing used and new powder, or elimination of all powder after each build. The more often the powder can be re-used, the more sustainable and cost-effective PBF is.

In this work, the authors report how thermal ageing in simulated printing conditions affects the properties of PPS that are critical to PBF processing. PBF-grade PPS powder was aged for up to 300 h at three process relevant temperatures: 200 , 230, and 278°C under nitrogen. The aged powders were then evaluated for properties relevant to the three PBF manufacturing process sub-functions: powder recoating, energy input, and coalescence and cooling. While there are no observable changes for recoating related properties, significant and systematic deviations from the as-received state of the powder are observed for energy input and coalescence and cooling related properties with increasing ageing time and temperature. These changes are most pronounced for properties pertaining to the coalescence and cooling sub-function (e.g., melt flow rate). These changes are interpreted both in terms of physical and chemical changes in PPS and in terms of how these changes may impact the PBF printing process.

References

(1) Chatham, C. A.; Long, T. E.; Williams, C. B. Powder Bed Fusion of Poly(Phenylene Sulfide) at Bed Temperatures Significantly below Melting. *Additive Manufacturing* **2019**, *28*, 506–516.

611347 Upcycling Waste Polyesters to Polymers for Additive Manufacturing

Abstract:

This year, <30% of poly(ethylene terephthalate) (PET) bottles are recycled, and PET textiles currently exhibit nearly zero recycling rates. Existing approaches for plastics recycling, such as mechanical reprocessing and pyrolysis, are effectively downcycling, as they lead tolower valued products. Chemical conversion of plastic wastes into materials of higher value, or upcycling, may economically incentivize a higher extent of plastics reclamation. To that end, this presentation will cover our recent efforts in demonstrating the conversion of reclaimed polyesters into polymers for light-based additive manufacturing. Thermomechanical properties of these upcycled polymers suggest they are drop-in replacements to petroleum-derived materials. Supply chain energy calculations predict that such plastics upcycling processes can significantly reduce both energy consumption and greenhouse gas emissions, relative to virgin material reproduction. Overall, waste plastics derived materials are demonstrated as new materials for 3D printing and an incentivizing recycling method toward a circular economy.

611387 One-Layer Alkylated Poly(olefin) Water Vapor Barriers As Flexible Substrate Coatings

Abstract:

The effect of one-layer poly(olefin) barrier coatings onto flexible and thin polyethylene terephthalate (PET) films will be discussed towards the blockage of water vapor. Results on well-known semi-crystalline polymers such as poly(ethylene) (PE), partially chlorinated PE (47% Cl-PE), poly(propylene) (PP), poly(vinylidene fluoride) (PVDF), and poly(vinylidene chloride) (PVDC) will be presented, as well as the effect of amorphous hydrophobic polymers such as poly(dimethyl siloxane) (PDMS) and commercial organo- and perhydro- poly(silazanes) on PET's water vapor transmission rate (WVTR) reduction. Results at different coating thicknesses will be highlighted. Finally, the water vapor barrier properties of synthetic alkylated poly(olefins) of controlled molecular weight and low dispersity, prepared by living cationic polymerization, will be presented as transparent polymeric coatings onto PET, and comparisons with metal sputtered PET (same substrate) will be realized. The incorporation of nanomaterials into high polymeric barrier coatings will be also discussed. Lastly, results on mechanical tests of PET/barrier coatings on elongation, along with coating-induced resistance towards cracking/ deformation will be presented.

607379 Numerical Modeling of a Fused Deposition Modeled Embedded Polymer Fiber in a Micro-Dispensed Cladding for Optical Fiber Interconnects

Abstract:

Optical fiber interconnects transmit information between optical chips on PCB boards at higher bandwidths and with less energy requirements then traditional copper interconnects. Creating a smooth round fiber that connects components in three-dimensions is required for today's fast data rates and tight packaging requirements. An additive manufacturing solution would need to satisfy these requirements to be a viable solution. However, fiber produced using fused deposition modeling (FDM) will sag under the force of gravity, non-homogeneous cooling, and surface tension when printed on a surface. Here we show by embedding a PMMA fiber produced using FDM printing, into a micro-dispensed optical urethane cladding, a smooth round fiber can be produced. We found in experimental studies that by controlling the processing conditions, we could produce round embedded fibers. Furthermore, we found in numerical simulation, that we could confirm those results at our processing conditions. Our results demonstrate that the roundness of a fiber can be measured at less than 1 micron when embedded within a cladding material and optical transmission rates of as high as 62%. We anticipate that embedding the fiber in the cladding material is one of the key steps in the larger laser enhanced-direct print additive manufacturing (LE-DPAM) process of producing smooth round plastic optical fibers. We anticipate that having the ability to produce a round fiber in the LE-DPAM process will be the foundation in the development of more sophisticated electro-optical computing devices.

602848 High T_g and Toughness Materials for Mask-Projection Stereolithography Via *in-Situ* Sequential Interpenetrating Network

Abstract:

Stereolithography is one of the most popular techniques used in additive manufacturing as it offers high dimensional accuracy. However, there are limited choices of materials for this process as the resins need to be photocurable. Acrylates and epoxies are commonly used, but their thermal and mechanical properties are not suitable for applications in which high-temperature performance is needed. In contrast, high-performance materials such as bismaleimide and cyanate ester are not popularly used in stereolithography as they are typically thermally processed. In order to utilize these materials in stereolithography, we developed a photocurable resin containing bismaleimide and cyanate ester that provides outstanding thermal and mechanical characteristics. In this system, an in-situ sequential interpenetrating network (IPN) is formed wherein bismaleimide and a reactive diluent copolymerize during printing resulting in cyanate ester swollen network with a sub room temperature T_g. During post-processing, cyanate ester is thermally cured. The resulting IPN has great dimensional stability with glass transition temperature above 250°C. Additionally, the IPNs show a 50% reduction in shrinkage, improved interlayer strength, and improved toughness relative to standard systems.

604156 Fused Filament Fabrication of Hot Melt Adhesives

Abstract:

Fused filament fabrication (FFF), a desktop form of material extrusion additive manufacturing (AM), is the most common AM technique for thermoplastics. In FFF, thermoplastic filaments are heated until molten, then extruded onto a print surface layer-by-layer. Broader adoption of FFF is limited by challenges such as narrow material range, anisotropic mechanical properties, and inferior mechanical properties as compared to traditional manufacturing techniques such as injection molding or compression molding. The crystalline regimes in semicrystalline polymers enhance their strength and chemical resistance; however, increased crystallinity can also lead to more brittle materials. As a result, in conventional plastics manufacturing, processing conditions and material selection are optimized to lead to the desired crystallinity and resulting properties for a given application. The rapid heating and cooling inherent to FFF leads to complex crystallization behavior, warpage, and poor weld strength.

In this work, we demonstrate FFF of a polyamide hot melt adhesive. This adhesive is a semicrystalline polymers with a sub-ambient glass transition temperature, intermediate melting temperatures, and low recrystallization termperature. Due to this combination, this is a room temperature flexible materials that exhibits high toughness and strength, and good adhesion. The effect of extruder temperature on print quality and the effect of raster angle on mechanical properties of prints was studied. The optimized print parameters, as well as low viscosity and high melt strength of the material, resulted in elimination of voids within the prints. Additionally, tensile properties comparable to or better than compression molded hot melt adhesives are observed. This work broadens the material window for FFF and demonstrates that hot melt adhesives can be used to additively manufacture optically transparent, flexible structures capable of >1200% strain to failure.

602264 Modeling and Nonlinear State Estimation for Advanced Process Control of the Enzymatic Conversion of Lactose into Value-Added Products

Abstract:

Due to increased concerns over sustainability of chemical processes, significant research has been conducted to minimize the amount of process waste generated. Despite advances in efficiency, selectivity, and conversion, waste management is still a concern for all processes, especially biological. The most effective method for waste disposal would be its conversion into a value-added product through profitable means. Many industries are attempting to achieve this goal, but it is challenging to design a novel catalyst, enzyme, or process. In particular, recent advances in enzymatic process modeling have created an opportunity for small-scale cheese manufactures to convert their waste into value-added products. The manufacturing of cheese results in the formation of lactose rich whey permeate that cannot be easily utilized and thus is often disposed of at cost to manufacturers [1].

A recent advance in enzymatic modeling corresponds to a new kinetic model to convert lactose waste into value-added galactooligosaccharide (GOS) products using β-galactosidase [2]. This specific reaction has been studied by several researchers in the past, but each proposed model presented challenges such as mass balance issues and overfitting due to the excessive number of parameters [3]. The proposed kinetic model contains no mass balance errors, while minimizing the number of parameters to accurately depict the process characteristics. Furthermore, this model uniquely distinguishes between purely galactose derived disaccharides, trisaccharides, and tetrasaccharides and those containing a glucose saccharide group. Due to the similarity of glucose and galactose, there are no online measuring techniques that can discern between the specific di, tri, or tetrasaccharides stereoisomers. A High Performance Liquid Chromatography (HPLC) method coupled with an ion-exchange column and refractive index detector can distinguish between glucose and galactose, but has no resolution for other isomers such as disaccharide, trisaccharide, and tetrasaccharide stereoisomers. Without the ability to measure specific component concentrations, state estimation must be used to predict the individual states using the online measurements and the proposed kinetic model.

This work addresses the state estimation and advanced control employing the developed enzymatic model for the first time. Unique estimation challenges are posed and addressed, with the specific novelty of using nonlinear state estimation to find stereoisomer concentrations from available online measurements and the nonlinear kinetic model. Both an Extended Kalman Filter (EKF) and a Moving Horizon Estimator (MHE) are analyzed for process implementation. The EKF is an unconstrained estimator that allows for rapid estimation but does not guarantee feasible and optimal results. The MHE is a constrained estimator that guarantees feasibility at the cost of increased computational time [4] [5]. Using these estimation techniques, a Biologically-Inspired Optimal Control Strategy (BIO-CS) algorithm is applied to optimize the feeding of lactose and enzyme into a semi-batch reactor [6][7]. The proposed estimation and advanced control framework allows the feasibility of this process to be examined and provides a guide for other bioprocesses to follow.

References:

[1] Illanes, A. (2011). Whey Upgrading by Enzyme Biocatalysis. *Electronic Journal of Biotechnology*.

[2] Schultz, G., Alexander, R., Lima, F.V., Giordano, R., & Ribeiro, M. (2019). Kinetic Model for the Enzymatic Synthesis of Galacto-Oligosaccharides: Describing Galactobiose Formation. *AIChE Annual Meeting.*

[3] Vera, C., Guerrero, C., Illanes, A., & Conejeros, R. (2011). A Pseudo Steady-State Model for Galacto-Oligosaccharides Synthesis With ß-Galactosidase From *Aspergillus oryzae*. *Biotechnology and Bioengineering*.2270-2279.

[4] Campani, G., Ribeiro, M., Zangirolami, T., Lima, F.V. (2019). A Hierarchical State Estimation and Control Framework for Monitoring and Dissolved Oxygen Regulation in Bioprocesses. *Bioprocess and Biosystems Engineering*.1467-1481.

[5] Lima, F.V., Rawlings, J. (2011). Nonlinear Stochastic Modeling to Improve State Estimation in Process Monitoring and Control. *Process Systems Engineering*.996-1007.

[6] Mirlekar, G., Gebreslassie, B., Diwekar, U., Lima, F.V. (2018). Biomimetic Model-Based Advanced Control Strategy Integrated with Multi-Agent Optimization for Nonlinear Chemical Processes. *Chemical Engineering Research and Design*.229-240.

[7] Mirlekar, G., Al-Sinbol, G., Perhinschi, M., Lima, F.V. (2018). A Biologically-Inspired Approach for Adaptive Control of Advanced Energy Systems. *Computers and Chemical Engineering*.378-390.

604108 Designing Data-Based and Model-Based Methods for Process Monitoring and Equipment Degradation Tracking

Abstract:

A wide variety of industrial processes utilize monitoring methods in order to ensure that safety and quality is maintained. Conventional model-based methods include dimensionality reduction techniques such as Principal Component Analysis (PCA) are often employed as they are computationally simple, and easy to implement in practice [1]. Unfortunately, these techniques assume that the process data is Gaussian, uncorrelated, and only contain a moderate level of noise. This work will demonstrate how wavelet-based representation of data can be utilized in order to efficiently handle data that violate these assumptions. Statistical hypothesis testing methods such as the generalized likelihood ratio chart have shown promise with respect to fault detection, and thus will be used to enhance monitoring capabilities [2], [3].

Additionally, this work will demonstrate the necessity to track deviations in the process models themselves in order to identify process drifts or equipment degradation. When a process operates under control, a controller continuously adjusts the level of manipulated variables in order to ensure that the controlled variables are being maintained within specific limits. Unfortunately, this may be at the expense of increased operating and additional costs. Therefore, this work will utilize a dynamic contour-based algorithm in order to illustrate how equipment degradation can be efficiently tracked in multiple operating regimes [4].

For both algorithms, illustrative examples using simulated synthetic data, and real data from different applications will be utilized in order to highlight the effectiveness of the developed algorithms.

References

[1] I. T. Joliffe, *Principal Component Analysis*, 2nd ed. New York, NY: Springer-Verlag, 2002.

[2] M. R. Reynolds and J. Y. Lou, "An Evaluation of a GLR Control Chart for Monitoring the Process Mean," *J. Qual. Technol.*, vol. 42, no. 3, pp. 287–310, 2010.

[3] M. Z. Sheriff, M. Mansouri, M. N. Karim, H. Nounou, and M. Nounou, "Fault detection using multiscale PCA-based moving window GLRT," *J. Process Control*, vol. 54, 2017, doi: 10.1016/j.jprocont.2017.03.004.

[4] M. Z. Sheriff, H. Nounou, M. Nounou, and M. N. Karim, "Monitoring process degradation through operating regime based process monitoring," in *AIChE Spring Meeting and Global Congress on Process Safety: Process Control Monitoring and Analytics*, 2019.

610824 Optimal Sensor Network Design and Nonlinear State Estimation for in-Situ Corrosion Monitoring in Coal-Fired Boilers

Abstract:

Existing coal fired power plants are increasingly being subjected to significant load-following due to the increased penetration of intermittent renewables to the grid. During load following operations maintaining the boiler environment at the desired conditions is significantly difficult. In particular, higher flue gas temperature than expected and higher concentration of corrosive sulfur species in the flue gas are causing external sulfidic corrosion especially in the waterwall (WW) tubes, which is one of the primary causes for the WW failure [1]. Experimental studies on corrosion rates of Ni-Cr alloys suggest that corrosion rate is a function of alloy composition, temperature, SO_2 and O_2 concentrations in the flue gas [2-3]. The effect of metal temperature on the corrosion rate is highly nonlinear with a "bell" shaped response between 600°C and 1000°C for most of the alloys used in the WW section. Unfortunately, as of now, there is no online commercial sensor for real-time measurement of corrosion rate due to the high operating temperature and challenging environment in the WW section. Recently a novel type of electrochemical sensor is being developed in-house for in-situ measurement of the corrosion rate [4]. As the operating temperature and concentration of sulfur species greatly vary spatially and temporally in the WW, the corrosion rate can vary significantly with space and time. However, it is infeasible to install such sensors in the entire WW. Motivated by these challenges, an optimal sensor network is designed, and a nonlinear state estimation technique is developed for real-time monitoring of corrosion rate at all desired locations inside the WW.

A first-principles dynamic process model of the WW is developed for calculating the spatial and temporal variation in the temperature and concentration of sulfur species. Models are also developed for the corrosion rate. Electrochemical sensor model has been previously developed at West Virginia University [4]. This is a high nonlinear differential algebraic equation (DAE) system. Therefore, an unscented Kalman filter algorithm for DAE system is developed for state estimation, particularly for estimating the corrosion rate given the data from the electrochemical sensor.

Typically, process information which can be obtained from a given sensor network and the reliability of sensor network improves as the number and/or the accuracy of the installed sensor increases. However, it is neither feasible nor necessary to install sensors at all desired locations. An optimal sensor network is designed by minimizing the posterior error covariance of the optimal estimator. The algorithm provides the optimal number, location and type of sensors.

The optimal sensor network is used to estimate the corrosion rate for a coal-fired boiler. A number of operating scenarios are simulated by changing the temperature in the WW and concentration of the sulfur species. Our work shows that even in the presence of large mismatch between the process and the model and high measurement uncertainty, the sensor network provides satisfactory spatial and temporal resolution of the corrosion rate under aggressive load-following operation.

References

1. North American Electric Reliability Council (NERC), "State of Reliability," 2017.

o

2. L. Luthra and D. A. Shores, "Mechanism of Na_2SO_4 Induced Corrosion at 600-900 C," *J. Electrochem Soc.*, no. 1971, pp. 2202–2210, 1978.
3. Viswanathan and C. J. Spengler, "Corrosion of 85 Ni-15 Cr Alloy at 1600 F in Controlled Atmospheres Containing O_2, SO_2, SO_3, H_2S, and N_2," *Corrosion*, vol. 26, no. 1, pp. 29–41, 1970.
4. N. Aung and X. Liu, "High temperature electrochemical sensor for in situ monitoring of hot corrosion," *Corros. Sci.*, vol. 65, pp. 1–4, 2012.
5. Kumari, S. K. Das, and P. K. Srivastava, "Modeling fireside corrosion rate in a coal fired boiler using adaptive neural network formalism," *Port. Electrochim. Acta*, vol. 34, no. 1, pp. 23–38, 2016.

609445 CO_2 Injection Monitoring with a Scalable, Automated Sparse Seismic Array

Abstract:

Successful applications of the 3D surface land seismic method to monitor CO_2 injection into a reservoir have been found in the literature. Reducing the operational effort to acquire 3D data and the turnaround time for the processing and interpretation of field data are highly desirable for more efficient seismic monitoring. The scalable, automated sparse seismic array (SASSA) method has been ideated as a cost-effective alternative or complementary to conventional 3D seismic methods. Based on a sparse array of receivers deployed strategically on Earth's surface and one or more fixed seismic sources, the SASSA method can be used to study specific portions of the subsurface to detect the movement of CO2 in a reservoir with lower environmental footprint and shorter turnaround time to deliver results than conventional 3D seismic methods.

The Energy & Environmental Research Center (EERC) carried out seismic monitoring of an oilfield undergoing CO_2 enhanced oil recovery in Montana, USA, using the SASSA method. Initial efforts in collaboration with the field operator were invested in designing an optimum distribution of two sources and 96 receivers not only to honor the subsurface geologic conditions and CO_2 injection patterns but also to minimize the impact of cultural noise sources. The signal-to-noise ratio was further improved in data acquisition by use of two fixed surface orbital vibrators that were remotely controlled by the EERC team in North Dakota. Weekly data acquisition was conducted on weekends from May 2019 through March 2020. Data-processing workflows were developed to preserve the true seismic amplitudes representing the time-lapse changes of the reservoir due to CO_2 injection and maximize the representation of the reservoir in the seismic sections from the distribution of sources and receivers on the surface. Encouraging results were obtained using the SASSA method in areas adjacent to the CO_2 injection wells. Two validation methods were considered: a time-lapse 2D seismic survey acquired at the beginning and the end of the monitoring activities and reservoir simulation.

597019 What Makes Sensors the Basis of Any Digital Transformation?

Abstract:

Whether a brownfield or greenfield facility, a company manufacturing chemicals or food supplements or a 10-person startup versus a Fortune 500 company, businesses of all sizes and sectors can achieve a simple entry into the world of digital transformation. There's the option to implement the full scope of the Digital Enterprise, simply pick individual digital applications that can get the digital journey started or to pick something in between. There's not a single path that can be applied to everyone.

One of the focal points of Industry 4.0 is the ability to increase the availability of plant assets and to give on-site personnel quick and meaningful decision-making tools for their work where and when they need it. The objective is to make processing plants more reliable through proven products and innovative services and to be able to more dependably predict the condition of assets. Sensor data plays a critical part in this effort as they are the eyes and ears of the process. Combined with the wide availability of simple data collection solutions and multi-variable sensors powered by cloud computing, process information can now be analyzed to provide insights that was difficult to get just a few years ago.

The overall goal of the presentation will be to provide a solid understanding of the power of Industry 4.0, how process sensors support asset management and provide example of real, tangible solutions available today. Some of the concepts to be discussed in more detail include:

- Discuss various ways of gathering data at the field level to be used for gaining insights into the process
- How to bring plant and tag data into the field and to make optimal use of remote service capabilities to enable faster and more efficient maintenance
- Provide specific examples of how to use existing sensor data to provide insights that provide value in terms of improved predictive maintenance, less downtime and faster maintenance
- Briefly touch on the digital twin, 3D and simulation and how it impacts automation and process control

611943 An Explainable Artificial Intelligence Based Approach for Interpretation of Fault Detection Results from Deep Neural Networks

Abstract:

Data Driven Fault Detection and Diagnosis methods can play an important role in Abnormal Event Management by timely detecting and diagnosing faults with the help of the enormous amount of data recorded in chemical process industries. Recently, deep learning techniques have been a promising area of research in this field. Models built using deep learning methods provide better predictive performance on benchmark datasets like the Tennessee Eastman Process than other statistical methods[1]. But these techniques do not possess good interpretability and we may be unable to determine how the model has reached a certain conclusion. Using these methods, it is difficult to diagnose how a fault has occurred and we cannot provide suitable recommendations to prevent their occurrence in the future.

eXplainable Artificial Intelligence (XAI) is a new field in machine learning and has been gaining popularity because of the need for the solutions provided by AI technologies to be understood by human experts. Research progress in XAI has been focussed in two areas - building interpretable and transparent AI Systems and post-hoc methods. Post-hoc methods make explanations from pre-developed model, treating the model as a black box and it does not impact performance of the model. [3] One way to obtain post modelling explanations is using the Shapley Value framework that is based on the Shapley concept from co-operative Game Theory. In Game Theory, Shapley Value provides a solution to fairly distributing the gains and costs to all the actors working in a coalition to achieve some outcome. It is essentially the average expected marginal contribution of one actor towards the outcome after all the possible combinations are considered. In machine learning, the Shapley Value is becoming a popular method to attribute the model's prediction(the outcome) on an input to its base features(the actors). The Integrated Gradients Technique is one of the multiple ways to operationalize Shapley Value for the feature attribution problem.[4] Integrated Gradients aggregate along the inputs that fall on the straight path between the chosen baseline to the input and the input. It is based on the Aumann-Shapley cost sharing technique which extends the discrete Shapley Value to the continuous context and quantifies the importance of each input feature in a prediction.[5]

In this work, we present a novel approach to build an interpretable AI system for detecting and diagnosing faults. In this system, a deep neural network is trained to solve a fault detection problem. Then, the Integrated Gradients Technique is used on the trained deep neural network model to understand the relative contributions of input features in detecting faults. This helps in identifying which factors have contributed to the occurrence of a fault. We also perform Key Variable Selection by choosing an optimal subset of measured variables by removing those input variables which have negligible contributions to the model predictions. In order to implement this approach, the software used is AI Platform Predictions by Google Cloud.

The performance of the approach is demonstrated on 3 synthetic datasets where each dataset contains 10000 samples of 4 input features and one output variable. The output variable is related to various combinations of the input variables through both linear and nonlinear functions. A neural network was trained for fault detection using the data in each case study and could correctly detect the faults with high accuracy. Next, using the proposed approach we determined the input variable(s) which influenced the output for each test sample. In all thress case studies, the XAI method could accurately determine the input variable(s) correctly. In this paper, we will present the proposed methodology as well as the results from the 3 synthetic data sets. The results for fault detection in the benchmark Tennessee Eastman process will also be presented.

REFERENCES

A novel process monitoring approach based on variational recurrent autoencoder - Feifan Chang, Peter B Qe, Jinsong Zhao. Link - https://doi.org/10.1016/j.compchemeng.2019.106515

Optimal variable selection for effective statistical process monitoring by Kaushik Ghosh, ManojkumarRamteke, Rajagopalan Srinivasan. Link - https://doi.org/10.1016/j.compchemeng.2013.09.014

Explainable artificial intelligence: A survey. By Dosilovic, F. K., Brcic, M., & Hlupic, N. Link - https://ieeexplore.ieee.org/abstract/document/8400040

The many Shapley values for model explanation by Mukund Sundararajan, Amir Najmi. Link - https://arxiv.org/abs/1908.08474

Axiomatic Attribution for Deep Networks by Mukund Sundararajan, Ankur Taly, Qiqi Yan. Link - https://arxiv.org/abs/1703.01365

599243 Low-Temperature Chemical Looping Oxidative Dehydrogenation of Ethane for Modular Ethane to Liquids

Abstract:

Of the natural gas liquids, ethane poses particular transport challenges for producers. The concentrations of ethane in wet-gas can exceed natural gas pipeline specifications; it is not easily liquefied for transport; and dedicated ethane pipeline capacity is limited. A modular scale process that can convert ethane into valuable, easily-transported liquids at gas processing facilities can capture significant value from this domestic resource. However, selective ethane conversion at modular scales is difficult; steam cracking to ethylene (the traditional industrial approach) requires very high temperatures (>1000 °C), and typical oxidative dehydrogenation approaches require energy intensive air separation and precise reactor control. Here we present a modular ethane-to-liquid fuel (m-ETL) system. This system is enabled by low-temperature (≤750 °C) redox-catalysts for the chemical-looping oxidative dehydrogenation of ethane (CL-ODH). In this approach, ethane is partially oxidized to ethylene and water with the lattice oxygen of the redox catalyst in a packed bed reactor. After the redox catalyst is depleted of oxygen, valve switching is used to introduce air to regenerate the lattice oxygen and provide heat for the net reaction. By bundling multiple packed beds together in parallel, heat can be integrated between reaction steps. The low-temperature parallel packed bed approach makes it well suited for modular implementations. The ability to use number-up manufacturing for the CL-ODH bundles, and the modular scale of the bundles make them well suited for low-cost distributed implementation. In the m- ETL system, the resulting ethylene rich product stream from the CL-ODH reactor would then be compressed and oligomerized to light liquid olefins suitable for use as gasoline. Gaseous hydrocarbon byproducts are then recycled with a slip stream burned in an engine to provide power for the system. Experimental CL-ODH redox catalyst is presented along with process and economic modeling studies.

601424 Ethane Oxidative Dehydrogenation in Microwave-Heated Fixed Bed and Structured Catalyst Reactors

Abstract:

Light olefins, including propylene and ethylene, are extensively used as chemical building blocks. The demand for light olefins has been growing and expected to continue. Steam cracking of naphtha has traditionally been the most common method for light olefins production. However, the intensive energy demand of the process and severe coking and the abundance of shale gas have changed the feedstock in the US. Specifically, catalytic dehydrogenation of light alkanes obtained from shale gas has attracted attention. Yet, this is a highly endothermic and high temperature reaction. Microwave heating is energy-efficient, rapid and selective and can lead to process intensification for modular and distributed ethylene production.

In this talk, we will present ethane oxidative dehydrogenation with carbon dioxide over Ga_2O_3/Al_2O_3 catalyst to provide insights into microwave reactor design. Our results showed that the heating rate with microwaves increases at least 10 times compared to conventional heating. Next, we compare the performance of fixed beds and structured beds of different characteristic features. The temperature uniformity is enhanced in the structured bed due to increased heat transfer. IR thermal camera shows a standard deviation in the temperature distribution of 5.0 °C in the structured bed and of 12 °C in the fixed bed. The structured bed is more stable as the reactivity dropped only 4% after 2 hr of reaction vs. 18% in the case of a fixed bed. Lastly, we will compare a merit index of microwave and conventional heating reactors to assess the potential of microwave technology for light olefins production.

604012 Gas to Liquid Flow Reactor for Conversion of Biogas or Waste Methane to Organic Acids with Hydrogel-Encapsulated Methanotrophs As Fixed Biocatalysts

Abstract:

Small-scale and disparate sources of methane continue to increase in abundance with the growth of biogas generated from waste processing, including digestion of solid wastes and waste-water treatment. Methane is expensive to transport and a greenhouse gas with a warning potential 25 times that of CO_2. Biological upgrading of methane and/or biogas by methanotrophic microorganisms is a promising approach for efficient conversion to value-added products. Metabolic engineering of methanotrophs like *Methylococcus capsulatus* has been successful in converting methane to a variety of organic acids including muconic, lactic, and succinic acids. However, liquid fermenters utilizing methanotrophs are fundamentally mass transfer limited due to the poor solubility of methane in water. Gas-sparged fermentation broths are ineffective for these bioconversions, and an alternative bioreactor design to overcome these limitations is presented herein.

Whole cell methanotroph biocatalysts are encapsulated within a fixed hydrogel support to contact a continuous gas phase in a flow reactor. This approach imagines hydrogel-encapsulated cells as a fixed catalyst bed similar to thermochemical gas flow reactors. Methanotrophs encapsulated in biocompatible hydrogels such as cross-linked PEGs are shown to be viable and function as effective biocatalysts for months. Here, hydrogel biocatalysts are mechanically supported within 3D printed lattice scaffolds, allowing for controlled catalyst geometries for diffusion and reaction. This gas flow reactor was utilized for the conversion of simulated and real biogas feedstocks with these hydrogel-encapsulated biocatalysts. Early work demonstrated that reducing feature sizes to sub-mm scale achieves more than an order of magnitude increase in volumetric productivity. Initial demonstrations at mL-scale were scaled up by several orders of magnitude in flow reactor operation. Wild type methanotrophs were utilized to study methane consumption and process parameters, and engineered strains were studied for organic acid productivity. Various 3D printed scaffold geometries were tested and tradeoffs between mass transfer improvements and overall biocatalyst loading were assessed. The effects of process variables including feed composition, pressure, and gas flow rates on flow bioreactor performance were evaluated. Finally, product recovery approaches and reactor configurations are addressed. This work demonstrates an flow bioreactor for flexible biogas upgrading incorporating advances in additive manufacturing, biocatalyst immobilization, methanotroph metabolic engineering, and reactor design.

610522 Techno-Economic Analysis of Direct Methane Aromatization for Natural Gas Upgrading at Isolated Sites

Abstract:

Inexpensive sources of natural gas have led to a boom in development of technologies taking advantage of C1 chemistry; however, a significant number of potential natural gas sources are stranded and lack the infrastructure for economically-feasible transportation. The resulting gas, which is normally flared, can, in theory, be recovered at the well-site, purified, and then processed to a product which is easier to transport. Gas-to-liquid (GTL) processes exist which can convert the feed to liquid products which can be more easily transported, although there are process constraints due to the need for deployment in remote locations, and several are being commercialized.[1] One such process, direct methane aromatization (DMA), produces significant amounts of benzene, a valuable commodity chemical, and hydrogen. The benzene can be purified on-site and then transported by truck; the hydrogen can be dealt with on-site as well by either transforming it to another transportable chemical, such as ammonia or cyclohexane, or using it to provide heating or electricity.

Previous economic studies of DMA have been reported which either invoke unrealistic assumptions in the process model,[2, 3] do not consider use of the gaseous hydrogen,[2, 3] or do not consider the limitations imposed on the process due to remote deployment.[4] With these issues in mind, a DMA mini-plant processing 1.0 MMSCFD of natural gas was designed with currently available technology to provide a platform for the investigation of process intensified alternatives. The reactor section of the plant uses a pre-carburized DMA catalyst developed at Texas Tech.[5] The unit was investigated for economic feasibility and sustainability. In particular, the analysis identifies four particular areas which can benefit from intensification: aromatics yield (i.e. one-pass conversion), coke selectivity, hydrogen recovery, and separation and purification of the aromatic products. Although intensified reactor technology for the DMA process is already being investigated for improving the yield with[6, 7] and without integrated hydrogen recovery[8, 9], the process model framework should provide opportunities for investigating additional intensification technologies, particularly in separations.

1. H. Fleisch. [White Paper], World Bank-Global Gas Flaring Reduction Partnership, 2015.
2. Pérez-Uresti, S.I. *et al.*, *Processes*, 2017, 5, 33.
3. Camilo Corredor, E., P. Chitta, and M. D. Deo. *Fuel Pro. Tech.*, 2019, 183, 55 – 61.
4. Huang, K. *et al.*, *Joule*, 2018, 2, 349 – 365.
5. Rahman, M., A. Sridhar, and S.J. Khatib. *Cat. A Gen.*, 2018, 558, 67 – 80.
6. Kee, B. *et al.*, *Eng. Chem. Res.*, 2017, 56, 3551 – 3559.
7. Morejundo, S.H. *et al.*, *Science*, 2016, 353(6299), 563 – 566.
8. Brady, C., B. Murphy, and B. Xu. *ACS Catal.*, 2017, 7(6), 3924 – 3928.
9. Cao, Z. *et al.*, *Chem. Int. Ed.*, 2013, 52, 13794 – 13797.

613984 Advanced Nanocomposite Membranes for Natural Gas Purification

Abstract:

This presentation discusses preparation and characterization of novel nanocomposite membranes based on recently discovered porous aromatic framework (PAF-1) material in a matrix of a polymer of intrinsic microporosity (PIM-1) for CO_2 removal from natural gas. Both pure and mixed gas permeation and separation properties will be reported, as will pure and mixed gas physical aging data over periods of 1,000 hours. Permeability coefficients decreased over time and selectivity increased. We also characterized the influence of temperature on gas permeability and selectivity properties. We characterized both thick films, as well as thin-film (~2 μm) composite membranes on a porous support. We find evidence of competitive sorption at short (e.g., 24 hours) aging times, with mixed gas CO_2/CH_4 selectivity values being higher than pure gas values. However, at longer aging times (e.g., >200 hours), the mixed gas selectivity is somewhat less than the pure gas selectivity, indicating the influence of CO_2-induced plasticization in the aged membranes. These mixed matrix membranes are, in some cases, more than 10x more permeable than cellulose acetate, a commonly used membrane for natural gas purification. Selectivity of the nanocomposites was somewhat lower than that of cellulose acetate, but it increased as the samples underwent physical aging.

600896 Iron Oxide Supported on Silicalite-1 As a Multifunctional Material for Biomass Chemical Looping Gasification and Syngas Upgrading

Abstract:

Iron oxide supported on silicatite-1 was synthesized as a novel oxygen carrier for biomass chemical looping gasification (BCLG) with additional functionalities of its reduced form as a catalyst for catalytic decomposition of tar and an adsorbent for capturing CO_2 in syngas. The temperature-programmed-reduction/oxidation profiles showed that the iron oxide supported on silicatite-1 had lower reduction and oxidation temperatures than the iron oxide supported on conventional silica. It was found the iron oxide supported on silicatite-1 was resistant to the formation of strong metal-support interaction and irreversible silicate at a high temperature and exhibited high thermal/chemical stability in a cyclic process. The experimental results showed that the novel oxygen carrier can gasify biomass into high value compounds of H_2 and CO at a moderate temperature. After reduction at 800°C, the iron supported on siliclaite-1 showed higher catalytic activity in tar decomposition than on a conventional silica support, due to the higher surface area and better metal dispersion. As an adsorbent for CO_2 capture, the reduced oxygen carrier with 15% Fe had a 1.01 mmol/g CO_2 adsorption capacity at 1 atmosphere pressure, which is comparable to other CO_2 adsorbents reported in literature. Therefore, the iron oxide supported on silicatite-1 is an effective looping material as an oxygen carrier for biomass gasification, a catalyst for tar decomposition and an adsorbent for CO_2 capturing.

EVALUATION OF THE IRON-BASED MATERIAL WITH Al_2O_3/ZrO_2 ADDITION AS OXYGEN CARRIER APPLIED IN CHEMICAL LOOPING PROCESS

C.T. Kao*, J.Y. Cheng, C.H. Shen and H.P. Wan

Green Energy and Environment Research Laboratories, Industrial Technology Research Institute, Taiwan

Abstract

Chemical looping process is one of the promising technologies for carbon capture. Oxygen carrier plays an important role in the chemical looping process. By the reaction of oxidation and reduction, iron could carry or release the oxygen by its phase change. In the present study, iron-based composite with alumina and zirconia addition were used for oxygen carriers in chemical looping. Material properties including relative density, microstructure, reactivity and mechanical ones were measured. Feasibility of this composite for chemical looping could therefore be evaluated.

The result indicated that all of the specimens formed a layer of Fe_2O_3 in the surface. Different ionic diffusion rate made this phase separation occurred. Moreover, the reactivity was then decreased. In the case of the addition of ZrO_2, an extra oxygen vacancy will produce due to the substitutional reaction between Zr and Fe. Although the specimen which only composited with ZrO_2 and Fe_2O_3 showed the best reactivity, the consuming of Fe_2O_3 might lead it not suitable to be the material for chemical looping.

Keywords: chemical looping, oxygen carrier, zirconia, microstructure, redox reaction

1. Introduction

There was more than 70% electricity generated by fossil fuel in Taiwan during 2010 to 2018. Moreover, over 90,000 kton CO2 are produced each year. According to the statics from IEA 2016 (International energy the global average CO_2 emission per capita was 4.35 tCO_2. While the CO_2 emission in Taiwan was 10.98 tCO_2 per capita, which was twice as much as global average.

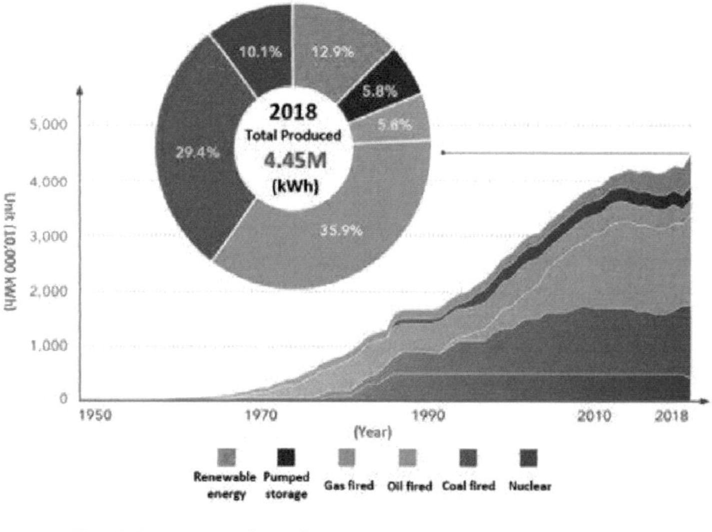

Fig.1 Resource for electricity generation in Taiwan

Among the types of fossil fuel using for electricity generation in Taiwan, coal occupied more than 60% of CO_2 emission. The problem of the CO_2 emission in Taiwan should indeed be solved. The greenhouse gas reduction and management act came into force in 2015. This act laid down GHG emission reduction targets in 2050 for Taiwan should reduce by 50% from 2005 levels.

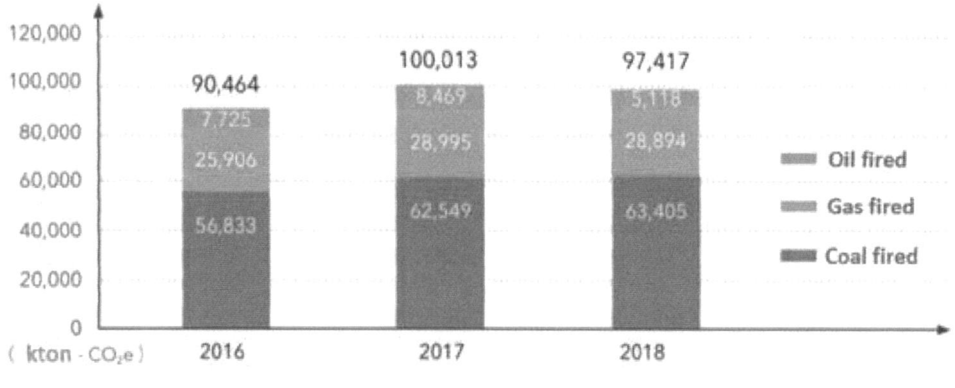

Fig.2 Amount of CO_2 emissions produced from different type of fossil fuel

Chemical looping is one of the most promising routes to solve the CO2 emission fundamentally. It doesn't need extra facility to separate CO2 and other gas.

The concept of chemical looping was shown in Fig. 4. Metal oxide (Me_xO_y) which is called oxygen carrier (O.C.) plays an important role for the chemical looping. The major function for oxygen carrier was to transport oxygen between oxidation and reduction reactor.

Fossil fuel will pass through the reduction reactor and then react with metal oxide (Me_xO_y). Only CO2 and H2O are produced during this reactor. Afterward, H2O could be removed by cooling process. High purity of CO2 was collected and could storage directly. Afterward, the oxygen depleted particle (Me_xO_z) will be transported to oxidation reactor and then react with air. The oxygen could be recovered again. Meanwhile, heat will also generate during this process.

Chemical looping combustion(CLC) was one of the applications in chemical looping process. The major purpose of CLC technology is a low-carbon alternative to the fossil fuel power generation. Fossil fuel such as coal, natural gas could be reacted with the oxygen carrier and then produce the heat and then generate the electricity.

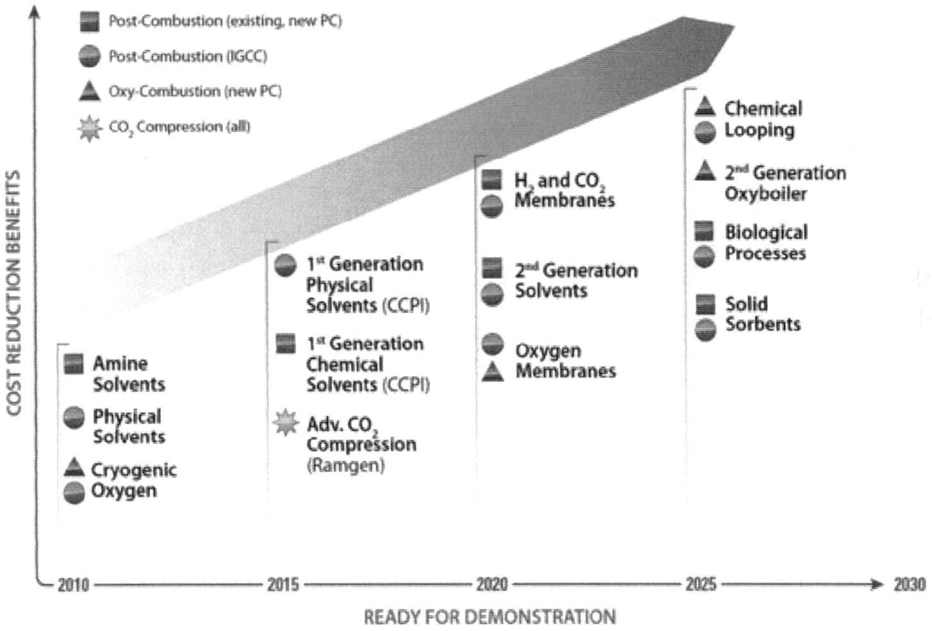

Fig.3 Progress and the future trend for CCS

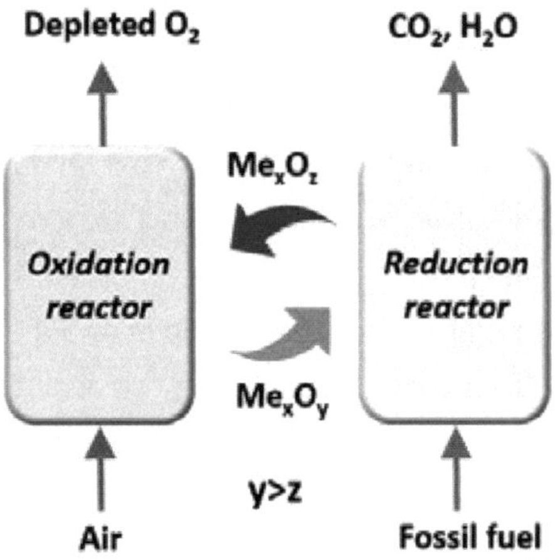

Fig.4 Concept for chemical looping process

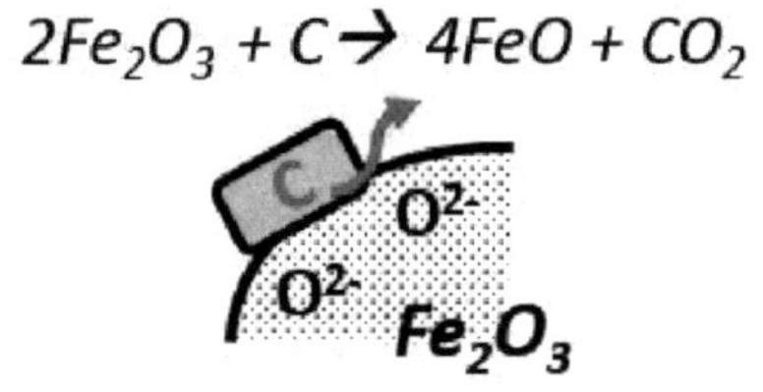

Fig.5 Concept for chemical looping combustion (CLC) process

The progress of chemical looping system developed in ITRI was exhibited in Fig. 6. Both lab. scale testing system including TGA (Thermogravimetric analyzer) and fixed bed system which combined with GC (Gas chromatograph) were used for the performance measurement. Moreover, a 30 kW_t pilot scale system was also established by our group.

For the initial stage (2013-2018), chemical was used as the material for oxygen carrier in chemical looping. Recently, we tried to search the material which was cheaper and environment friendly as the oxygen carrier.

Progress of chemical looping in ITRI, Taiwan

2009-2012 Fundamental research		2013-2014 Gas fuel combustion		2015-2017 Solid fuel combustion

	TGA	Fixed bed	Cold model moving bed	30 kW$_t$ moving bed system	30 kW$_t$ coal direct chemical looping
Reactor System					
Oxygen carrier loading	0.1 g	100 g	20 kg	60 kg	60 kg
Function	Material property	Reaction behavior analysis (CO_2 and H_2 yield)	OC circulation test	Methane combustion and H_2 generation	Coal direct combustion and H_2 generation

Fig.6 Progress of chemical looping system developed in ITRI

Progress of oxygen carrier in ITRI

2013-2015 Outsourcing	2015-2018 Chemical prepared	2019-2020 Recycled material

Material	FeAl	FeAlZr	FeAlCe FeAlZr	Recycled material
Shape	Pellet	Ball	Ball	Ball

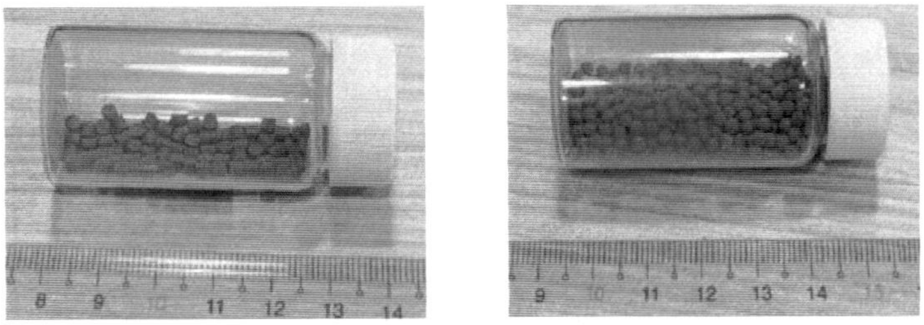

Pellet shape **Ball shape**

Fig.7 Progress of oxygen carrier developed in ITRI

2. Experiment procedure

The preparation procedure for the composite of Al_2O_3/ZrO_2 addition to iron-based showed in Fig. 8. Starting powders including Fe_2O_3, Al_2O_3, ZrO_2 and pore former were mixing in dilute water by mechanical method for 15 h. Drying, granulation and sieving process were then conducted. Particle size around 270 to 500 um was chosen for the following pressing process. Pellet shape with 3 mm in diameter was pressed by die-pressing machine. De-binder and sintering process were then conducted in order to remove the pore former and sinter the specimens.

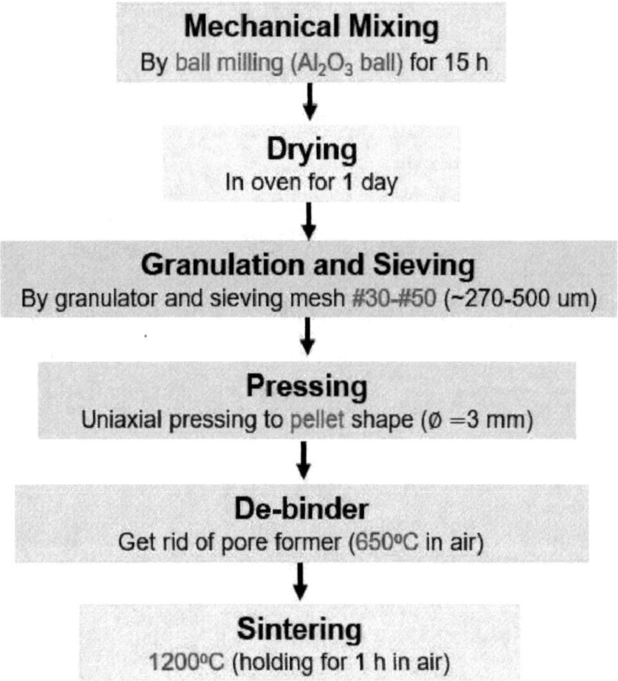

Fig. 8 Flow chart of the specimens prepared in present study

Image of the as-sintered specimens could be found in Fig. 9. For the reactivity test, 50 redox reaction cycle was performed for all of the as-sintered specimens. The parameter of reactivity test for each cycle was shown in Fig. 10. The definition of reactivity could be found in the following equation.

$$X_{OC} = \frac{w_O - w_{(t)}}{w_O - w_R}$$

W_0 is the weight of the composite being fully oxidized
W_R is the weight of the composite being fully reduced
$W_{(t)}$ is the weight of the composite at specific time point

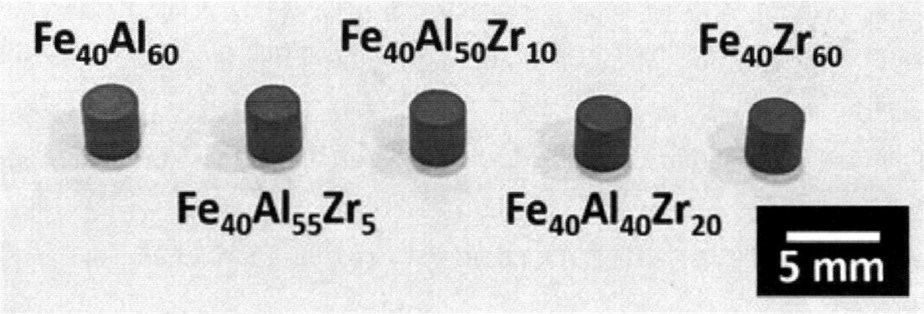

Fig. 9 Image of the as-sintered specimens

Atmosphere	Composition and purity	Flow rate	Holding time
Oxidation	Air 99.9%	200 cc/min	10 min
Purge gas	N_2 99.9%	200 cc/min	5 min
Reduction	H_2 20% with N_2 compensation	200 cc/min	40 min
Purge gas	N_2 99.9%	200 cc/min	5 min
Ramp 10°C/min from room temperature to 900°C			

Fig. 10 Parameter of the reactivity test for single redox reaction cycle

Image for the specimens reacted with 50 redox reaction cycle was shown in Fig. 11. There was a red layer attached in the surface of specimens. $Fe_{40}Zr_{60}$ was broken after reactivity test.

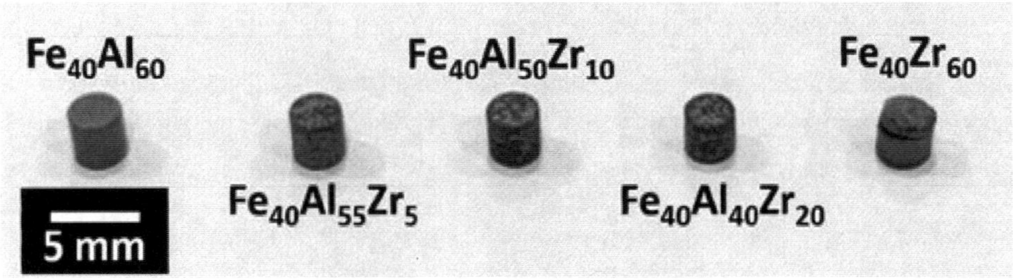

Fig. 11 Image of the as-sintered specimens reacted with 50 redox reaction cycle

One of the bottlenecks for the chemical looping was its cost. A thousand MW_t chemical looping hydrogen generation system (CLHG) was applied for simulation the economic analysis in by Aspen plus in ITRI. The result showed that nearly 40% of the cost for CLHG was derived from the consuming of oxygen carrier material. For both moving bed or fluidized bed reactor, the particle of oxygen carrier should withstand such as impact or abrasion during circulation. Afterward, the particle will loss or break, and need to supplement again.

To elevate the reactivity and durability for oxygen carrier was the major purpose. Both Al_2O_3 and ZrO_2 were applied as the inner support material in the iron-based material. The feasibility of Al_2O_3/ZrO_2 addition in iron-based for chemical looping was then revealed.

3. Result and discussion

The reactivity of the Al_2O_3/ZrO_2 addition in iron-based composition was studied in present study. Microstructure variance was introduced to explain how it affect the reactivity for as-sintered specimens.

Microstructure observation:

The microstructure of the as-sintered were shown in Fig. 12. All of the specimens exhibited porous structure which was attributed by the elimination of pore former during de-bindering process. High porosity could provide more area to oxygen carrier and fuel gas to react.

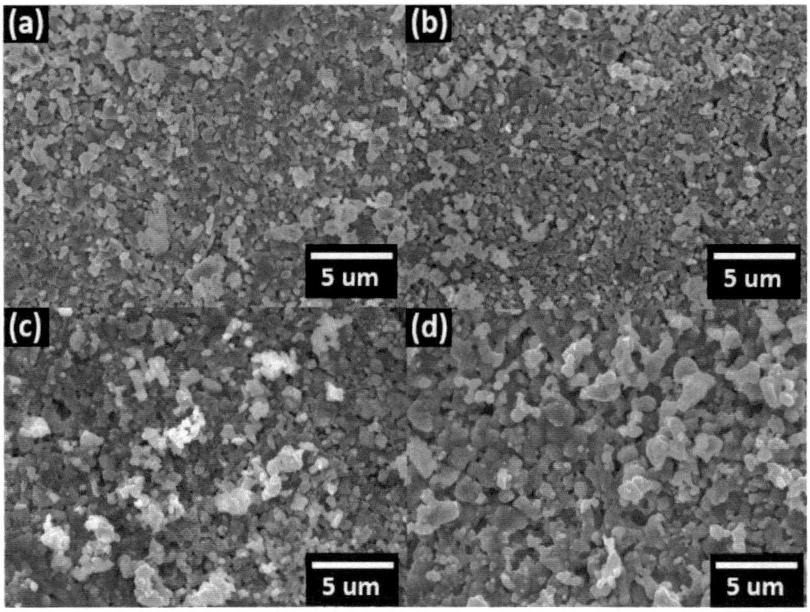

Fig. 12 Microstructure of the as-sintered specimens (a)$Fe_{40}Al_{60}$; (b)$Fe_{40}Al_{50}Zr_{10}$; (c)$Fe_{40}Al_{40}Zr_{20}$ and (d)$Fe_{40}Zr_{60}$.

The specimens after 50 cycle redox reaction was mounted into the epoxy in order to observe its cross-section view. The corresponded microstructure was shown in Fig. 13. And the brighter particle was identified as the Fe_2O_3. Along with the addition of

ZrO_2, the thickness of Fe_2O_3 layer formed in the surface was also increased. Moreover, this layer will detach from the surface in the specimen of $Fe_{40}Zr_{60}$.

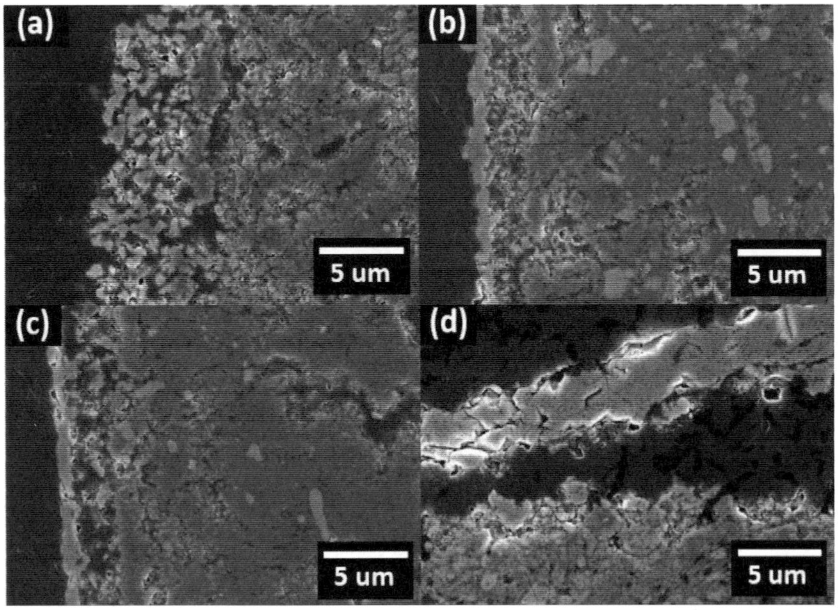

Fig. 13 Microstructure of the as-sintered specimens reacted with 50 redox reaction cycle (a)$Fe_{40}Al_{60}$; (b)$Fe_{40}Al_{50}Zr_{10}$; (c)$Fe_{40}Al_{40}Zr_{20}$ and (d)$Fe_{40}Zr_{60}$.

Layer formation mechanism:

The formation of Fe_2O_3 layer could be regard as the phase separation. By Ellingham diagram, only Fe_2O_3 has phase change during redox reaction. Fe_2O_3 will migrate to the surface in order to react and then segregate in the surface.

Defect in this composite could explain the result which exhibited in Fig. 12. Oxygen vacancy is deem as the most critical defect which could accelerate the ion diffusion in ceramic material. It provides the lower energy gap and easier route for ion diffusion in ceramic material.

For the addition of Al_2O_3 into Fe_2O_3. There was no any extrinsic oxygen vacancy produced.

$$Fe_2O_3 \xrightarrow{Al_2O_3} Fe_{Al}^{\times} + O_O^{\times}$$

However, the addition of ZrO_2 into Fe_2O_3 will produce extra oxygen vacancy. The diffusion rate in this composite was then accelerated.

$$\frac{1}{2}Fe_2O_3 \xrightarrow{ZrO_2} Fe_{Zr}' + \frac{1}{2}V_O^{\cdot\cdot} + O_O^{\times}$$

Therefore, the thickness of the Fe_2O_3 layer for ZrO_2 addition was thicker than that in Al_2O_3 addition.

The reactivity for 50 cycle redox test was shown in Fig. 14. $Fe_{40}Zr_{60}$ showed the best reactivity with around 70%. Other as-sintered specimens showed the similar reactivity with around 45%.

The Fe_2O_3 layer formed in surface might play as the barrier which inhibited the reaction inside the specimen. And the reaction area was thus only occurred in surface.

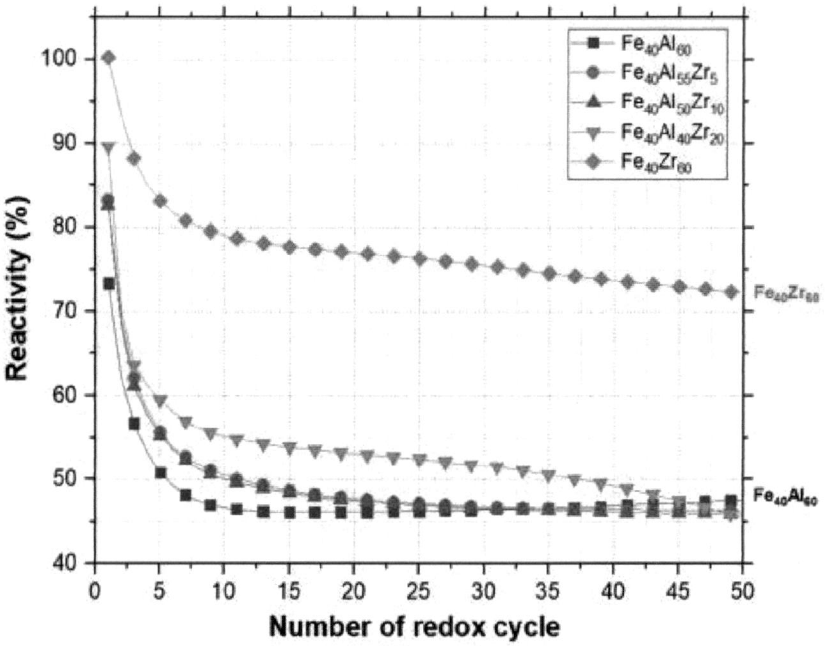

Fig. 14 Reactivity of the as-sintered specimens reacted with 50 redox reaction cycle.

In case of $Fe_{40}Zr_{60}$, because of the Fe_2O_3 layer was too thick to detach, Fig. 13(d), the inner material could thus expose and react with fuel gas again. The highest reactivity for $Fe_{40}Zr_{60}$ to all of specimens was therefore exhibited.

Although the specimen of $Fe_{40}Zr_{60}$ showed the best reactivity, the expected reactivity of $Fe_{40}Zr_{60}$ which over 50th redox cycles might be gradually decreased due to the consuming of Fe_2O_3. Fe_2O_3 layer will formed and then detach to the mother body repeated.

The addition of ZrO_2 in Fe_2O_3 showed the negative effect to the mechanical properties, Fig. 15. The major reason could be caused by the phase transformation from zirconia and it decreased the relative density.

Composition	Relative density (%)	Crash strength (kgf)	Attrition (%)
$Fe_{40}Al_{60}$	66.42 ± 1.96	57.17 ± 11.75	0.75
$Fe_{40}Al_{50}Zr_{10}$	62.43 ± 5.49	40.93 ± 7.60	1.22
$Fe_{40}Al_{40}Zr_{20}$	65.39 ± 4.61	54.45 ± 7.86	0.70
$Fe_{40}Zr_{60}$	54.45 ± 5.81	27.81 ± 5.42	2.03

Fig. 15 Mechanical property of crash strength and attrition.

4. Conclusion

The feasibility of the addition of Al_2O_3 and ZrO_2 in iron-based oxygen carrier was revealed in the present study. The corresponded microstructure before and after 50 redox reaction cycle were also exhibited.

For the reactivity test conducted in present study, all of the as-sintered specimens showed a drastic decrease within the first ten redox cycle. The Fe_2O_3 layer formed in the surface played as the barrier to inhibit the deeply reaction to the inner layer material.

Due to the substitution reaction between Zr and Fe, additional oxygen vacancy was produced and then promoted the ion diffusion. The thickness of the Fe_2O_3 layer which formed in the surface was thus increased.

$Fe_{40}Zr_{60}$ showed the best reactivity among the as-sintered specimens in present study. However, the worst mechanical property and the consuming of Fe_2O_3 during redox reaction make it be excluded the candidates for being the material for chemical looping.

Acknowledgements

This project was supported by Bureau of Energy, Ministry of Economic Affairs, project number 108-E0216.

610444 Feasibility Analysis of Chemical Looping Ammonia Synthesis from Stranded Natural Gas.

Abstract:

Stranded Natural Gas (SNG) flaring is intermittent flaring of natural gas at a volume or location which makes it uneconomical to recover by traditional methods (compression to liquid natural gas, or combustion for electricity). SNG flaring releases a substantial amount of greenhouse gas emissions, prompting public and private entities to focus efforts on developing technologies to convert SNG to a transportable value-added product.[1] Ammonia production from SNG is a practical solution, considering it is industrially produced from natural gas today through the Haber Bosch process (1).

N_2+$H_2$$NH_3$, *T=450°C, P=200atm*

While the high temperature and pressure of the Haber Bosch process is needed to push the reaction equilibrium forward at an industrially acceptable rate, it makes the process uneconomical at small scales [2]. Recently attention has been given to chemical looping for ammonia synthesis (CLAS), because of its ability to produce ammonia at atmospheric pressures and favorable small-scale economics [2]. A literature review identified previously proposed chemical loops for ammonia synthesis with varying levels of technical success. Therefore, the objective of this work is to provide a techno-economic assessment of ammonia production from SNG using chemical looping synthesis.

Chemical looping is capable of circumventing the high pressures of the Haber Bosch process by separating the reaction into individual steps mediated by a material carrier. Generally, in CLAS a material carrier is used to fix nitrogen, then a hydrogen source is used to reduce the carrier and evolve ammonia leaving the material carrier to complete the chemical loop. Recently, a chemical loop using alkali metals as the material carrier was demonstrated to produce ammonia at atmospheric pressures and 100 [3]:

$xN_2 + 4AH_x\ 2xA_{2/x}NH + xH_2$

$2xA_{2/x}NH + 2xH_2 xNH_3 + 2AH_x$

(A denotes an alkali metal and x is the valence of A)

A different chemical loop was proposed in [4] using water as the hydrogen source mediated by a calcium carrier. Although, the hydrolysis of calcium nitride is thermodynamically favored to produce ammonia it also produces a stable calcium oxide. This results in an energy intensive process to regenerate calcium and complete the chemical loop:

$3Ca + N_2\ Ca_3N_2$

$Ca_3N_2 + 3H_2O\ 2NH_3 + 3CaO$

$CaO + R\ Ca + RO$

(R denotes suitable reducing agent)

Here, to evaluate previously proposed chemical loops for mild condition synthesis of ammonia, focus is given to their thermodynamic feasibility. Thermodynamic feasibility is determined as a negative Gibbs free energy of reaction at a temperature in which the constituents of the reaction do not phase change. Along with thermodynamic feasibility, process safety, reaction yield and other criteria are used to create a decision tree to rationally evaluate chemical loops presented in this work. For each material carrier and chemical loop combination, the temperature that minimizes the Gibbs free energy of reaction at a constant pressure of 1 atm was found, assuming ideal states. Aspen plus was used to simulate the equilibrium reactions and to assess economic viability of the process using its economic analyzer. From this work it was discovered that modifying a previously proposed unsuccessful chemical loop, creates a new feasible chemical loop, shown in Fig.1. This new chemical loop along with other techno-economically feasible chemical loops are studied and presented in this work. Further studies are recommended to study the kinetic feasibility of the proposed chemical loops for industrial applications.

Acknowledgements

This work was partially sponsored by the United Technologies Corporation Institute for Advanced Systems Engineering (UTC-IASE) of the University of Connecticut. Any opinions expressed herein are those of the authors and do not represent those of the sponsor.

References

[1] Natural Gas Flaring and Venting: State and Federal Regulatory Overview, Trends, and Impacts, 2019.

[2] R. Michalsky, B.J. Parman, V. Amanor-Boadu, P.H. Pfromm, Solar thermochemical production of ammonia from water, air and sunlight: Thermodynamic and economic analyses, Energy. 42 (2012) 251–260. doi:10.1016/j.energy.2012.03.062.

[3] W. Gao, J. Guo, P. Wang, Q. Wang, F. Chang, Q. Pei, W. Zhang, L. Liu, P. Chen, Production of ammonia via a chemical looping process based on metal imides as nitrogen carriers, Nat. Energy. 3 (2018) 1067–1075. doi:10.1038/s41560-018-0268-z.

[4] R. Michalsky, Thermochemical production of ammonia using sunligth, air, water and biomass, Kansas State University, 2012.

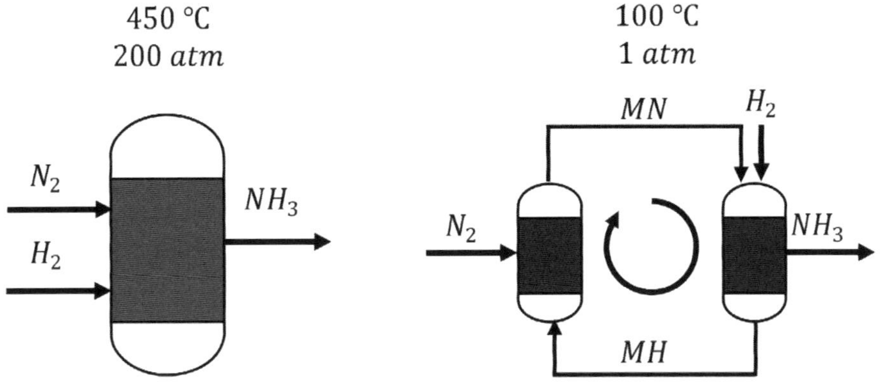

Figure 1. Ammonia production via the Haber Bosch process (left) compared to chemical looping process (right). Where M is the metal carrier, MN is the metal nitride and MH is the metal hydride.

596792 Oxygen Vacancy Healing for Desorption of Stored Ammonia from a Metal/Zeolite Nanocomposite

Abstract:

Recently, the US Department of Energy (DOE) has set targets for the storage and delivery of hydrogen for light passenger vehicles which require a sorbent material with reversible storage, low cost, fast kinetics, and stability in ambient conditions. In this regard, gaseous ammonia has gained major attention for its potential as a high density hydrogen carrier. Storage of ammonia in porous materials could address many of the DOE target metrics and offer additional benefits including hazard mitigation, minimal pressurization and cooling, and facile material handling.

Zeolites offer highly tunable pore chemistries and geometries, which allows precise control of the chemical environment of the adsorbate. The intrinsic acidity of zeolites promotes strong gas-solid interactions but simultaneously hinders the effective release of the stored gas due to chemisorption. To overcome this strong binding energy, temperature or pressure swing is typically employed to remove adsorbed ammonia.

In the current work, the partial reduction of a nickel-oxide/zeolite nanocomposite led to a remarkable heat release upon exposure to air, which was attributed to the exothermic healing of oxygen vacancies in the NiO_{1-x} lattice. This heating phenomenon was shown to be repeatable over 10 at least cycles. Temperatures as high at 694 °C were measured in the catalyst bed on exposure to air. By applying the redox properties of the nickel-oxide/zeolite nanocomposite, significant desorption of ammonia was promoted. The removal efficiency was found to be 38%, an improvement in comparison to vacuum swing desorption (30%). Future outlooks are proposed to improve the total ammonia uptake and to improve the cycling process.

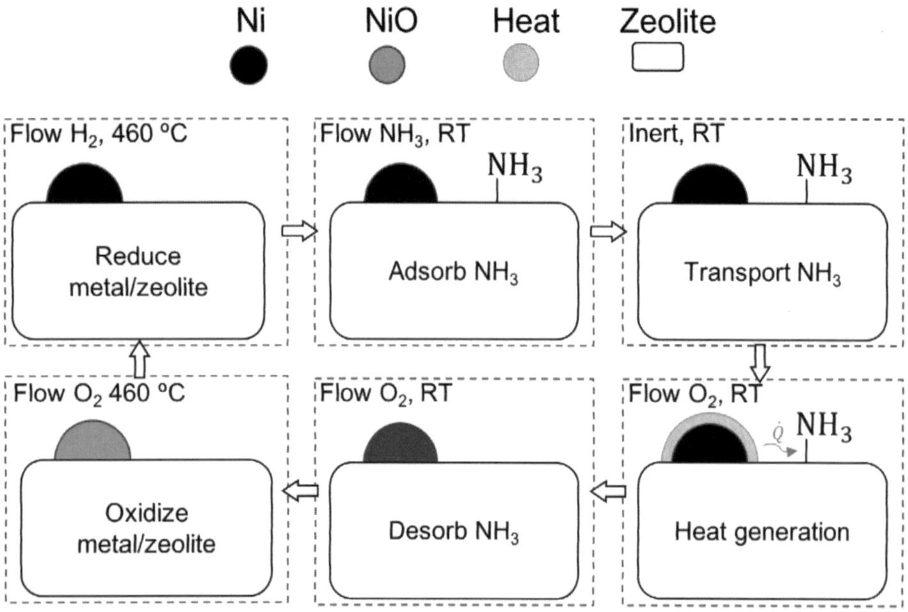

Scheme 1. Ammonia storage on metal/zeolite sorbents using the redox properties of the reducible metal sites

601361 A Molten Carbonate Shell Modified Perovskite Redox Catalyst for Anaerobic Oxidative Dehydrogenation of Light Alkanes

Abstract:

Acceptor doped, redox-active perovskite oxides such as $La_{0.8}Sr_{0.2}FeO_3$ (LSF), $La_{0.8}Sr_{0.2}MnO_3$ (LSM) and $La_{0.8}Sr_{0.2}Co_{0.2}Fe_{0.8}O_3$ (LSCF) are active for light alkane oxidation to CO_x. However, they show poor selectivity to olefins. This study investigates molten alkali metal salts as effective "promoters" to modify the surface of perovskite oxide for chemical looping–oxidative dehydrogenation (CL-ODH) of light alkanes including ethane, propane and butane. Li_2CO_3 coated LSF (LSF@Li_2CO_3) is an efficient redox catalyst for CL-ODH of ethane [1]. Unlike previously reported redox catalysts which are typically endothermic during the lattice oxygen assisted ethane ODH step, the labile oxygen species in LSF@Li_2CO_3 facilitate exothermic operation in both redox steps as confirmed by TGA-DSC measurements (-69.5 kJ/mol [O] during the ODH step and -40 kJ/mol [O] during the re-oxidation step). Up to 92.2% ethylene selectivity and 63.6% ethane conversion were obtained. TEM, XPS, XRD characterizations and DSC measurement indicate that the redox catalyst is composed of a layer of molten Li_2CO_3 covering the solid LSF substrate. $^{18}O_2$-exchange experiments and electrochemical impedance spectroscopy indicate that the molten Li_2CO_3 layer facilitates oxygen shuttling from LSF bulk to the molten carbonate layer surface while blocking the non-selective sites for ethane oxidation. Further investigations of the potential reaction pathways indicates that peroxide species (O_2^{2-}) are the most likely active species for CL-ODH. TGA measurements, *in-situ* XRD and Mössbauer spectroscopy indicate that Fe^{4+} species reduction to Fe^{3+} is responsible for the formation of the active peroxide, which are subsequently transported to the outer surface of the molten Li_2CO_3 layer for the ODH reaction (**Fig. 1a**). The formation of active peroxide via Fe^{4+} to Fe^{3+} transition is further supported by DFT calculation. With other molten alkali metal salt promoters, the perovskite oxide can also be modified for CL-ODH of propane and butane. More than 50% of C_{2+} olefin yield (ethylene + propylene) at 600 °C for CL-ODH of propane was obtained. In a similar manner, >40% C_4 olefin yield at 500 °C for CL-ODH of butane was obtained (**Fig. 1b**). With high olefin selectivity, high olefin yield and favorable heat of reactions, the core-shell redox catalyst has excellent potential to be effective for intensified light alkane conversion. The mechanistic findings also provide a generalized approach for designing CL-ODH redox catalysts.

Reference

[1] Gao et al., *Science Advanced*, Accepted in 2020, **In Press**

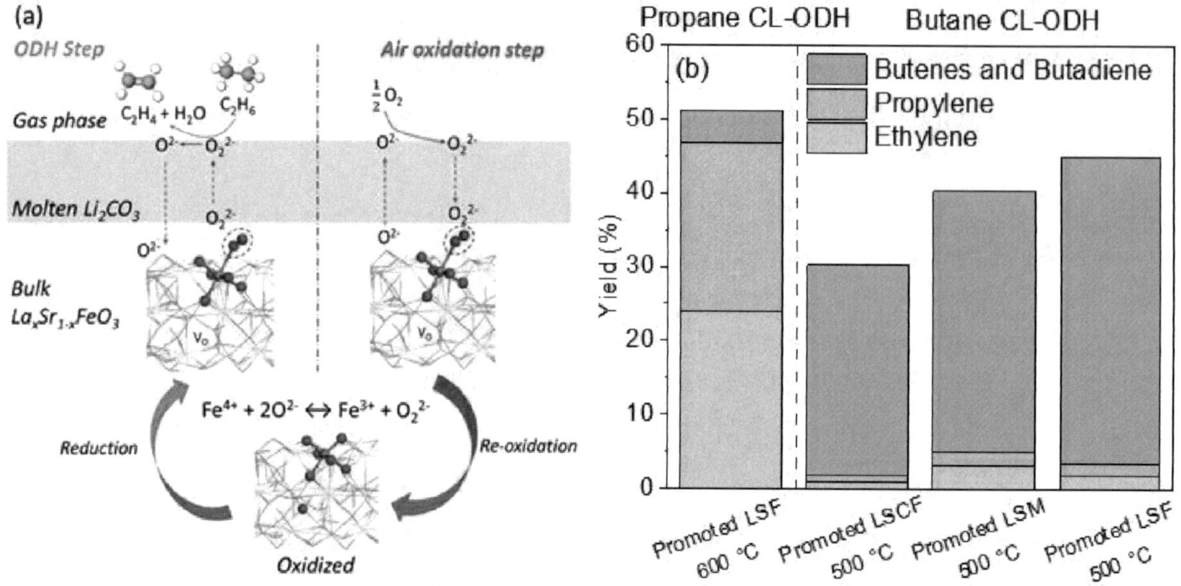

Figure 1. (a) Mechanistic drawing of ethane CL-ODH on Li_2CO_3 promoted LSF; (b) Butane and propane CL-ODH on perovskite oxides promoted with Li-containing salts.

609815 Solar Driven Iron Oxide Based Thermochemical Methane Reforming and Water Splitting for H_2 and Syngas Production

Abstract:

Due to the continuous increase in the population of world and drastic depletion of the fossil fuel reservoirs, it is highly essential to invest towards renewable energy technologies such as solar energy (storage, conversion and utilization). A two-step solar thermochemical H_2O and/or CO_2 splitting process which utilizes metal oxide based redox reactions is one of the promising ways of producing solar fuels such as solar H_2 or renewable precursors for fuels such as solar syngas (a mixture of H_2 and CO). This process includes the endothermic reduction of metal oxides at elevated temperatures by releasing O_2 and oxidation of the reduced metal oxide by H_2O, by CO_2, or by the mixture of the two producing H_2, CO or syngas. One of the major limitations associated with the MO based cycles is the higher thermal reduction temperature. To solve this issue associated with the higher reduction temperatures, methane can be used as the reducing agent. In this investigation, thermodynamic analysis of the two-step iron oxide based solar thermochemical methane reforming and water-splitting process was investigated in two sections with the help of commercial thermodynamic software's such as HSC Chemistry and FactSage. In the first section, the thermodynamic equilibrium compositions were determined as a function of temperature and other operating parameters at different experimental conditions. In the second section, the second law thermodynamic analysis of iron oxide based solar thermochemical methane reforming and water-splitting process was performed. Once the process flow diagram was set, effects of operating parameters on solar absorption efficiency of the solar reactor (η_{abs}), solar energy input to the solar reactor (Q_{solar}), rediation heat losses from the solar reactor (Q_{re-rad}), net energy absorbed in the solar reactor ($Q_{reactor-net}$), rates of entropy produced in the solar reactor ($I_{rr,reactor}$) and cooling unit ($I_{rr,cooling}$) were calculated and plotted. At the end, the solar-to-fuel conversion efficiency ($\eta_{solar-to-fuel}$) for this process was estimated and compared with the other thermochemical fuel production cycles.

598068 Kinetic Studies on Chemical Looping – Oxidative Dehydrogenation of Ethane Using Surface Modified Mixed Oxides

Abstract:

Ethylene is commercially produced from ethane via high-temperature pyrolysis in the presence of diluting steam a.k.a. steam cracking. The process requires high temperature (>1100 K) and a large steam generation load, making it energy intensive with an energy consumption of 17-21 GJ/tonne ethylene. Alternatively, Chemical Looping – Oxidative Dehydrogenation (CL-ODH) has the potential to substantially reduce the energy demand and to increase the single-pass ethane conversion. The ability of a redox catalyst for selective hydrogen combustion (SHC), i.e. selectively oxidizing hydrogen co-product from ethane dehydrogenation, represents an effective strategy for CL-ODH because it can shift the reaction equilibrium and facilitate autothermal operation.

In this work, redox kinetics of unpromoted and Na_2WO_4-promoted $CaMnO_3$ based redox catalysts were investigated in the context of CL-ODH. In the case of unpromoted $CaMn_{0.9}Fe_{0.1}O_3$, the reduction kinetics follow reaction-order models. The activation energy for C_2H_4 reduction is approximately three times greater than that for H_2 reduction. The greater dependence on active lattice oxygen concentration and the larger energy barrier under C_2H_4 reduction result in an order-of-magnitude decrease in the reduction rate. Despite of its favorable SHC properties, $CaMn_{0.9}Fe_{0.1}O_3$ redox catalyst leads to significant CO_2 formation in CL-ODH of ethane. To further improve the redox catalyst's selectivity towards hydrogen combustion, a Na_2WO_4 promoter was impregnated on $CaMn_{0.9}Fe_{0.1}O_3$. The reduction of Na_2WO_4-promoted $CaMn_{0.9}Fe_{0.1}O_3$ follows an Avrami-Erofe'ev nucleation and nuclei growth model. The addition of Na_2WO_4 more significantly suppressed C_2H_4 combustion relative to H_2 oxidation. As a result, the reduction rate of Na_2WO_4-promoted $CaMn_{0.9}Fe_{0.1}O_3$ under H_2 was three orders of magnitude greater than that under C_2H_4, demonstrating its excellent SHC properties.

In addition, a number of other mixed oxide redox catalyst particles with different alkali metal salt promoters were also investigated, showing promising performance in terms of long-term stability and single-pass ethylene yield at different operating temperature ranges. Both the kinetic studies and long-term experiments validate the feasibility and attractiveness of the CL-ODH approach.

610492 Catalytic Pyrolysis of Waste Automobile Tires to Low Sulfur Fuels

- Download/View Extended Abstract

Abstract:

Energy production and utilization is one of the main drivers of human civilization and until we master how to produce and distribute it efficiently and sustainably, the human civilization will always be threaten for survival. Currently, there are millions of tons of organic materials, which are classified as waste because we do not know how to efficiently utilize them to benefit humankind. A classic example is the so-called waste automobile tires, which are ubiquitous in every country. Disposal of these tires are increasingly becoming a challenge because of limited landfills and fire problems. Improperly disposed waste tires can potentially become the sources of diseases such as malaria fever and zika virus because they become breeding grounds for mosquitoes that carry the disease. The traditional disposal methods such as land filling, road applications, playground applications, direct pyrolysis to produce high sulfur heating fuels, and combustion as tire-derived fuels are becoming inadequate because of pollution problems associated with them and the low cost petroleum crude oil prices currently prevailing. In the USA about 300 million tires are disposed of annually either in landfills, combustion, and other disposal methods. Tire pyrolysis is extensively practiced in several countries especially, China and India where they are used to produce tire pyrolysis oils for heating. These oils contain 1-1.35 wt% sulfur, which make them unsuitable for transportation fuels where the sulfur content is typically below 0.5 %. The sulfur content of these oils can be reduced through various sweetening or desulfurization processes, but these unit operations increase the cost of fuel and thus noncompetitive relative to petroleum derived heating fuels. At the USTAR Bioenergy Center we have developed a sulfur-tolerant catalyst for catalytic pyrolysis that effectively removes over 60 wt% of the sulfur in the pyrolysis oil and also removes the chlorine in the butyl rubber liner. Thus, the pyrolysis oil can be easily upgraded to higher value fuel. In general, most waste tire pyrolysis oils reported in literature contain 0.80 – 1.35 wt% sulfur for atmospheric pyrolysis oil compared to 0.38 wt% that we measured for the oils produced from the tire crumbs used for these studies. The oxygen content of the oil was less than 1%, which meant that the oil will need minimal hydrogenation to remove the residual oxygen. The hydrodeoxygenation of the tire pyrolysis oils also converted the residual sulfur to hydrogen sulfide and hence making the fuel sulfur free. The higher heating value (HHV) of the pyrolysis oil was 42.7 Mg/kg and after hydrotreatment increased to 45.7 MJ/kg. The ASTM D86 distillation of the pyrolysis showed 95% diesel fraction while after hydrotreating the diesel fraction was 55% and gasoline fraction was 35%. Thus, the catalytic tire pyrolysis oil would be suitable a marine diesel fuel because of its low sulfur content and high energy content.

599918 Probabilistic Analysis of Converting Marine-Borne Plastics into Usable Fuels

Abstract:

Oceanic plastics represent a major global environmental pollution problem. Every year between 4.8 and 12.7 million tons of plastic enter the ocean and as a result a potentially dangerously high amount of plastic has already accumulated in the oceans. This plastic is highly concentrated in the naturally occurring gyre's in the ocean which could aide in any clean-up efforts, but the question remains what should be done with the plastic once it is collected? We have proposed a process that would involve a ship collecting plastic in the ocean and converting it into fuel in an onboard hydrothermal liquefaction reactor to fuel itself. Before starting experimental work, we wanted to determine whether the afore mentioned process could in fact make more energy than it consumes. To do this in the presence of uncertainty a probabilistic exergy analysis was carried out on a theoretical system for three different and common plastics, polystyrene, polyethylene and polypropylene, as well as for a mixed plastic stream. The theoretical process consisted of a plastic shredder, reverse osmosis system, reactor, separator and an ultraviolet & hydrogen peroxide oxidation system. The exergy model accounted for the energy requirements of the equipment, heating up the reactants to the reaction temperature and the chemical exergy of the streams before and after reaction. This analysis allowed us to determine the break even point (when the net exergy was equal to zero) as well as the point at which there was a 50% chance of having net exergy production and create probability curves as seen in Figure 1. The analysis showed that with less than 10% by volume plastic in the feed (ocean) enough energy could be created to run the equipment and fuel a boat large enough to hold it. The work also allowed for the creation of a decision-making tool that allows one to assess the risk associated with a project.

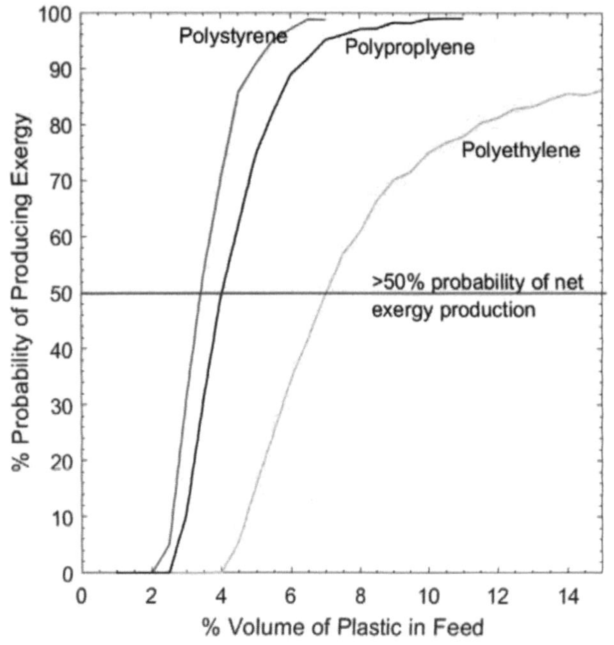

Figure 1 Exergy analysis of polystyrene, polypropylene and polyethylene including the energy required for the ship's engine.

611928 Struvite Recovery from Manure By Way of Hydrothermal Carbonization

Abstract:

Dairies and other animal feed lots have grown in size in recent years. Traditional practices of manure disposal were focused on land application, but this is problematic for large operations. The manure is either applied heavily in nearby fields, leading to overapplication and deleterious effects on environment, or the cost of hauling to spread on wider spaced fields is prohibitive. At the same time, there is an opportunity to recover valuable nutrients in manure, primarily phosphorous but also nitrogen, into a valuable form and reduce the costs associated with manure management.

Hydrothermal carbonization has been established as a promising treatment technology for wet biomass. A recent study has shown its potential in solubilization of nutrient content of cow manure into the hydrothermal aqueous product specially when its combined with a strong organic acid such as citric acid. In the current study dairy manure was treated by HTC process at 170C for 10 minutes with different concentration of citric acid. The resulting liquid phase was then used as a supernatant for struvite crystallization at pH range of 9-10. We show experimentally effects of HTC reaction conditions on recovery of phosphorous in struvite. Different ratios of N:P and Mg:P were studied to obtain the optimum precipitation condition. The effect of different concentrations of the organic acid on the crystallization mechanism and the quality of the final struvite were investigated.

603049 Conversion of Recycled Drywall into Nitrogen Efficient Fertilizers

Abstract:

New, cost-effective and sustainable methods are needed to decrease the reactive N-losses into the environment from mineral fertilizers. For this purpose, we demonstrate a successful mechanochemical synthesis of urea cocrystals containing secondary nutrients calcium (Ca) and sulfur (S) using widely abundant drywall gypsum waste. Resulting $CaSO_4 \times 4$urea cocrystal possesses unique reactive properties. In particular, it possesses a higher urea decomposition temperature of 195 ^{o}C as it foregoes the melting of urea at 137 ^{o}C as inferred from thermogravimetric measurements. A combination of Dynamic Vapor Sorption (DVS) and *in situ* Raman spectroscopy measurements suggested a much lower $CaSO_4 \times 4$urea propensity to adsorb and react with water as relative humidity than urea alone. *Ab initio* thermodynamics framework was devised and showed a large apparent thermodynamic barrier of crystalline $CaSO_4 \times 4$urea to transform into urea and the resulting reactive gases, NH_3 and CO_2. Detailed N and S measurements suggested their enhanced availability when using cocrystals especially during the early stages of the experiments. The synthesis process described using waste drywall gypsum to form $CaSO_4 \cdot 4$urea cocrystal mechanochemically thus can provide a new value proposition for this abundant waste and contributing to sustainable use as it can decrease the use of various enzyme inhibiting synthetic organophosphorus compounds.

604232 Synthesis and Characterization of Improved Energy Materials from Spent Lithium-Ion Battery Cathodes

Abstract:

Lithium-ion batteries (LiBs) are widely used as electrochemical power sources in energy applications because of their high energy and power density. With the global market for electric vehicles estimated to reach $150 billion by 2025, an efficient and sustainable metal recovery process has become essential to minimize pollution created by disposing of LiBs in landfills. In this work, a closed-loop, sustainable process for recovery of Li and Co from $LiCoO_2$ (the most common cathode of LiBs) is presented using oxalate chemistry. Oxalic acid has the unique property of separating Li and Co directly by precipitating cobalt oxalate dihydrate ($CoC_2O_4 \cdot 2H_2O$) during the digestion while lithium oxalate ($Li_2C_2O_4$) remains in solution. The addition of hydrogen peroxide as an external reducing agent during the metal extraction process reduces the amount of oxalic acid required for digestion and results in the precipitation of cobalt oxalate dihydrate having a unique micro-rod structure. The rod-like structure is maintained after calcination to form unique cobalt oxide materials (Co_3O_4 and CoO) that can be used in synthesizing high-performance $LiCoO_2$ cathode materials. This presentation will describe the experimental and mechanistic details of the synthesis and characterization of the improved energy materials along with their applications.

602611 Upgrading Biochar Via Co-Pyrolysis of Agricultural Biomass and Spent Tires/Sulfur Wastes for Environmental Remediation

Abstract:

This study was carried out to use 5 biochars (woochip (WC) biochar, raw sulfur/rice straw (RS) biochar, modified sulfur/RS biochar, spent tire/RS biochar, and CO_2-activated biochar) as sorbents for explosive (2,4-dinitrotoluene (DNT)), halogenated phenol (2,4-dichlorophenol (DCP)), toxic metals (Pb and Ba) and toxic anions (Cr and Se). RS biochar was also evaluated as control material. Spent tire/RS biochar was the most effective sorbent for DNT, showing the maximum sorption capacities of 23.4 mg/g. Statistical analysis indicates that BET surface area and hydrophobic may be involved in the sorption of DNT to biochar. Compared to RS biochar, WC biochar samples effectively remove DCP considering maximum sorption capacity and properties of sorbents and sorbates, it appears that the sorption capacity of biochar for DCP is related to the pH, pK_a of DCP to biochar. Raw sulfur/RS biochar and CO_2-activated RS biochar had the highest sorption capacities for Pb, showing 126.6 and 132.2 mg/g, respectively. In the case of Ba, CO_2-activated RS biochar was the most effective (23.98 mg/g). Correlation analysis between biochar properties and maximum sorption capacities suggests that surface functional groups and CEC may affect the sorption of cationic heavy metals to biochar. Microtox® bioassay and TCLP tests did not show any toxic effects of the synthesized biochar. Our results suggest that co-pyrolysis of spent tires with biomass wastes may be a viable option to dispose spent tires by decreasing mobility and toxicity of metals as well as increasing sorption capacity of biochar for some contaminants.

597288 Synthesis of Nitrogen-Containing Carbon Dots As Fe (III) Probe after Extracting Humic Acid from Compost of Sludge and Leaf

Abstract:

Sewage sludge is generated as a by-product during wastewater treatment and needs to be efficient and environmentally friendly disposal. In the past decades, the excess sludge in municipal wastewater treatment plants was mainly treated environmentally-unfriendly such as landfill, incineration. Therefore, recycling sludge into high value-added products, such as humic acid, arising great attention. Composting is an ideal aerobic transformation process for biowaste stabilization. Especially, humic acid contains conjugated double bonds and aromatic rings, the similar structure with natural biomass which were successfully used to hydrothermally synthesize carbon dots. Therefore, it's promising to utilize humic acid as a precursor for hydrothermal oxidation to synthesize N-CDs.

In this work, blue fluorescent N-CDs was synthesized for the first time by the use of humic acid. The schematic illustration on the synthesis of fluorescent N-CDs is in Figure 1. Firstly, the humic acid was extracted from the compost of sludge and leaf for recycling solid waste. Then hydrothermal treatment of humic acid using hydrogen peroxide as an oxidant, blue fluorescent N-CDs was synthesized within short time. A series of characterization methods including TEM, XPS, FTIR and Raman analyses were used to conform the successful synthesis of N-CDs. As shown in Figure 2, the N-CDs has an average diameter of 1.88 nm and exhibits a well-defined spherical shape, and the distances between the lattice fringes are further characterized to be 0.218 nm and 0.355 nm, corresponding to the (010) in-plane lattice spacing and inter-layer distance of graphite[1]. Furthermore, Raman analysis of N-CDs revealed the existence of D and G bands at ~ 1390 cm−1 and ~ 1600 cm−1, respectively, confirming their architectures belonging to the carbon nanostructure family. Notably, the observed higher intensity of the G band than the D band verified the high quality of the synthesized products. Moreover, XPS measurements suggested that the synthesized N-CDs contains C, O and N elements with atom ratio of C:O:N = 67.56: 27.61: 4.82.

Nowadays, a large amount of iron ions-polluted wastewater is discharged into nature due to the widespread application and imperfect post-processing system in industrialized countries, leading to an increased accumulation of Fe^{3+} in the water system. Therefore, it's an urgent task to probe and monitor the Fe^{3+} in the water system. Among the various method developed to probe Fe^{3+}, fluorescence detection is a favourable method due to high sensitivity, selectivity and fast detection. The aqueous solution of the obtained N-CDs with excitation-dependent properties showed a strong blue emission with 8.8% quantum yield (QY) under the excitation of 340 nm, which indicates the N-CDs a promising sensing probes for Fe^{3+} in water. With the addition of Fe^{3+}, the fluorescence intensity of the N-CDs gradually decreased. This research indicates an extremely sensitive method for detecting Fe^{3+} at a lower 1.9 nM level by using the synthesized N-CDs as fluorescent probes without a complicated procedure. The detectability is lower than the concentration of Fe^{3+} in the specified drinking water standard of 5.36mM[2]. Furthermore, it is lower than the limit concentration of Fe^{3+} in reverse osmosis influent, which assure the N-CDs can be used as a sensor to prevent the reverse osmose membrane from being contaminated by excess Fe^{3+}. On the other hand, the practical feasibility of the synthesized N-CDs for probing Fe^{3+} in the actual sample of water has also been investigated. The reduction rate of fluorescence intensity of the N-CDs had a good linear relationship with the added water volume of the Youth Lake when the amount of lake water added was changed from 0 to 300 μL. The concentration of Fe^{3+} in the lake water was calculated according to the standard curve to be 20.4 mM, which is consistent with the result detected by ICP-MS, confirming the accuracy of the N-CDs method for detecting Fe^{3+} in actual environment.

Moreover, the mechanism of the N-CDs sensing technique was investigated in detail. It can be explained by the fluorescence static quenching effect and a ground-state complex formed in this process. The formation of the ground-state complex ensures the specificity of the N-CDs for the detection of Fe^{3+}, that is, the N-CDs will have no quenching effect with other metal ions.

Acknowledgments

The authors are grateful to the financial support of National Natural Science Foundation of China (Nos. 21776203, 21576187 and 51478308).

References

[1]Tang, L , et al. Energy-level structure of nitrogen-doped graphene quantum dots. Journal of Materials Chemistry C 2013, 32: p. 4908-4915.

[2]Organization, World Health. "Guidelines for Drinking-water Quality 4th Ed". (2011).

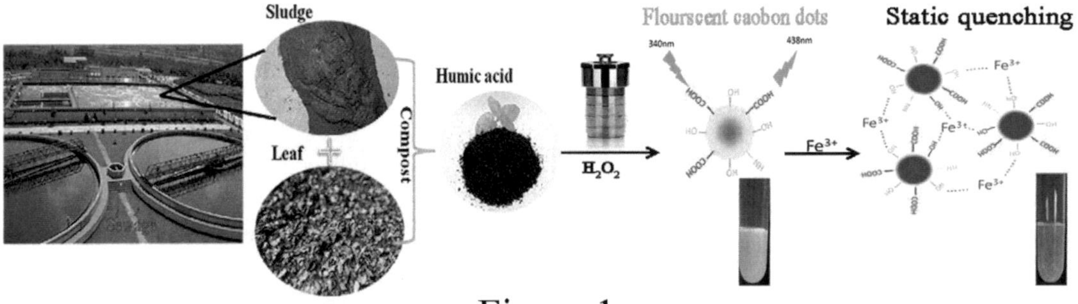

Figure 1

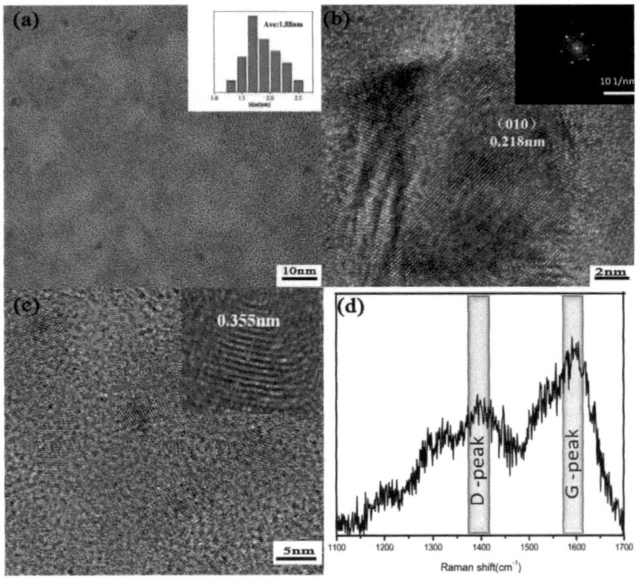

Figure 2

608441 Structural Characterization and Molecular Representation of Refined Unburned Carbon Recovered from Fly Ash

Abstract:

In order to improve and efficient utilization of fly ash derived from coal-based power plants, a structural study of refined unburned carbon (RUC) recovered from fly ash was carried out with multiple analytical tools, including X-ray diffraction (XRD), Raman spectroscopy, solid-state ^{13}C nuclear magnetic resonance (^{13}C-NMR) spectroscopy, Fourier transform infrared (FT-IR) spectroscopy, X-ray photoelectron spectroscopy (XPS), Laser desorption ionization time-of-flight mass spectroscopy (LD-TOF-MS), and high-resolution transmission electron microscopy (HRTEM). These analytical data was used to construct a molecular representation of GUC by using a built-in Amorphous Builder module in Materials Studio.

In the present study, GUC was recovered by a two-step process including refining of fly ash by demineralization with HF/HCl/HF process and then graphitization and crystallization by heat treatment up to 2500 ^{o}C.

The crystalline dimensions such as interlayer spacing (d_{002}), stacking height (L_a), and average lateral size (L_c) of GUC was calculated from XRD spectrum and found to be 0.338, 19.60, and 13.15 nm respectively. Moreover, the increase in peak intensity of D and G band was clearly observed during the heat treatment for carbon graphitization. The aromaticity (f_a) of RUC was found in the range of 0.45-0.51 by the multiple techniques. The relative intensity ratio of D and G band of GUC was 0.47.

The –OH, aliphatic and aromatic C–H, C=O, C-C, and C-O functional groups was clearly observed in FTIR and XPS spectrums. The quantitative analysis of carbon and oxygen functionalities such as –OH, aliphatic and aromatic C–H, C=O, C-C, C-O groups was found to be 12.1, 38.7, 10.4, 28.5, and 10.3% respectively from C1s and O1s XPS profile. The molar content of different carbon types such as aliphatic, aromatic, and carbonyl carbons of GUC was 37.5, 57.1, and 5.1% calculated from ^{13}C-NMR spectrum. Four aromatic rings per clusters with 3-4 substitute side chains on each aromatic cluster was calculated from ^{13}C-NMR spectrum. The aromatic layers was clearly observed in the HRTEM image in the form of fringes with a consistent crystalline dimension from XRD.

250 aromatic fringes was selected from the HRTEM image, and the heteroatoms and aliphatic side chains obtained from different analytical tools was used to construct 250 individual RUC fragments. A random cross-linking of individual fragments with different combinations of aromatic, aliphatic, and oxygen linkages till observing a consistent distribution of molecular weight observed in LD-TOF MS. A final molecular model was presented with 52 unique individual molecules with a molecular formula $C_{7754}H_{4576}O_{109}N_{48}S_{32}$. The structural and molecular properties obtained from analytical tools are consistent with the constructed model. This is the first molecular representation of unburned carbon recovered from fly ash ever proposed, which may lead to further understand the unburned carbon from fly ash and improve their utilization.

References:

1. Jaiswal, Y., Pal, S.L., 2020. Structural Characterization of Indian Vitrinite-Rich Bituminous Karharbari Coal. ACS Omega 5, 6336–6347. https://doi.org/10.1021/acsomega.9b03674.
2. Niekerk, D. Van, Mathews, J.P., 2010. Molecular representations of Permian-aged vitrinite-rich and inertinite-rich South African coals. Fuel 89, 73–82. https://doi.org/10.1016/j.fuel.2009.07.020.
3. Sharma, A., Kyotani, T., Tomita, A., 2001. Quantitative evaluation of structural transformations in raw coals on heat-treatment using HRTEM technique. Fuel 80, 1467–1473. https://doi.org/10.1016/S0016-2361(01)00018-7.

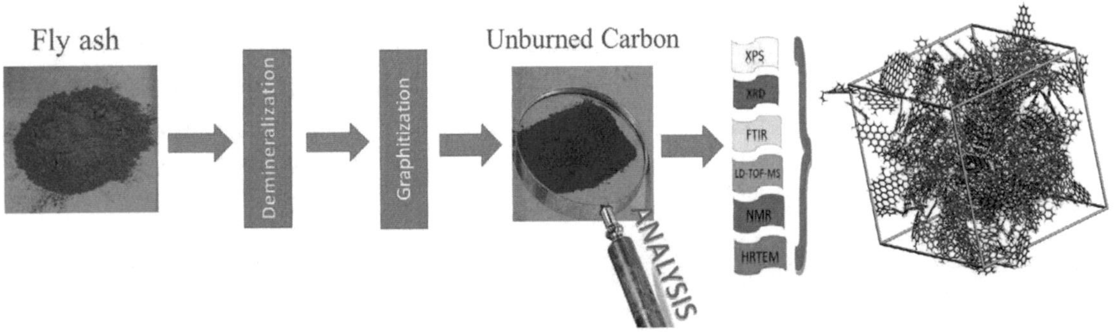

603070 Planning and Optimization of a Regional Hydrogen Supply Chain: An ASEAN Case Study

Abstract:

The world is seeking scalable and affordable low carbon and renewable energy sources to reduce atmospheric CO_2 concentration levels while meeting rising global energy demand. Hydrogen is receiving significant attention worldwide as a versatile energy carrier and feedstock that can achieve deep decarbonization across energy and industrial sectors. Researchers have proposed various approaches to evaluate hydrogen supply chains. For example, Almansoori and Shah[1] developed a multi-period hydrogen supply chain model, based on the hydrogen demand of fuel cell vehicles in the Great Britain from 2005 to 2034. Reuß et al.[2] carried out a techno-economic comparison between different transport and storage infrastructure options for scenarios with 25%, 50% and 75% fleet penetration of fuel cell electrical vehicles in Germany. Seo et al.[3] developed a novel hydrogen supply chain model to compare centralized storage and decentralized storage modes. However, most previous work has focused on country-specific supply chains, such as for the UK, Germany, or South Korea, and considered the transport sector as the only hydrogen demand, neglecting that hydrogen can also be utilized in power and industrial sectors. There is no study on regional hydrogen supply chains with multiple H_2 demand sectors. A regional supply chain can capture the benefits of integrating energy-abundant areas with high energy-consuming areas.

In this research, a multi-period regional hydrogen supply chain model is developed for a better understanding of different pathways for a hydrogen economy in Singapore and the ASEAN countries through 2050. A comprehensive superstructure including hydrogen production, storage, delivery, and utilization is proposed. Both renewable and non-renewable hydrogen production technologies with applicable carbon capture and sequestration are included, reflecting the diversity of energy sources available in the region for hydrogen production. To compare different options for an ASEAN hydrogen supply chain through 2050, we carry out techno-economic analyses of hydrogen production taking into account location-dependent attributes, such as cost and carbon emissions. The variation of hydrogen demand over a long-term planning horizon is considered, as well as uncertainty of energy source prices and technological progress, to find the competitive production technologies in each year. Since the volumetric energy density of hydrogen is low, hydrogen must be transported and stored at high pressure or in liquid form, or via liquid organic hydrogen carriers or ammonia. Consequently, different transmission-storage-distribution pathways are also considered and evaluated to address the trade-off between the energy density, energy consumption, and facility cost. Based on the superstructure, a mixed-integer non-linear programming (MINLP) model is formulated to optimize both the total cost and carbon emissions of the hydrogen supply chain. The most effective pathways for producing, transporting, and storing hydrogen are being identified. Several what-if scenarios are explored to derive insight on relative emissions and costs associated with various hydrogen pathways for the ASEAN region.

1. Almansoori, A., Shah, N., 2009. Design and operation of a future hydrogen supply chain: Multi-period model. *International Journal of Hydrogen Energy*, 34(19): 7883-7897.

2. Reuß, M., Grube, T., Robinius, M., Stolten, D., 2019. A hydrogen supply chain with spatial resolution: Comparative analysis of infrastructure technologies in Germany. *Applied Energy*, 247: 438-453.

3. Seo, S.-K., Yun, D.-Y., Lee, C.-J., 2020. Design and optimization of a hydrogen supply chain using a centralized storage model. *Applied Energy*, 262.

611098 Predicting Reaction Mechanism of Electrochemical CO_2 Reduction Using Machine Learning

Abstract:

Understanding the mechanism of an electrochemical reaction could, in principle, be one of the important factors in the accelerated discovery of highly efficient electrocatalysts. Cyclic voltammetry (CV) is one of the techniques providing important mechanistic insight into an electrochemical reaction by the virtue of the change in current density when a reactant or an intermediate species is adsorbed/desorbed on the electrode surface. However, interpreting a CV can be influenced by subjectivity from the experience of the electrochemist. Moreover, quantifiable confidence in such interpretations is wanting. Machine Learning (ML) algorithms can be applied in such cases to circumvent the heuristics in CV interpretations and user subjectivity to present robust, quantifiable confidence in predicting electrochemical reaction mechanisms. One of the most attractive electrochemical reactions in recent years has been the electrochemical CO_2 reduction (CO2R) to value-added chemicals due to its potentially high impact on the industries, society, and the environment. This work shows the implementation of an ML model (Neural networks) to predict the reaction mechanism of CO2R by learning from different data sets of cyclic voltammograms. The simulated and experimental cyclic voltammograms are used to train the neural network to identify the dependence on the parameters like CV scan rates and uncompensated resistance. For each individual CV, the voltammogram serves as an input to the model. The expected output predicts the likelihood of the occurrence of either simple reversible electron transfer reactions or irreversible product formation after an electron transfer reaction. This ML model confirms that the predictions can be made even by simple models as a substitute for the conventional subjective approach.

610876 Revealing the Mechanism of Intriguing Predominant Urea Formation from Thermal Degradation of CO_2-Loaded Aqueous Ethylenediamine

Abstract:

Amine scrubbing using aqueous amine solutions remains to be the most efficient near-term solution for post-combustion CO_2 capture from flue gas at coal-fired power plants.[1,2] Monoethanolamine (MEA) is one of the most extensively tested alkanolamines, but its commercial application tends to be limited because of degradation and corrosion issues and enormous energy requirement for solvent regeneration.[3–6] About (~10 %) of total capital cost of CO_2 capture is attributed to makeup of amine solvent loss.[7–10] Moreover, some degradation products are considered to be dangerous if exposed to the environment.

Ethylenediamine (EDA) has been reported to have several advantages over MEA for its higher CO_2 capacity, faster CO_2 absorption rate, easier producible implementation from renewable resources, and slightly lower degradation rates at moderate CO_2 loading.[11–16] EDA and MEA have a similar molecular structure as a primary amine; the former contains two primary amine groups and the latter has one primary amine and one hydroxyl group, while the end groups are separated by two -CH_2- groups in both molecules. Despite their structural similarity, the major thermal degradation products of CO_2-loaded EDA and MEA solutions have been found to be distinctly different.[11,17]

Thermal degradation of amines is generally understood to occur primarily in the stripper due to high temperature (~ 100 - 120 °C) and is known to increase with CO_2 loading.[18] Previous experimental studies proposed that thermal degradation of primary and secondary amines would be initiated by polymerization of carbamates, which can ultimately degrade to urea and imidazolidinones.[19,20] The MEA degradation is thought to proceed via 2-oxazolidinone (OZD) formation that may undergo further reaction with amines to form more stable products.[21] Our previous study predicted isocyanate formation from dehydration of carbamic acid to be highly probable, and also showed that cyclization of intermediate isocyanate to OZD would be kinetically more facile than reaction with MEA to form urea in aqueous MEA solution. The formation of isocyanate, albeit very small quantity, has also been experimentally reported.[22,23] In contrast to the case of MEA, the major thermal degradation products of EDA have been identified to be 1,3-Bis(2-aminoethyl)urea and 2-imidazolidone from mass spectrometry and high performance liquid chromatography measurements, while urea formation appears to be preferred.[11,17]

In this talk, we will present molecular mechanisms underlying thermal degradation of EDA in CO_2-loaded solution, especially the underlying reason for the preferential formation of urea, rather than cyclic compounds, in contrast to the case of MEA. The thermodynamic favorability of previously proposed reaction pathways, including the relative stabilities of intermediates and products, were evaluated using quantum mechanical calculations with an implicit solvent model. The free-energy barriers for key elementary reactions involved in the EDA degradation were determined using *ab initio* molecular dynamics simulations combined with metadynamics. Besides the simulation results, further analysis of the interactions between short-lived intermediates and surrounding solvent molecules provide an explanation as to why EDA and MEA show different degradation behaviors. This work highlights the importance of kinetic effects associated with the local structure and dynamics of solvent molecules around the intermediates that may significantly influence the degradation process of amine solvents.

References

[1] G. T. Rochelle, *Science*, 2009, **325**, 1652–4.

[2] P. Luis, *Desalination*, 2016, **380**, 93–99.

[3] M. R. M. Abu-Zahra, L. H. J. Schneiders, J. P. M. Niederer, P. H. M. Feron and G. F. Versteeg, *Int. J. Greenh. Gas Control*, 2007, **1**, 37–46.

[4] M. R. M. Abu-Zahra, J. P. M. Niederer, P. H. M. Feron and G. F. Versteeg, *Int. J. Greenh. Gas Control*, 2007, **1**, 135–142.

[5] D. M. D'Alessandro, B. Smit and J. R. Long, *Angew. Chem. Int. Ed. Engl.*, 2010, **49**, 6058–82.

[6] H. M. Stowe and G. S. Hwang, *Ind. Eng. Chem. Res.*, 2017, **56**, 6887–6899.

[7] A. B. Rao and E. S. Rubin, *Environ. Sci. Technol.*, 2002, **36**, 4467–4475.

[8] A. J. Reynolds, T. V. Verheyen, S. B. Adeloju, A. L. Chaffee and E. Meuleman, *Int. J. Greenh. Gas Control*, 2015.

[9] G. S. Goff and G. T. Rochelle, *Ind. Eng. Chem. Res.*, 2006, **45**, 2513–2521.

[10] S. A. Mazari, B. Si Ali, B. M. Jan, I. M. Saeed and S. Nizamuddin, *Int. J. Greenh. Gas Control*, 2015.

[11] S. Zhou, X. Chen, T. Nguyen, A. K. Voice and G. T. Rochelle, *ChemSusChem*, 2010, **3**, 913–918.

[12] M. Rabensteiner, G. Kinger, M. Koller, G. Gronald and C. Hochenauer, *Int. J. Greenh. Gas Control*, 2014, **27**, 1–14.

[13] S. Kadiwala, A. V. Rayer and A. Henni, *Chem. Eng. J.*, 2012, **179**, 262–271.

[14] J. Li, A. Henni and P. Tontiwachwuthikul, *Ind. Eng. Chem. Res.*, 2007, **46**, 4426–4434.

[15] Z. Wang, M. Fang, Y. Pan, S. Yan and Z. Luo, *Chem. Eng. Sci.*, 2013.

[16] A. P. Salvi, P. D. Vaidya and E. Y. Kenig, *Can. J. Chem. Eng.*, 2014, **92**, 2021–2028.

[17] J. Thompson, H. Richburg and K. Liu, in *Energy Procedia*, 2017.

[18] D. Hatchell, O. Namjoshi, K. Fischer and G. T. Rochelle, *Energy Procedia*, 2014, **63**, 1558–1568.

[19] H. Lepaumier, D. Picq and P. L. Carrette, in *Energy Procedia*, 2009.

[20] H. Lepaumier, E. F. Da Silva, A. Einbu, A. Grimstvedt, J. N. Knudsen, K. Zahlsen and H. F. Svendsen, in *Energy Procedia*, 2011.

[21] J. Davis and G. Rochelle, *Energy Procedia*, 2009, **1**, 327–333.

[22] T. Sakakura, J. C. Choi and H. Yasuda, *Chem. Rev.*, 2007, **107**, 2365–2387.

[23] T. E. Waldman and W. D. McGhee, *J. Chem. Soc. Chem. Commun.*, 1994, 957.

611409 Techno-Economic and Sustainability Analysis of the Production of Acetic and from Methane and Carbon Dioxide

Abstract:

Methane and carbon dioxide together represent 90% of greenhouse gases, and one way of diminishing their emissions is by converting them into added value chemicals. Acetic acid is one of such chemicals, and efforts at the laboratory scale have been carried out to prove it can be directly and indirectly synthesized from CH4 and CO2. However, there are very few studies that evaluate the possibility of implementing any of these processes at a larger scale. In this thesis, techno-economic and sustainability analyses were used to assess the feasibility of upscaling two pathways to produce acetic acid: (1) by direct conversion of methane and carbon dioxide using a plasma-assisted reactor and (2) by indirect catalytic conversion, using syngas (obtained from dry methane reforming) as an intermediate. ASPEN Plus was used for simulation, and with literature values and plant cost estimation techniques, the capital and operational expenses were estimated and compared to those of the commercial route.

The indirect and direct pathways attained return on investment (ROI) values of 1.5% and 5.5%, respectively, and unless credit is given to the reduction in the carbon footprint, neither process can economically compete with the commercial route, which attained an ROI of 19%. In terms of sustainability, the indirect and direct processes captured 0.7 and 2.5 times more carbon dioxide compared to the emitted amount by the commercial pathway. For the direct pathway it was found through a sensitivity analysis the carbon credit value can be reduce down to a value of \$50/tonne to render the process economically viable.

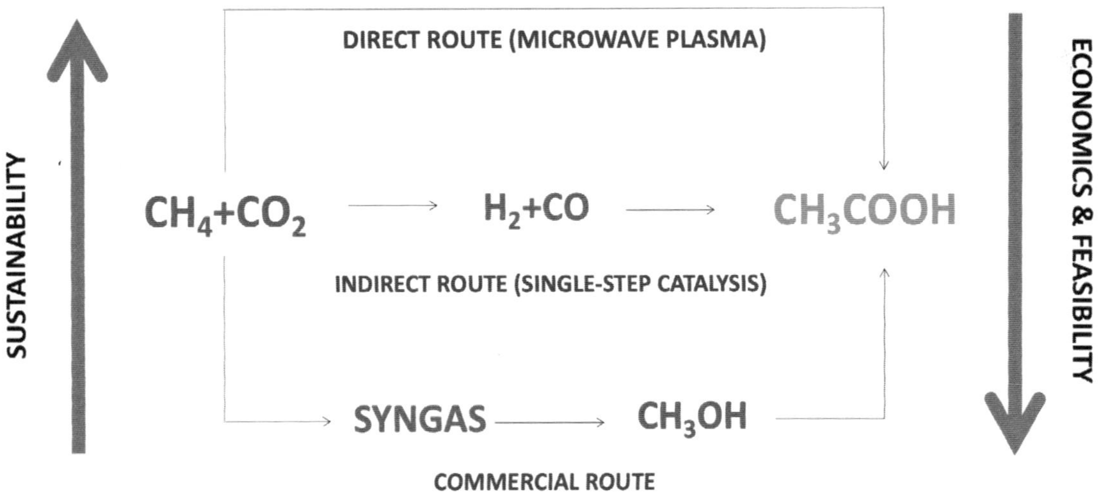

THE CARBONSAFE INITIATIVE: LAUNCHING THE CONVERGENCE OF CARBON CAPTURE AND STORAGE PROJECTS BY 2025

Mary Sullivan[1], Kanwal Mahajan[2], Darin Damiani[3], and Tomas Mora[4]

[1]National Energy Technology Laboratory, Pittsburgh, PA 15236-0940, U.S.A

[2]National Energy Technology Laboratory, Morgantown, WV 26507-0880, U.S.A

[3]U.S. Department of Energy, Washington, DC 20585, U.S.A.

[4]KeyLogic Systems, Inc., Morgantown, WV 26505, U.S.A.

Abstract

The Carbon Storage Program implemented by the U.S. Department of Energy (DOE) Office of Fossil Energy and managed by the National Energy Technology Laboratory (NETL) is helping to develop technologies that safely and permanently store carbon dioxide (CO_2) without impacting human health and the environment or hindering economic growth. Building upon almost two decades of knowledge and experience gained from the field lab testing projects and Regional Carbon Sequestration Partnership (RCSP) Initiative efforts, the Program initiated the Carbon Storage Assurance Facility Enterprise (CarbonSAFE) Initiative. The CarbonSAFE Initiative addresses key gaps on the critical path toward carbon capture, utilization and storage (CCUS) deployment by reducing technical risk, uncertainty, and cost of a geologic storage complex with capacity to store 50+ million metric tons of CO_2 from industrial sources over a 30-year period. The CarbonSAFE Initiative is taking a phased approach: (I) Integrated CCS Prefeasibility, (II) Storage Complex Feasibility, and (III) Site Characterization, Permitting and Carbon Capture Assessment. Subject to the availability of funding, a fourth phase would include construction and completion of wells and infrastructure leading to authorization to begin injection operations. A total of 13 projects were completed under Phase I, which provided high-level evaluations of potential CCS scenarios in regions throughout the United States. Six projects were awarded under Phase II to confirm the adequacy of storage complexes through initial site characterization. Phase III projects commenced in 2020 and will involve the entire integrated process, including a capture assessment, detailed site characterization, and preparing an application for a Class VI Underground Injection Control (UIC) permit to construct. These efforts are designed to help launch the industry by 2025 and contribute to increasing confidence that geologic storage of captured CO_2 can be a safe enterprise on a broad scale.

Keywords: U.S. Carbon Storage DOE Program; National Energy Technology Laboratory (NETL); carbon capture, utilization and storage (CCUS) research; geologic storage technologies; Regional Carbon Sequestration Partnerships (RCSPs); field projects; accomplishments; CarbonSAFE

1. United States Leadership in Commercial Deployment of CCUS

One of the great challenges facing the world today is the need to reduce global greenhouse gas (GHG) emissions and thereby mitigate the potential impacts of climate change without provoking undue detriment to global economies. In the United States (U.S.), energy-related CO_2 emissions have declined over the last decade in part due to a trend in fuel-switching from coal-based to natural gas-based sources in power generation, as well as increased production of renewable electricity, and improved energy efficiencies of vehicles. However, the U.S. Energy Information Administration (EIA) projects that U.S. energy-related emissions will remain essentially unchanged for the next 30 years unless carbon reduction programs and policies are adopted [1, 2]. Numerous international organizations, from the Intergovernmental Panel on Climate Change to the International Energy Agency (IEA), recognize that CCUS is an essential clean energy technology if a low-carbon economy is to be achieved at the lowest possible cost. Geologic Carbon storage includes the capturing of carbon dioxide (CO_2) from a stationary source; compressing it to a liquid-like state; transporting it via pipeline, ship, truck, or other means; and injecting it underground into deep reservoirs safely and securely. Although not part of CarbonSAFE, CO_2 recycling to produce economically valuable products and services such as improved materials, processes, fuels, and chemicals can be added to the CCUS value chain.

The United States is ideally situated for commercial CCUS projects because of its abundant carbon storage resources in proximity to many industrial CO_2 sources such as coal- and gas-fired power plants, gas-processing plants, ethanol plants, cement manufacturers, hydrogen production facilities and a host of others.

The United States also has over 50 years of CO_2-enhanced oil recovery experience and more than 5,000 miles of CO_2 pipeline infrastructure. Moreover, enhancements to federal and state policies are providing the needed incentives for advancing commercial CCUS projects. According to the IEA, 45Q tax credits could trigger new capital investments of up to $1 billion (B) for CCUS technologies over the next 6 years, resulting in an additional 10 to 30 million (M) metric tons (MT) of CO_2 capture capacity [3]. In the U.S power sector alone, 45Q credits could incentivize retrofit technologies in coal and natural gas power plants and capture of 49 MMT of CO_2/year (yr) by 2030 [4]. To date, there are over 30 carbon capture projects in various stages of project development in the U.S. [5, 6, 7]. With respect to regulations, recent U.S. Environmental Protection Agency (EPA) approval of primacy for North Dakota and Wyoming for Underground Injection Control (UIC) Class VI wells (used for geologic saline storage of CO_2) has provided developers an opportunity to work more closely with state jurisdictions, which might help facilitate the process and expedite permitting in these two states.

Over the past two decades, the U.S. Department of Energy (DOE) Office of Fossil Energy (FE) and its partners have reduced the cost of carbon capture from power and industrial sources and advanced the integration of CCUS systems in storage complexes with different geologic settings. One example of such an integrated system was NRG Energy's, Inc. Petra Nova project, the world's largest coal-fired post-combustion CO_2 capture system. The captured CO_2 was sent for use in enhanced oil recovery (EOR) in a partially depleted oil field in Jackson County, Texas. Lessons learned from Petra Nova and other industrial projects, such as the Air Products and Chemicals, Inc. hydrogen production facility in Port Arthur, TX and the Archer Daniels Midland Company ethanol facility in Decatur, IL, will result in lower costs for future projects. The next large-scale coal-fired post combustion CO_2 capture and compression project is expected to reduce costs by nearly 30% compared to Petra Nova's cost [8].

The knowledge gaps impeding growth in commercial CCUS deployment are both technical and non-technical. FE's Carbon Storage Program is working to close these remaining gaps by supporting a portfolio of research and development (R&D) projects and initiatives that investigate the technical gaps and aid in addressing the non-technical aspects of CCUS, such as the regulatory process, finance and markets, and public perception. Perhaps one of the best-known and most successful endeavours is the Regional Carbon Sequestration Partnerships (RCSPs). Over the last 17 years the RCSPs have characterized the CO_2 storage resource potential in their respective regions, completed pilot and large-scale CO_2 injection projects that cumulatively stored more than 11 MMT of CO_2, demonstrated the effectiveness of various technologies to verify storage containment in multiple reservoir types, and established a broad national network among companies and professionals supporting the deployment of commercial-scale CO_2 storage projects throughout the United States. Perhaps most importantly, the RCSPs have helped to validate storage security by virtue of its injection projects showing no evidence of impacts to human health or the environment. Through these successes the RCSP Initiative has laid the foundation for the CCUS industry, with key findings and recommendations captured in a series of Best Practice Manuals [9] made available to the public.

2. Convergence of DOE-FE's Carbon Capture and Carbon Storage programs

At present, the current state of CCUS technology is sufficient to develop and execute a CCUS project. However, cost reduction and performance optimization are among the biggest challenges to an industry launch by 2025. Increasing the number of CCUS projects at commercial scale enables development of the following:

- Cost-effective and accurate approaches for appraising site-specific storage resources and supporting investment-level decisions.
- Fully optimized and cost-effective long-term monitoring technologies and tools for commercial-scale onshore storage sites.
- Information and approaches for more targeted data acquisition and understanding of risk(s) in support of the permitting process.

DOE-FE's Carbon Storage Program is addressing these knowledge gaps primarily through its Carbon Storage Assurance Facility Enterprise (CarbonSAFE) Initiative, as described in more detail below. In parallel, DOE-FE's Carbon Capture Program is advancing R&D post-combustion carbon capture technologies that have the potential to reduce capture cost to $30/MT (at 95% purity). Both DOE-FE programs are planned to converge in 2025 with the intent to deploy integrated CCUS projects that would link commercial-scale storage complexes with state-of-the-art capture facilities.

3. CarbonSAFE Initiative

The Carbon Storage Program envisioned CarbonSAFE as the natural next step following the success of the Regional Carbon Sequestration Partnership (RCSP) efforts [10, 11], which provided the technical and non-technical foundations for scaling up CCUS operations to commercial scale. RCSP successes have already been demonstrated by contributions to major demonstration projects (Archer-Daniels-Midland [ADM], Air Products, Petra Nova), which cumulatively have stored more than 11 MMT of captured CO_2. The CarbonSAFE Initiative was created to ensure carbon storage complexes with capacities of 50+ MMT of CO_2 will be ready for integrated CCUS system deployment starting in 2025. The vision of the CarbonSAFE Initiative is to provide lessons learned and examples of the development of a CCUS storage complex from the feasibility study up until the point of injection. The progress made by this effort will instil greater confidence that commercial-scale CCUS projects are technically feasible and can be integrated in an economically feasible manner. The CarbonSAFE Initiative was first envisioned in 2016 as consisting of four phases:

- Phase I: Integrated CCS Prefeasibility – 13 projects awarded in 2016
- Phase II: Storage Complex Feasibility – 6 projects awarded: 3 in 2016, 3 in 2018
- Phase III: Site Characterization and CO_2 Capture Assessment – 5 projects awarded in 2020
- Phase IV: Permitting and Construction – potential future phase, dependent on appropriations

Thirteen prefeasibility studies were conducted in ten states and two offshore locations in the lower 48 states. Prefeasibility projects conducted economic feasibility analyses for targeted sites under varying scenarios. In addition, extensive regional data was collected, analyzed, and modeled; multiple storage sites and infrastructure were evaluated; and business plans for a variety of CCUS scenarios were developed. Currently, the program is implementing the Storage Complex Feasibility and Site Characterization and CO_2 Capture Assessment efforts. Research conducted within the six Storage Complex Feasibility projects focused on one or more specific reservoirs within the defined storage complex, encompassing (a) data collection; (b) geologic analysis; (c) identification of contractual and regulatory requirements; (d) subsurface modeling to support geologic characterization, risk assessment, and monitoring; and (e) public outreach.

The Site Characterization, Permitting and CO_2 Capture Assessment phase (Phase III) began in 2020. These projects will converge with Front-End Engineering Design (FEED) studies awarded by DOE's National Energy Technology Laboratory (NETL) Carbon Capture Program for commercial-scale post-combustion capture technologies implemented in the power sector.* The key objectives of this phase are to (a) complete a detailed site characterization of a commercial-scale CO_2 storage site (50+ MMT of captured CO_2 within a 30 yr period); (b) apply for and obtain UIC Class VI permit(s) to construct injection well(s); (c) complete a CO_2 capture assessment; (d) perform all work required to support a National Environmental Policy Act determination for each proposed project; and (e) address all other permitting processes and requirements, liability relief, and finance agreements in support of the business model for eventual commercial operations. Once the UIC Class VI permit applications to construct are submitted, additional activities will include working with the regulators to satisfy their requirements until construction permits are granted. Figure 1 shows locations of the CarbonSAFE Phase III projects with FEED studies across the U.S. The characterization projects in Phase III will converge with five of nine FEED studies awarded by DOE's NETL Carbon Capture Program for commercial-scale carbon capture systems on coal and natural gas power plants.

Subject to availability of funding, the final phase of CarbonSAFE, Construction of the Storage Complex, will involve the drilling and completion of the injection and monitoring wells and development of risk and mitigation plans to obtain authorization by the regulator to begin injection operations. Issued final reports for all project phases are located at the DOE's Office of Scientific and Technical Information site (https://www.osti.gov/search). Project-specific data produced by the CarbonSAFE effort are available on the Energy Data Exchange (edx.netl.doe.gov).

---

* The Funding Opportunity Announcements (FOAs) for the CarbonSAFE Phase III projects and Carbon Capture Program FEED studies were run independently with no dependence between them in terms of project selection. CO_2 capture assessments are being completed under CarbonSAFE for sources that did not receive FEED funding from the Carbon Capture Program.

Figure 1. Map of the contiguous U.S. showing locations of CarbonSAFE Phase III projects and Carbon Capture Program FEED studies

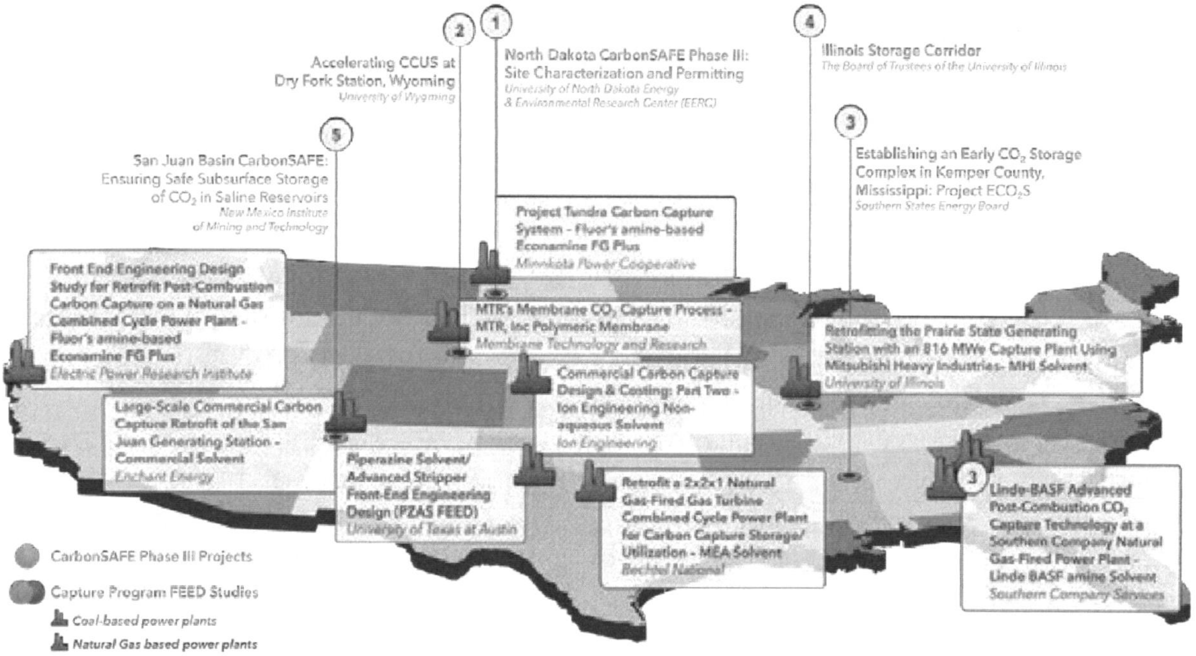

3.1 Integrated CCS prefeasibility projects

While not all prefeasibility projects were able to receive funding to continue their investigations into the next phase, a significant amount of useful and actionable research was produced. Extensive regional data was collected, analyzed, and modeled; multiple storage sites and infrastructure were evaluated; and business plans for a variety of CCUS scenarios were developed. A summary of each project's sources and storage estimates were provided in a previous publication [12]. Further characterization of the subsurface is needed at each site to verify the general initial findings. Common themes identified for making an integrated CCUS project successful from the prefeasibility projects include the following:

- Collaboration with oil and gas companies and strong host site support is essential.
- Uncertainty in federal policy results in lack of commitment from industry and the public for CCUS projects.
- Capture costs will need to decrease to make integrated CCUS projects economical.

- Reservoir heterogeneity requires improved models and data to increase plume behavior accuracy.
- Pressure management may be necessary to meet target rates and volumes in applicable geologic settings.
- Delineating the area-of-review for a geologic storage project is a critical component of permitting and the technical risk assessment.
- Regional state-of-stress and geomechanical characterization are important for site selection.

3.2 Financial and economic analysis

The prefeasibility projects performed a regional storage resource assessment to explore economic strategies for capturing, transporting, and storing large-scale anthropogenic sources in potential 50 MMT storage sites. Revenue from 45Q tax credit incentives for dedicated and associated storage scenarios and the potential sale of CO_2 to EOR fields were considered to offset the costs of these projects. Associated storage revenues were calculated at $20-33/MT of CO_2 sold for EOR at an oil market price range of $55-60/barrel (bbl). Table 1 includes the total cost of several fully integrated systems for CO_2 capture, compression, and transportation from sources to injection sites in the Onshore U.S. Prefeasibility projects with primary storage sites in the U.S. Offshore and Coastal regions in the Gulf of Mexico, Coastal Louisiana, and Offshore Washington are not summarized in this section.

Table 1. Total cost and revenue (shown in parentheses) for prefeasibility projects in onshore U.S.

Key	Project Name	Source(s) by Location	Econ. Cases	Revenue Type	Total Cost ($/MT)
1	Integrated Mid-Continent Stacked Carbon Storage Hub	Coal-fired power plant in Lincoln County, NE Ethanol plant corridor in eastern NE	3	Dedicated storage	31 to 67
			3	Oil+Net CO_2 stored for EOR	(1.72) to 57
2	Nebraska Integrated Carbon Capture and Storage Prefeasibility	Coal-fired power plant in Lincoln County, NE	1	Dedicated storage	20
			1	CO_2 sold for EOR	32
3	Integrated Carbon Capture and Storage for Kansas (ICKan)	Two coal-fired power plants in Finney and Pottawatomie County, KS	6	Dedicated storage	2 to 27
			6	CO_2 sold for EOR	14 to 39

		Ethanol plant corridor in northeast KS			
4	Integrated CCS Prefeasibility Study at Dry Fork Station	Coal-fired power plant in Campbell County,WY	1	CO_2 sold for EOR and dedicated storage	(0.14)
5	CarbonSAFE Illinois East Basin[1]	Chemical plant in Vigo County IL	1	CO_2 sold for EOR and dedicated storage	-
6	CAB-CS: Central Appalachian Basin CarbonSAFE Integrated Prefeasibility Project	Coal-fired power plant, a Gas-fired power plant, and Chemical plant in eastern Ohio. Potential sources: steel mill and gas-fired power plant in eastern and southern Ohio	6	dedicated storage	9 to 66
			6	CO_2 sold for EOR and dedicated storage	9 to 68
			6	CO_2 sold for EOR	1 to 67
7	Northern Michigan Basin CarbonSAFE Integrated Prefeasibility Project	Coal-fired power plant in Otsego Co., MI. Potential sources: steel mill and gas-fired power plant in northern and central MI	4	Dedicated storage	21 to 116
			4	CO_2 sold for EOR and dedicated storage	17 to 114
			4	CO_2 sold for EOR	12 to 108
8	California CO_2 Storage Assurance Facility Enterprise	Three integrated energy corridors in Southern San Joaquin Valley, CA	1	CO_2 sold for EOR and dedicated storage	(40)
9	CarbonSAFE Rocky Mountain Phase I: Ensuring Safe Subsurface Storage in the Intermountain West	Two coal-fired power plants in Emery County, Utah.	4	Dedicated storage [NA]	50 to 51
			9	CO_2 sold for EOR [NA]	17 to 27
10	Integrated Commercial CCS Prefeasibility Study at Rock Springs Uplifts, Wyoming[2]	Coal-fired power plant in Sweetwater Co., WY.	1	CO_2 sold for EOR and Dedicated storage	-

[1]Capital investment was estimated at $58.7 M. Annual operating costs were estimated at $12.7 M to supply 1.7 MMT/yr of CO_2.

[2]Capital investment was estimated at $787 M. Annual operating costs were estimated at $54 M to supply 2 MMT/yr of CO_2.

[NA] Economic analysis did not include 45Q tax credit incentives.

The economic analysis for each prefeasibility study can be categorized into three types of scenarios: coal-fired power plant point-to-point transport (Scenario 1), multi-source point-to-point transport (Scenario 2), and long-range energy corridor transport (Scenario 3). The scenarios are meant to offer potential commercial synergies for reducing the unit cost of CO_2 storage and the risk of investment. One representative prefeasibility study for each scenario is summarized in the following sections.

3.2.1 Scenario 1: Coal-fired power plant point-to-point transport

The University of Utah performed a prefeasibility study in central Utah (Table 1, Key 9). The PacifiCorp's Hunter Power Plant and Huntington Power Plant are located in Emery county near a 7,000-foot deep brine-bearing sedimentary rock formation. A commercial amine-based capture technology (Mitsubishi Heavy Industries (MHI) KM-CR Process® with KS-1™ solvent) was evaluated to estimate the capture and compression costs. Approximately 1.84 MMT and 7.2 MMT of CO_2 per yr can be captured from each plant at $50 and $61/MT, respectively. The proposed locations of two storage sites named Buzzards Bench and Drunkards Wash have estimated storage capacities of 201 MMT and 380 MMT of CO_2, respectively.

Table 2 lists the point-to-point economic scenarios for each power plant. Cost analysis supports the primary core scenario consisting of 65% capture (equivalent to 1.84 MMT of CO_2 per year) from each coal-fired power plant and transporting and injecting it into a sandstone formation at the Buzzards Bench or Drunkards Wash area. Simulation studies suggest that one well is sufficient to store 50 MMT of CO_2 over 30 years in each storage complex. The second scenario involves selling CO_2 to an EOR field located in the Uinta Basin in Utah and a natural source of CO_2 in the McElmo Dome in southwestern Colorado. Findings suggest that larger-scale infrastructure deployments to oil fields in the Uinta Basin for EOR applications could result in a cost savings of up for 67%/MT of CO_2. None of the scenarios considered splitting tax credit incentives from storing CO_2 in both dedicated and associated storage sites. Detailed discussion of the results of this study are provided by McPherson [13].

Table 1. Economic scenarios without 45Q tax credit incentives

Case	Source(s) and Sink(s)	Pipeline (Miles)	Total Cost ($M/yr)	Capture ($M/yr)	Trans. ($M/yr)	Storage ($M/yr)	Total Unit[1] ($/MT)

1	Hunter – Buzzards	22	86	78	0.8	6.6	50.40
2	Hunter – Drunkards	11	87	78	2.4	6.7	51.37
3	Huntington – Buzzards	9	86	78	1.3	6.6	50.67
4	Huntington – Drunkards	7	87	78	1.8	6.7	50.98
5	Hunter – Uinta EOR	100 >	29	78	1.5	(51)	16.87
6	Hunter – McElmo Dome	300 >	42	78	14.5	(51)	24.57

Note: Numbers in parentheses indicate negative costs or profit derived from 45Q tax credit.
[1] Total Unit cost includes storage and monitoring costs for 30 years.

3.2.2 Scenario 2: Multi-source point-to-point transport

Battelle performed a prefeasibility study in northern Michigan (Table 1, Key 7). Multiple CO_2 sources, rather than a single source, were considered in the analysis. These include: 1) the existing Wolverine Power Cooperative-owned natural gas-fired power station in Otsego County that will be retrofitted to a combined cycle facility with CO_2 capture supplemented by CO_2 from the DCP Midstream-owned natural-gas separation facility, also located in Otsego County; 2) a potential new source (PNS) assumed to be located close to the saline storage site(s) and the EOR fields that is a conventional natural gas combined cycle (NGCC) facility with capture; 3) a second variation of the PNS that incorporates the emerging natural gas-fired technology being developed and built by NET Power, LLC, based on the Allam Cycle; and 4) Project TIM, a proposed steel manufacturing facility by New Steel, Inc. in Shiawassee County. The saline storage sites, Saint Peter Sandstone (SPSS) Site 2 and SPSS Site 7, were down selected for detailed economic analysis. Site 2 is located in Montmorency County and Site 7 is located in Kalkaska and Missaukee counties in Michigan. Simulation studies suggest that Site 2 requires six injection wells to inject 50 MMT of CO_2 and has a projected 50 MMT CO_2 plume area of 18,200 acres. Site 7 requires only three injection wells and has a projected plume area of 14,000 acres.

Table 4 lists the multi-source point-to-point economic scenarios evaluated. Interest costs during construction and escalation were included for each of the storage, pipeline, and capture facilities during the construction period prior to the commencement of operations in 2025. Economic analysis of Net Power, an emerging gas-power technology, was not included. The three types of scenarios include: 50 MMT of CO_2 stored at Site 2 for each of the identified sources, 50 MMT of CO_2 supplied to the EOR fields (i.e., carbonate reefs), and a hybrid case in which the CO_2 is divided between saline storage (50%) and CO_2-EOR (50%). In the hybrid case, only one group of EOR reefs is needed to store 25 MMT CO_2. The availability of EOR (ongoing and expanded) as an option for stacked storage with saline storage is an attractive feature for the Northern Michigan Basin area, as this can help defray some of the capital and operating costs for saline storage, increasing the feasibility of the project. Results show that either a 20% reduction in the cost of capture or a $30/MT of CO_2 sale price is necessary for a new natural gas-fired combined cycle with 100% CO_2-EOR storage to achieve a breakeven net revenue requirement. Core Energy, a key member of the Battelle Carbon Storage – Northern Michigan Basin project team, operates in EOR fields and could benefit from this project. Detailed discussion of the results of this study are provided by Kelly [14] in Table 4.

Table 2. Economic scenarios with 45Q tax credit incentives

Case	Source(s) and Sink(s)	Pipeline (Miles)	Total Costs ($M)	Capture ($M)	Trans. ($M)	Storage ($M)	Total Unit Cost[1] ($/MT)
1	Wolverine – SPSS-2 site	21	1,056	691	40	325	54-74
2	Wolverine – EOR and SPSS-2 site	74	1,017	691	131	195	55-73
3	Wolverine – Niagaran Reefs	105	881	691	190	-	47-60
4	New NGCC – SPSS-2 site	17	1,016	658	33	325	44-58
5	New NGCC – EOR and SPSS-2 site	46	937	658	84	195	43-56
6	New NGCC – Niagaran Reefs	83	808	658	150	-	38-48
7	Project Tim – SPSS-2 site	154	1,450	848	277	325	89-116
8	Project Tim – EOR and SPSS-2 site	183	1,370	848	327	195	89-114
9	Project Tim – Niagaran Reefs	225	1,250	848	402	-	85-118

[1]Cost components and anticipated amount of revenue for a scenario on a 30-year levelized basis, expressed in 2018 $/tonne (the project-start date) on a low or high cost of capital basis.

3.2.3 Scenario 3: Large-scale energy corridor transport

The University of Kansas, along with the Kansas Geological Survey, conducted a preliminary feasibility study of two coal-fired power plants in Kansas connecting to large storage corridors in Nebraska (Table 1, Key 3). The Westar Energy's Jeffrey Energy Center (JEC) and S.E. Holcomb station were selected to evaluate multiple capture levels using a commercially available Linde-BASF system as the basis for the capture technology. The estimated capture costs for JEC (2.5 MMT/yr) and S.E. Holcomb Station (1.7 MMT/yr) were $45–67/MT and $46–$72/MT, respectively. In addition, about 34 ethanol plants within the region could supply CO_2 to EOR and storage sites located in southwest Kansas and the Permian basin in Texas. Capture and compression costs from these ethanol plants were estimated at $19/MT and represent a yearly production 10.9 MMT of CO_2 from 3.6 B gallons of ethanol.

Saline aquifers that are overlain by oil reservoirs were studied in the state of Kansas. As a primary site option, the Patterson site, located in southwest Kansas, has an estimated storage capacity of 60 MMT. Previous studies in the region estimated that four of these fields in the NHSC have an estimated associated storage potential for EOR of 1.8-2.3 MMT/yr [15]. To date, 5.5 M barrels (bbls) of oil have been produced from the combined Patterson and Heinitz fields before and after unitization, and an additional 1.2 Mbbls of oil are expected from the waterflood (6.7 Mbbls ultimate recovery). As a secondary site option, the Davis Ranch and John Creek fields, located in the Forest City Basin in Wabaunsee county, have an estimated storage capacity of 24 MMT.

Table 5 lists the scenarios presented by the University of Kansas. All cost estimates include both capital costs and operating expenses over a 2-yr construction period followed by 20 years of operation, and a 10% cost of capital. The most feasible option, as outlined in Case 4, consists of delivering 4.3 MMT/yr of CO_2 from ethanol plants in Kansas and Nebraska to several EOR fields located in southwest Kansas. The total project cost can be further reduced by supplying 2-2.5 MMT to dedicated storage and selling 1.8-2.3 MMT of CO_2 to EOR in the nearby EOR field. Profits from the sale of CO_2 for EOR could offset transportation. Detailed discussion of the results of this study are provided by Holubnyak [16].

Table 3. Economic scenarios with 45Q tax incentives

Case	Source(s) and	Pipeline	45 Q	Pipeline	Capture	Total

	Sink(s)	(miles)	Revenue Type	Costs ($/MT)	Costs[1] ($/MT)	Cost ($/MT)
1	Holcomb – Patterson Site	25	EOR	(23.3)	46-72	23
			Saline	(35.1)	46-72	11
2	JEC – Davis Ranch/John Creek	50	EOR	(22.9)	45-67	22
			Saline	(34.8)	45-67	10
3	34 ethanol plants – Permian Basin	1,546	EOR	9.0	10	19
			Saline	0.8	7	8
4	15 ethanol plants – SW Kansas	737	EOR	5.0	9	14
			Saline	(2.5)	5	2

Note: Numbers in parentheses indicate negative costs or profit derived from 45Q tax credit.
[1]The derived costs for scenarios 1 and 2 used lowest capture costs from pre-FEED studies

3.3 CarbonSAFE storage complex feasibility projects

CarbonSAFE feasibility projects focused on the next phase of advancing carbon storage toward commercial scale. Work conducted under the feasibility projects included activities such as completing at least one characterization well, updating of static earth models, updating of dynamic and geomechanical modeling, 3D seismic analysis, core and log analysis updates and well testing updates. Initial characterization results demonstrated the likely viability of multiple 50+ MMT geologic storage complexes in deep saline formations. A risk assessment exercise was also conducted to identify and assess technical and nontechnical risks that could prevent potential candidate storage complexes within the study area from serving as commercial storage sites. The assessment identified and evaluated six technical risk categories: 1) CO_2 injectivity, 2) storage capacity, 3) lateral migration of CO_2, 4) lateral pressure propagation, 5) vertical migration of CO_2 or formation brine, and 6) induced seismicity. The risk assessment results can be used to complete a site characterization plan (e.g., future site characterization, modeling and simulation, monitoring activities) and management strategies for the project. Three feasibility projects that have not yet advanced to the characterization phase are briefly described below.

3.3.1 Wabash CarbonSAFE

The Wabash CarbonSAFE project, led by the University of Illinois at Urbana-Champaign, determined the feasibility of commercial-scale CO_2 storage in the Mt. Simon sandstone formation located beneath the future Wabash Valley Resources

(WVR) ammonia production plant. Targeted to begin operation in 2021, the ammonia plant is under construction to repurpose and expand the former integrated gasification combined cycle plant located near Terre Haute, Indiana. Approximately 1.65 MMT of CO_2/yr will be produced as a by-product, which can then be captured through a process using Rectisol.

The selected storage site, located on the eastern side of the Illinois Basin, is on the WVR plant site. The primary target reservoir is the Mt. Simon sandstone formation. This formation is estimated to have a storage capacity of 12-172 billion metric tons of CO_2 in the Illinois Basin [17]. Secondary storage targets include units in the Knox Group, and the St. Peter Sandstone. A stratigraphic well was drilled in the WVR facility for the acquisition of 265 (ft) feet of core and to obtain geologic data to update models and confirm reservoir properties. Core samples were collected from the Maquoketa Group (secondary seal), the Eau Claire Formation (primary seal), and the Upper Mt. Simon Sandstone. An extensive suite of geophysical logs was completed to evaluate CO_2 storage potential, hydrocarbon potential, and general lithological and petrophysical characteristics of the sedimentary column. In addition, 20 kilometers (km) of 2D seismic data is being processed to assist with subsurface geologic characterization. Future vertical seismic profiles (VSPs) can be performed to improve delineation of geologic structures. Final reservoir simulation, based on the stratigraphic well's core and log analyses and drill stem test data, would be performed to assess the CO_2 injection potential of the Mt. Simon sandstone.

3.3.2 CarbonSAFE Illinois Macon County

The CarbonSAFE Illinois Macon County project, led by the Illinois State Geological Survey, assessed the feasibility of developing a CO_2 storage complex that can store more than 50 MMT of CO_2 in the central Illinois Basin. The closest potential CO_2 sources are the Abbot Power Plant located in Champaign County, Illinois; City Water, Light, and Power located in Springfield County, Illinois; and ADM located in Decatur County, Illinois. These three sources emit a total of approximately 3.0 MMT/yr of CO_2. A fourth, more distant potential source is the 1,600-megawatt (MW) coal-fired Prairie States Energy Campus, which began operations in 2012 and emits greater than 10 MMT of CO_2/yr.

The original project site in Macon County was switched to the Podolsky Oil Company's field in Christian County, Illinois to accommodate the larger potential sources of CO_2. The target reservoir was the Lower Mt. Simon sandstone formation.

Initial characterization work included drilling the T.R. McMillen #2 (TRM) stratigraphic well for the acquisition of 148 ft of core and to obtain geologic data to update models and confirm reservoir properties. The well was drilled through the targeted storage interval into the Precambrian basement to a total depth of 6,490 ft. Preliminary results indicated that the Mt. Simon sandstone at the location of the TRM #2 well was suitable for commercial-scale CO_2 storage. Initial modeling showed that 30 years of injection in three wells with 80 acre-spacing projected a 190 MMT CO_2 plume extension of 2.7 miles with little interaction with the adjacent Illinois Basin–Decatur Project (IBDP). A 30 km 2D seismic survey and a VSP were performed to obtain data needed to interpret the subsurface structure of the site. Seismic data is being processed and analyzed. Additionally, stacked storage in oil-bearing carbonate oil reservoirs in the Mt. Auburn trend of the Sangamon Arch in the Illinois Basin are being studied for CO_2-EOR potential.

Previous characterization efforts at the IBDP site suggested that induced microseismicity migrated from sediment into Precambrian basement. At the site, the microseismicity was monitored using deep borehole sensors. About 1400 events were recorded, with moment magnitudes between 1.1 and -1.7. [18, 19]. Induced seismicity risk was assessed at the Macon site through the characterization of geomechanical properties of rocks and fault analysis, and the stress state field measurements in the basement. Well log cross sections displaying the geomechanical properties at the Macon and IBDP sites were prepared to illustrate the differences between the sites. Laboratory tests performed on the cores from the TRM well were also performed on the basement rock. In situ stress direction characterization at the Macon site was completed using indicators observed in a Formation Micro-Imager log and well tests (e.g., Diagnostic Fracture Injection Test). Future characterization work would include the integration of these data and well site studies with improved geomechanical models and dynamic reservoir models.

3.3.3 Integrated Midcontinent Stacked Carbon Storage Hub

The Integrated Midcontinent Stacked Carbon Storage Hub project, led by Battelle Memorial Institute, is a combination of three prefeasibility projects located in Kansas and Nebraska. The intent of the project is to develop a source corridor that can run southwest from eastern and central Nebraska toward Red Willow County, Nebraska. Detailed discussion of the Phase I results of this study are provided by Duguid [20]. The feasible pipeline routes connecting sources along the source corridor to potential saline sinks (i.e., dedicated storage) or CO_2-EOR sinks (i.e., associated storage) in the storage corridor account for environmentally and culturally sensitive areas. The team identified 17 technically and economically feasible CO_2-EOR oilfields and three saline reservoir fields in the storage corridor. The designated oilfields have estimated storage of 577.4 MMT of CO_2 with the potential to produce 181.9 Mbbls of oil via EOR, equivalent to an estimated gross revenue for stacked storage and EOR of $30.9 B.

Three saline sinks were evaluated: Madrid field in southwest Nebraska, Patterson-Heinitz-Hartland field (PHH) in southwestern Kansas, and the Sleepy Hollow Field (SHF) in southwest-central Nebraska. One stratigraphic well was drilled at the SHF site to obtain geologic data through well logging and 110 ft of core for analysis. Two additional stratigraphic wells were drilled at the PHH site to collect geologic data through well logging and obtaining 778 ft of core for analysis. The effort also included the acquisition of 26 square miles (mi^2) of new 3D seismic data to interpret the subsurface structure in the PHH site. Results from seismic data processing revealed two major reverse faults that offset the reservoir and seal intervals and constitute an uplifted block in the Patterson area. In addition, three- and four-way structural closures were identified that can assist with trapping CO_2 in the Arbuckle-Osage reservoirs. Updated models indicated that the original 50 MMT target injection amount was not viable in the SHF site. Numerical simulation for the PHH site indicated that three wells would be required to store 50 MMT of CO_2 in deep stacked saline Paleozoic formations over 30 years. Future recommended characterization work would include fault analysis at the PHH site and conducting a 3D seismic survey and drilling a characterization well at the Madrid site.

3.4 CarbonSAFE site characterization and CO_2 capture assessment projects

Three of the five Site Characterization and CO_2 Capture Assessment projects build upon previous phases that indicated the commercial viability of storage complexes located in North Dakota, Mississippi, and Wyoming. The two remaining projects, located in New Mexico and Illinois, demonstrated the viability of their respective sites through previous work conducted, unrelated to the CarbonSAFE effort. The site characterization efforts conducted by these five projects, and the association of these projects with capture assessment projects, are discussed below.

3.4.1 Illinois Storage Corridor (Phase III)

The Illinois Storage Corridor is a region with proven geologic storage performance and numerous industrial carbon sources. Numerous DOE-FE funded projects in the Illinois Basin are already addressing the challenges posed by CCUS deployment at scale through the field validation of storage complexes and new technologies. These projects include two pilot carbon capture technology projects [21] two commercial-scale CCUS projects led by RCSPs [22], and Wabash and Macon CarbonSAFE phase II projects. As part of the CarbonSAFE phase III work, the Board of Trustees of the University of Illinois (Champaign, IL) will determine the feasibility of developing a storage complex near two facilities: One Earth Energy ethanol facility in Ford County, Illinois and Prairie State Energy Campus (PSEC) facility, a 1,200-MW coal-fired electricity generation plant in St. Clair County, Illinois. Annually, these two sources have the potential to capture 0.45 and 6 MMT of CO_2, respectively. A Carbon Capture Program funded FEED study at PSEC will evaluate the use of the commercial MHI's KM CDR Process™ CO_2 capture technology, and a capture assessment at the One Earth Energy ethanol facility will be completed under CarbonSAFE. Future characterization work at each site will include the drilling and logging of one characterization well, coring new samples, and acquiring 26 miles of 2D seismic data and 4-6 mi^2 of 3D seismic data.

3.4.2 San Juan Basin CarbonSAFE Phase III: Ensuring safe subsurface storage of CO_2 in saline reservoirs

New Mexico Institute of Mining and Technology (Socorro, NM) is performing a comprehensive site characterization study of a mature oil and gas field in the San Juan Basin, northwest New Mexico. The field, operated by Hilcorp Energy, has great stacked storage potential in alternating sequences of deep saline formations and oil-bearing reservoirs. Saline storage targets include the Salt Wash Member, Bluff sandstone, and Slick Rock Member.

The selected site is approximately 20 miles from the San Juan Generating Station (SJGS), an 847-MW coal-fired electricity generation plant. Annually, this plant has the potential to capture 6-7 MMT of CO_2. The CCUS project will consist of supplying 2 MMT of CO_2/yr to the storage complex and transporting the remaining CO_2 via the Cortez pipeline network to EOR fields in the Permian Basin in Texas. A Carbon Capture Program funded FEED study at SJGS will be completed using MHI's commercial amine-based solvent CO_2 capture technology. Ongoing and upcoming efforts at the storage complex site include the acquisition and evaluation of existing field data (e.g., 3D seismic, logging and coring samples), the drilling and logging of one characterization well, and coring of new samples.

3.4.3 Establishing an early CO_2 storage complex in Kemper County, Mississippi: Project ECO_2S

Southern States Energy Board (Peachtree Corner, GA) will fully characterize the storage complex facility in Kemper County, Mississippi. Previous characterization efforts recognized Kemper County as a world class geologic setting capable of securely storing over 1,200 MMT of CO_2 over a 30,000-acre area. Detailed discussion of the Phase II results of this study are provided by Reistenberg [23]. Three high-quality sandstone reservoirs include the lower Cretaceous Paluxy Formation, Washita-Fredericksburg interval, and the Upper Cretaceous Lower Tuscaloosa group. The initial proposed CCUS scenario comprised capture from the Kemper County Energy Facility with minimal pipelines required to transport the CO_2 to the storage complex. Due to operational changes at the facility, the project now includes two additional sources totaling 22.5 MMT of CO_2/yr for the development of a regional hub. A Carbon Capture Program funded FEED study will be completed using Linde-BASF's OASE blue Advanced Amine solvent-based capture technology at Southern Company's Plant Daniel in Moss Point, Mississippi. Additional CO_2 capture assessments of Plant Ratcliffe in Kemper County, Mississippi and Plant Miller in Jefferson County, Alabama will be completed under this project. Ongoing and upcoming efforts at the storage complex site include the drilling and logging of three characterization wells, coring new samples, and acquisition of 4-6 mi^2 of 3D seismic data.

3.4.4 North Dakota CarbonSAFE Phase III: Site characterization and permitting

The University of North Dakota Energy & Environmental Research Center (Grand Forks, ND) will fully characterize a storage complex near Minnkota's Milton R. Young Station (MRYS) one of two storage sites considered in earlier feasibility efforts. Detailed discussion of the Phase II results of this study are provided by

Peck [24]. The site is capable of storing over 3 MMT of CO_2/yr from the plant's 455-MW Unit 2. The project will investigate three storage reservoirs for further study: Broom Creek, Inyan Kara, and Deadwood sand reservoirs. Previous work suggests that Broom Creek formation is capable of securely storing over 563 MMT of CO_2. Coordinating with this CarbonSAFE Initiative is the Project Tundra initiative - an integrated CO_2 capture and storage project led by Minnkota Power Cooperative Targeted to begin operation in 2022-2023, the MRYS Unit 2 will be using a commercial amine-based Econamine FG Plus^{SM} capture technology from Fluor Corporation. A FEED study of this technology is being studied under the Carbon Capture program. Ongoing effort at the MRYS site will facilitate the completion of all necessary regulatory requirements for two Class VI well permits. Future characterization work will include the drilling and logging of two characterization wells, coring new samples, and 11 mi^2 of 3D seismic acquisition.

3.4.5 Wyoming CarbonSAFE: Accelerating CCUS commercialization and deployment at Dry Fork Power Station and Wyoming Integrated Test Center

The University of Wyoming (Laramie, WY) is finalizing the characterization of the storage complex in Campbell County, Wyoming. Detailed discussion of the Phase I results of this study are provided by Quillinan [25]. The selected storage site is centrally located in the Powder River Basin, which produces more than 40% of the coal mined in the United States [26], and is accessible to existing CO_2 pipeline infrastructure and potential CO_2-EOR fields. The project will investigate four storage reservoirs for further study: the Minnelusa, Lower Sundance, Lakota/Fall River Group, and the Muddy Sandstone. Initial analysis suggests a P50 storage volume of 3.4 MMT/mi^2 for the combined reservoirs. The project would capture CO_2 from Basin Electric's Dry Fork Station (DFS), one of the newest 385-MW commercial-scale coal-fired power plants in the United States. The power station is also home to the Wyoming Integrated Test Center (https://www.wyomingitc.org/), which is used to study the capture, storage, and management of carbon emissions from DFS. A Carbon Capture Program funded FEED study at the DFS will be completed for the commercial Polaris^{TM} membrane CO_2 capture technology from Membrane Technology and Research, Inc. Future characterization work will include analysis of previous phase II acquisitions (legacy 2D seismic, newly acquired 3D seismic, and coring new samples), re-entering one well to gauge for monitoring, constructing a second well for data collection and injection tests, and a new VSP seismic data acquisition.

4. Summary

CCUS is becoming widely accepted as a viable and necessary option for fossil-based energy sources, such as coal- or gas-fired power plants and other industrial sources to lower their carbon dioxide (CO_2) emissions. Over the past two decades DOE research efforts have provided supporting evidence that CO_2 can be stored in geologic formations safely and securely across multiple regions of the U.S. Efforts now must shift focus on the justification of the business case for widespread CCUS deployment. Research will continue to focus on technologies and methods to reduce cost, reduce operational risk and increase efficiencies to make commercial-scale CCUS economical.

The CarbonSAFE Initiative is providing this understanding of how to safely, efficiently and cost-effectively characterize and permit project site(s) at the commercial-scale throughout the U.S. Phase I and Phase II projects have investigated potential integrated CCUS project scenarios and identified both technical and non-technical research gaps. Each Phase III CarbonSAFE project now works to fill those gaps. Projects will need to consider and negotiate various aspects and business arrangements as part of the process to evaluate and secure CO_2 source(s), develop a plan for pore space ownership, and complete a UIC Class VI permit application(s) to construct the injection wells. To achieve wide adoption of CCUS, it is imperative that we build upon these learnings to test, mature and prove CCUS technologies at commercial scale.

5. Disclaimer

This report was prepared as an account of work sponsored by an agency of the United States Government. Neither the United States Government nor any agency thereof, nor any of their employees, makes any warranty, express or implied, or assumes any legal liability or responsibility for the accuracy, completeness, or usefulness of any information, apparatus, product, or process disclosed, or represents that its use would not infringe privately owned rights. Reference herein to any specific commercial product, process, or service by trade name, trademark, manufacturer, or otherwise does not necessarily constitute or imply its endorsement, recommendation, or favoring by the United States Government or any agency thereof. The views and opinions of authors expressed herein do not necessarily state or reflect those of the United States Government or any agency thereof.

6. Acknowledgement

KeyLogic's contributions to this work were funded by the National Energy Technology Laboratory under the Mission Execution and Strategic Analysis contract (DE-FE0025912) for support services.

7. References

[1] National Petroleum Council, "Meeting the Dual Challenge: A Roadmap to At-Scale Deployment of Carbon Capture, Use, and Storage, Chapter One: The Role of CCUS in the Future Energy Mix," 2019. [Online]. Available: https://dualchallenge.npc.org/files/CCUS-Chap_1-042320.pdf. [Accessed November 2020].

[2] EIA, "U.S. Energy-Related CO2 Emissions in AEO2019 Reference Case, 1990 to 2050 (figure 1-10)," EIA, 2019.

[3] S. Bennett and T. Stanley, "US budget bill may help carbon capture get back on track," 12 March 2018. [Online]. Available: https://www.iea.org/commentaries/us-budget-bill-may-help-carbon-capture-get-back-on-track. [Accessed November 2020].

[4] D. Nagabhushan and J. Thompson, "Carbon capture and storage in the US power sector - The impact of 45Q Federal Tax Credits," Clean Air Task Force, The impact of 45Q Federal 2019. [Online]. Available: https://www.catf.us/wp-content/uploads/2019/02/CATF_CCS_United_States_Power_Sector.pdf. [Accessed November 2020].

[5] Carbon Capture Coalition, "Carbon Capture Jobs and Project," June 2020.

[Online]. Available: https://carboncapturecoalition.org/wp-content/uploads/2020/06/Carbon-Capture-Jobs-and-Projects-1.pdf. [Accessed November 2020].

[6] Clean Air Task Force, "Potential Carbon Capture Projects Database," undated. [Online]. Available: https://docs.google.com/spreadsheets/d/115hsADg3ymy3lKBy4PBQRXz_MBknptqlRtlfuv79XV8/edit#gid=1540463113. [Accessed November 2020].

[7] Clean Air Task Force, "Interactive Map of CCUS Projects in Development in the U.S.," 27 July 2020. [Online]. Available: https://www.catf.us/2020/07/ccus-interactive-map/. [Accessed November 2020].

[8] H. Tanakaa, T. Tsujiuchi, T. Kamijo, S. Kishimoto, Y.-J. Lin, S. Kawasaki, Y. Nakagami and T. Nojo, "Advanced KM CDR Process using New Solvent," October 2018. [Online]. Available: https://www.cfaenm.org/wp-content/uploads/2019/03/GHGT14_manuscript_20180913Clean-version.pdf. [Accessed November 2020].

[9] NETL, "BEST PRACTICES MANUALS," undated. [Online]. Available: https://netl.doe.gov/coal/carbon-storage/strategic-program-support/best-practices-manuals. [Accessed November 2020].

[10] T. Rodosta, M. Ackiewicz and A. Albenze, "Status Update and Results from the U.S. Department of Energy Regional Carbon Sequestration Partnership Initiative, Proceedings GHGT-12," *Energy Procedia,* vol. 63, pp. 6039-6052, 2014.

[11] T. Rodosta, W. Aljoe, G. Bromhal and D. Damiani, "U.S. DOE Regional Carbon Sequestration Partnership Initiative: New Insights and Lessons Learned, Proceedings GHGT-13," *Energy Procedia,* vol. 114, pp. 5580-5592, 2017.

[12] M. Sullivan, T. Rodosta, K. Mahajan and D. Damiani, "An overview of the Department of Energy's CarbonSAFE Initiative: Moving CCUS toward commercialization," *Separations: Materials, Devices and Processes,* 6 November 2019.

[13] B. McPherson, M. Cather, R. Middleton, T. Chidsey, J. Heath, M. Saunders and S.-Y. Lee, "CarbonSAFE Rocky Mountain Phase I: Ensuring Safe Subsurface Storage of Carbon Dioxide in the Intermountain West," 31 August 2018. [Online]. Available: https://www.osti.gov/biblio/1559990. [Accessed November 2020].

[14] M. Kelley, A. Haagsma, P. Champagne, J. Glier, N. Gupta, M. H. Yugulis, J. Hawkins, J. Main, A. Pasumarti, J. Sminchak, S. Weber, B. Mannes, R.

Pardini, K. Sanders and W. Harrison, "Integrated Pre-Feasibility Assessment for a Northern Michigan Basin CarbonSAFE CO 2 Storage Complex," 2 September 2018. [Online]. Available: https://doi.org/10.2172/1469190. [Accessed November 2020].

[15] M. K. Dubois, E. T. Williams, J. C. Youle and D. E. Hedke, "Potential for CO2 storage and enhanced oil recovery in four southwest Kansas oil fields, an extended abstract; in W. L. Watney, PI, Final Report for DOE Award Number: DE-FE0002056," DOE, 2015.

[16] Y. Holubnyak, M. Dubois, T. Bidgoli, D. Wreath, L. Watney, S. Stover, D. Newell, F. Fazelalavi, A. Hollenbach, J. Jennings, C. Steincamp, J. Schremmer, B. Jordan and B. Crabtree, "Integrated CCS for Kansas (ICKan) Final Technical Report," December 31, 2018.

[17] United States Department of Energy, "Carbon Storage Atlas: 5th edition," DOE, Washington, D.C., 2015.

[18] B. Goertz-Allmann, S. Gibbons, V. Oye, R. Bauer and R. Will, "Characterization of induced seismicity patterns derived from internal structure in event clusters," *Journal of Geophysical Research: Solid Earth,* vol. 122, p. 3875–3894, 2017.

[19] R. Bauer, R. Will, S. Greenberg and S. Whittaker, "Chapter 19: Illinois Basin-Decatur project," in *Geophysics and Geosequestration*, T. M. Davis, M. Landrø and M. Wilson, Eds., Cambridge University Press, 2019, pp. 339-369.

[20] D. H. Bacon, D. Blankenau, D. Divine, A. Duguid, I. Fukai, J. Glier, J. Hawkins, M. Jimenez, R. Joeckel, S. McDonald, R. Middleton, R. Peterson and M. Scharenberg, "Integrated mid-continent stacked carbon storage hub, Phase I Final Report," 12 October 2018. [Online]. Available: https://doi.org/10.2172/1478726. [Accessed November 2020].

[21] NETL, "Large Pilot Testing of Linde-BASF Advanced Post-Combustion Carbon Dioxide Capture Technology at a Coal-Fired Power Plant," 2018. [Online]. Available: https://www.netl.doe.gov/projects/project-information.aspx?k=FE0031581. [Accessed November 2020].

[22] S. McDonald, "Illinois Industrial Carbon Capture & Storage Project: Eliminating CO2 Emissions from the Production of Bio Fuels - A 'Green' Carbon Process," 11 July 2017. [Online]. Available: https://www.energy.gov/sites/prod/files/2017/10/f38/mcdonald_bioecono my_2017.pdf. [Accessed November 2020].

[23] D. Riestenberg, R. Esposito and K. Gray, "Establishing an Early CO2 Storage Complex in Kemper County, Mississippi: Project ECO2S (FE0029465)," 28 August 2019. [Online]. Available: https://www.osti.gov/biblio/1617036.

[Accessed November 2020].

[24] W. Peck, K. Glazewski, D. Nakles, N. Azzolina, C. Crocker, R. Klapperich, B. Oster, D. Daly, A. Livers-Douglas, S. Butler, S. Smith, B. Botnen, I. Feole, J. He, N. Dotzenrod, A. Salazar, S. Patil and J. Crossland, "North Dakota Integrated Carbon Storage Complex Feasibility Study," 24 March 2020. [Online]. Available: https://doi.org/10.2172/1606011. [Accessed November 2020].

[25] S. Quillinan and K. Coddington, "Integrated Commercial Carbon Capture and Storage (CCS) Prefeasibility Study at Dry Fork Station, Wyoming," 30 May 2019. [Online]. Available: https://www.osti.gov/biblio/1557442/. [Accessed November 2020].

[26] U.S. Department of the Interior, U.S. Geological Survey, "Assessment of Coal Geology, Resources, and Reserve Base in the Powder River Basin, Wyoming and Montana," February 2013. [Online]. Available: https://pubs.usgs.gov/fs/2012/3143/fs-2012-3143. [Accessed November 2020].

602459 Towards the Development of Digital Twin Framework for Continuous Pharmaceutical Manufacturing: Real-Time Model Maintenance and System Analyses

Abstract:

Emerging digitization technologies within Industry 4.0 framework are bringing revolutionary transformations in the chemical manufacturing sector, including pharmaceutical manufacturing. Industry 4.0 in pharmaceutical manufacturing comprises efficient data collection and implementation strategies through the use of process digitization, Internet of Things (IoT), advanced data analytics using big data, machine learning, cyber-physical systems and cloud computing. These data-intensive strategies highlighted in Industry 4.0 are recently being applied towards the development of a digital twin framework [1,2]. For continuous pharmaceutical manufacturing, a data-centric digital-twin framework consists of a physical manufacturing plant inter-connected with its virtual, model-based representation, allowing seamless communication between the two entities. Within this framework, real-time data pertaining to process operation are recorded from the physical plant and sent to update the virtual model for advanced system analyses, and to update the plant operation when needed.

For the efficient implementation of digital-twin framework, real-time modeling strategies concerning virtual model and plant updates need to be developed. The presented work focuses on developing these strategies, where the major aspects involve: 1) model maintenance based on real-time data obtained from plant operation; 2) algorithms for real-time system analyses; 3) determination of frequency criterion for model and plant updates. Model maintenance of the virtual plant is performed using model adaptation strategies, including recursive adaptation and ensemble-based methods [3,4]. Unlike commonly used methods where models are updated and maintained off-line, these strategies allow real-time model updates in the virtual plant during manufacturing plant operation. After developing real-time model updating strategies within the digital twin framework, the model updates are used for dynamic system analyses, like sensitivity and feasibility analysis [5-7], providing uninterrupted monitoring of process performance. Based on these analyses, corrective process improvements are suggested for real plant operation, in situations where, the current process settings can potentially lead to violations of drug's quality specifications. Along with this capability of real-time update of virtual and real plant, frequency criterion for these updates need to be investigated for an efficient update strategy in the digital twin framework. These updates are generally performed at the same frequency associated with plant measurements [8] but can be modified based on a user-defined plant-model mismatch criterion [9] to reduce required computational resources.

In this presentation, the results of such a framework for a continuous direct compaction line, together with Industry 4.0 will be illustrated. This will allow for a computationally efficient and robust implementation of digital-twin framework in pharmaceutical manufacturing for scientific, risk-based regulation compliant process operation.

Reference:

[1] M. Grieves, J. Vickers, Digital Twin: Mitigating Unpredictable, Undesirable Emergent Behavior in Complex Systems, in: Transdisciplinary Perspectives on Complex Systems, Springer, Cham, 2017: pp. 85–113.

[2] A. Belhadi, K. Zkik, A. Cherrafi, S.M. Yusof, S. El Fezazi, Understanding Big Data Analytics for

Manufacturing Processes: Insights from Literature Review and Multiple Case Studies, Computers & Industrial Engineering. 137 (2019).

[3] P. Kadlec, R. Grbic, B. Gabrys, Review of adaptation mechanisms for data-driven soft sensors, Computers & Chemical Engineering. 35 (2011) 1–24.

[4] K. Hazama, M. Kano, Covariance-based locally weighted partial least squares for high-performance adaptive modeling, Chemometrics and Intelligent Laboratory Systems. 146 (2015) 55–62.

[5] G. Bano, Z. Wang, P. Facco, F. Bezzo, M. Barolo, M. Ierapetritou, A novel and systematic approach to identify the design space of pharmaceutical processes, Computers & Chemical Engineering. 115 (2018) 309–322.

[6] Z. Wang, M.S. Escotet-Espinoza, M. Ierapetritou, Process Analysis and optimization of continuous pharmaceutical manufacturing using flowsheet models, Computers and Chemical Engineering. 107 (2017) 77–91.

[7] F. Boukouvala, V. Niotis, R. Ramachandran, F.J. Muzzio, M. Ierapetritou, An integrated approach for dynamic flowsheet modeling and sensitivity analysis of a continuous tablet manufacturing process, Computers & Chemical Engineering. 42 (2012) 30–47.

[8] G. Bano, P. Facco, M. Ierapetritou, F. Bezzo, M. Barolo, Design space maintenance by online model adaptation in pharmaceutical manufacturing, Computers & Chemical Engineering. 127 (2019) 254–271.

[9] A.S. Badwe, R.D. Gudi, R.S. Patwardhan, S.L. Shah, S.C. Patwardhan, Detection of model-plant mismatch in MPC applications, Journal of Process Control. 19 (2009) 1305–1313.

610062 Digital Transformation of Unit Operations Laboratory

Abstract:

The Unit Operations Laboratory at Rose-Hulman Institute of Technology recently underwent its digital transformation and the facilities are now being used in multiple courses as well as research projects. The fundamental shift to allowing open and efficient access to data has allowed this rich resource to influence not only the traditional Unit Operations lab course, but to supplement other core courses, serve as the backbone for our new *Process Analytics* course, and act as a hub for multidisciplinary research projects.

603143 Development of a Joint Infrastructure for Chemical Engineering Applications Related to Industry 4.0 and Digital Transformation

Abstract:

This work provides an overview on the opportunities created when a developed Industry 4.0 infrastructure is applied to chemical engineering processes. Industry 4.0 defines a technology framework designed to revolutionize the world process industries. In this context, research groups have been exploring such a revolution by bringing to this environment improvements to link chemical engineering research, applications and teaching[1,2,3]. In particular, in this presentation, the cooperation between research groups at West Virginia University (WVU, USA) and Federal University of Campina Grande (UFCG, Brazil) in this area is described. Such cooperation aims to lead to a complete IIoT (Industrial Internet of Things) infrastructure connected to several dynamic process simulators (e.g., Aspen, CHEMCAD, SimCentral). With data availability from this infrastructure (at different data request points for a simulated process and in real time), Industry 4.0 aspects can be explored, such as: digital twin, machine learning, on-line data analysis, event frames and notifications, as well as topics directly linked to automatic and statistical process control, process optimization and economic analysis.

The current infrastructure envisioned for analysis involves the use of a simulation platform that represents a chemical or manufacturing plant, data processing/storage and data request points from client devices and applications. The particularly investigated simulations include a subcritical coal-fired power plant model (from WVU), an industrial process from a large Latin America Refining Company (from UFCG), as well as chemical processes from the literature. Central to this infrastructure is a database to store real-time simulation data that represent plant measurements. The database used corresponds to the OSIsoft PI historian (which is part of the PI System)[1]. A Supervisory Control and Data Acquisition (SCADA) system is employed to remotely monitor and control the studied simulation through Open Platform Communications (OPC). In an analogy between the developed infrastructure and the Automation Pyramid, levels 0 (sensors and actuators) and 1 (unit automation) are represented by the dynamic simulations, level 2 (SCADA) provides the remote access to process variables and specifications by customized HMIs (Human Machine Interfaces), and level 3 (Plant Information Management System – PIMS) contextualizes the data from an asset standpoint allowing the management of analytics, event frames, notifications and system integrations with a tree view of the plant structure. The PIMS makes the data available for custom applications, thus interpreting the information and adding value to process related techniques such as predictive process control, optimization and economics analysis. Therefore, this infrastructure can provide a more realistic environment for investigating process automation and implementations under development, using desktop or web-based custom reports and dashboards for decision making at all different levels (operators, engineers or executives).

This work thus explores aspects of Industry 4.0 such as Digital Twins, Simulation, System Integration and Data Analytics using a developed infrastructure at WVU and UFCG that can be used for operator training, research and engineering classes. Additionally, it corresponds to the starting point for future research on more robust data analytics, machine learning, big data and business intelligence studies.

References

[1] E. Ruiz-Ramos, J. M. Romero-García, F. Espínola, I. Romero, V. Hernández and E. Castro, "Learning and researching based on local experience and simulation software for graduate and undergraduate courses in chemical and environmental engineering", Education for Chemical Engineers, 21, pp 50 – 61, 2017;

[2] J. Uhlemann, R. Costa and J. Cl. Charpentier, "Product design and engineering — past, present, future trends in teaching, research and practices: academic and industry points of view", Current Opinion in Chemical Engineering, 27, pp 10–21, 2020;

[3] M. Teles dos Santos, A. S. Vianna Jr. and G. A. C. Le Roux, "Programming skills in the industry 4.0: are chemical engineering students able to face new problems?", Education for Chemical Engineers, 22, pp 69–76, 2018.

[4] PI Server. (2018 SP3). San Leandro, CA, USA. OSIsoft.

602636 Robust Large-Scale State-Estimate Prediction

Abstract:

Effective monitoring and control of a process require online measurements of all state variables of the process. Measurement of all state variables are typically not available due to the lack of reliable online sensors and/or high costs of online sensors. These unmeasured variables are often directly related to product properties, which are important to monitor and control. Their estimates are usually obtained using state estimators. As a result of the need for and the importance of state estimation, the development of accurate state estimators that are robust with respect to model uncertainties and unmeasured inputs, has been the subject of research for decades [1-5]. There have been a few reports on the design of large-scale state estimators [6-11]. Although the implementation of centralized estimators for large-scale systems seems to be an optimal approach, these estimators are not scaleable to complex, large-scale systems, as they suffer from the curse of dimensionality.

In this paper, we address the problem of large-scale robust state-estimate prediction; that is, the prediction of future values of state estimates robustly at every time instant in large-scale processes. To this end, we decompose a large-scale state-estimate prediction problem into a set of smaller-scale state-estimate prediction problems so that the resulting set of the smaller-scale state-estimate predictors are more robust, easier to design and implement, and more computationally efficient. We also study tuning of the smaller-scale state-estimate predictors to maximize the robustness of the predictors set while ensuring the stability of the error dynamics of the entire set. In addition, we address the problem of implementing the smaller-scale state-estimate predictors in parallel (using a computer with parallel processors) to improve computational efficiency of the state estimate prediction while preserving the accuracy and robustness of state-estimate predictions.

References

[1] J.M. Ali, N.H. Hoang, M.A. Hussain, D. Dochain, Review and classification of recent observers applied in chemical process systems, Computers & Chemical Engineering, 76 (2015) 27-41.

[2] A.K. Jana, A nonlinear exponential observer for a batch distillation, in: 2010 11th International Conference on Control Automation Robotics & Vision, IEEE, 2010, pp. 1393-1396.

[3] M. Soroush, State and parameter estimations and their applications in process control, Computers & Chemical Engineering, 23 (1998) 229-245.

[4] S. Tatiraju, M. Soroush, Nonlinear state estimation in a polymerization reactor, Industrial & engineering chemistry research, 36 (1997) 2679-2690.

[5] N. Zambare, M. Soroush, B.A. Ogunnaike, Robustness improvement in multi-rate state estimation, in: Proceedings of the 2001 American Control Conference.(Cat. No. 01CH37148), IEEE, 2001, pp. 993-998.

[6] N. Abdel-Jabbar, C. Kravaris, B. Carnahan, A partially decentralized state observer and its parallel computer implementation, Industrial & engineering chemistry research, 37 (1998) 2741-2760.

[7] N. Abdel-Jabbar, C. Kravaris, B. Carnahan, Structural analysis and partitioning of dynamic process models for parallel state estimation, in: Proceedings of the 1998 American Control

Conference. ACC (IEEE Cat. No. 98CH36207), IEEE, 1998, pp. 3170-3176.

[8] J.B. Carvalho, F.M. Barbosa, Parallel and distributed processing in state estimation of power system energy, in: MELECON'98. 9th Mediterranean Electrotechnical Conference. Proceedings (Cat. No. 98CH36056), IEEE, 1998, pp. 969-973.

[9] R. Ebrahimian, R. Baldick, State estimation distributed processing [for power systems], IEEE Transactions on Power Systems, 15 (2000) 1240-1246.

[10] D.M. Falcao, F.F. Wu, L. Murphy, Parallel and distributed state estimation, IEEE Transactions on Power Systems, 10 (1995) 724-730.

[11] D.B. Pourkargar, M. Moharir, A. Almansoori, P. Daoutidis, Distributed estimation and nonlinear model predictive control using community detection, Industrial & Engineering Chemistry Research, 58 (2019) 13495-13507.

601991 The Digital Transformation of the Chemicals Industry - and How to Prepare

Abstract:

Over the course of the next 20 years, the chemicals industry will see more change than it has in the last 50, in large part due to digital technologies. Digital transformation is affecting business processes from R&D, to manufacturing, to supply chain, to sales and go-to-market -- and the data from all these digital efforts is tying those diverse functions to optimize the chemical enterprise and create new business models and sources of value. This presentation looks at how the industry will change through 2040 and examines what the new sources of competitive advantage will be for both companies and individuals.

613114 Driving the Value from Digitalization

Abstract:

Many opportunities exist to utilize the technologies of Industry 4.0 and digitalization within the process industries, to improve reliability, productivity, profitability, and safety. However, organizations may become enamored with the new tool or capability, looking for problems to fit the technology rather than focusing on the true business needs. Only by being technology agnostic and using a strategy that places balance sheet impact and long term competitive advantage as selection criteria to the myriad of solutions available in the market will an organization focus on the proper challenges. This presentation will examine potential pitfalls to technology selection and implementation.

613028 Combining AI with Chemical Engineering First Principles Will Take Process Performance, Safety and Sustainability to New Levels

Abstract:

Digitalization is now in the spotlight in terms of creating improved agility, resilience, and ability to operate process industry assets remotely and safely.

Sensors, connectivity, and cloud computing are making process data more available for use in optimization and prediction. High performance, edge computing, and new algorithms are enabling placing process models online for monitoring, optimization, and operational advice.
Most process and chemical engineers in industry have observed the influx of data science and data scientists into their organizations. This session will cover an area called "hybrid modeling." This combines and embeds AI and machine learning within first-principles rigorous models. You may think of this as putting chemical engineering guardrails around machine learning. Rigorous, first principles models, brings to bear decades of chemical engineering knowledge. Machine learning and AI brings to bear ability to rapidly analyze, and progressively learn from empirical operating data. When combined, it is now possible to achieve a new level of modeling that will further leverage both types of analysis. We have been validating this breakthrough combination of AI and chemical engineering with over 100 companies in industry, and will cover several high value and immediately practical use cases during this session.

601992 Designing Materials and Processes in the Digital World

Abstract:

3M has been known as a materials innovation company for many years with thousands of new products launched annually, many of those enabling and empowering everyday products. This innovation model is based on leveraging deep technologies into novel combinations to develop product platforms.

This talk will address the challenges and progress in transforming this culture and enabling it with digital tools created internally and with outside partners to accelerate the speed of materials innovation dramatically.

607940 Digitalization of Biopharmaceutical Manufacturing

Abstract:

This presentation describes ways to leverage and implement digitalization technologies in biopharmaceutical manufacturing. An accelerated process development workflow is described that employs micro-scale technologies, modular unit operations with integrated process control and monitoring systems, systems integration, and full plant automation. The presentation describes how to best develop and transfer knowledge between steps in the workflow via first-principles models, data analytics, and machine learning. Case studies are described that illustrate the application of each of the above methods, including where process equipment was designed digitally by using first-principles models first, then the process equipment was constructed and implemented experimentally, with the experimental results confirming model predictions.

608784 Stepping Towards the Industrial Sixth Sense

Abstract:

Industry 4.0 is transforming chemical processes into complex, smart cyber-physical systems, by the addition of elements such as smart sensors, Internet of Things, big data analytics or cloud computing. Modern engineering systems and manufacturing processes are operating in highly dynamic environments, and exhibiting scale, structure and behaviour complexity. Under these conditions, plant operators find it extremely difficult to manage all the information available, infer the desired conditions of the plant and take timely decisions to handle abnormal operation[1]. Human beings acquire information from the surroundings through sensory receptors for vision, sound, smell, touch, and taste, the Five Senses. The sensory stimulus is converted to electrical signals as nerve impulse data communicated with the brain. When one or more senses fail, the humans are able to re-establish communication and improve the other senses to protect from incoming dangers. Furthermore, a mechanism of 'reasoning' has been developed during evolution, which enable analysis of present data and generation of a vision of the future, which might be called the Sixth Sense. As industrial processes are already equipped with five senses: 'hearing' from acoustic sensors, 'smelling' from gas and liquid sensors, 'seeing' from camera, 'touching' from vibration sensors and 'tasting' from composition monitors, the Sixth Sense could be achieved by forming a sensing network which is self-adaptive and self-repairing, carrying out deep-thinking analysis with even limited data, and predicting the sequence of events via integrated system modelling.

This contribution introduces the development of an intelligent monitoring and control framework for chemical process, integrating the advantages of Industry 4.0 technologies, cooperative control and fault detection via wireless sensor networks. The framework consists of four main components. The first one is a wireless sensors network, transmitting over a 5G communication network, that facilitates data management for improved fault detection. The second component is an efficient fault detection algorithm that can analyse the data and classify it in faulty or normal. The third component is a knowledge-based and a model-based fault detection monitoring system. For the fault-detection, a two-stage method based on a hybrid learning approach is applied, which utilises supervised and unsupervised learning. Finally, the fourth component is a cooperative model predictive control system that takes the required measures to ensure stable process information. Under the cooperative framework, the distributed controllers share information about their state with other controllers, thus interaction between processing units is not lost. Using information on the process' structure and behaviour, equipment information, and expert knowledge, the system is able to detect faults. The integration with the monitoring system facilitates the detection and optimises the controller's actions. The chemical process is a mini-plant that produces sodium ion solution as sodium chloride for the fine chemical, the pharmaceutical and the food industry. The dataset used for simulations consists of over 10 million samples, with each sample having 43 variable measurements, including temperature, flows, pressures and levels, collected by deployed wireless sensors set up on the process units. The results indicate that the proposed approach achieve high fault detection accuracy based on plant measurements, while the cooperative controllers improve the control of the process.

1. Natarajan, S., & Srinivasan, R., 2014, Computers and Chemical Engineering 60, pp. 182-196

AUTHOR INDEX

AIChE
120 Wall Street, FL 23
New York, NY 10005-4020

ISBN 978-1-7138-2291-2
9 781713 822912